CHARACTERIZATION OF POLYMERS

CHARACTERIZATION OF POLYMERS

Encyclopedia Reprints

Edited by

NORBERT M. BIKALES

Consulting Chemist
Livingston, New Jersey

WILEY-INTERSCIENCE

a Division of John Wiley & Sons, Inc.
New York • London • Sydney • Toronto

CHARACTERIZATION, reprinted from *Encyclopedia of Polymer Science and Technology*, Vol. 3, pp. 611–631

FRACTIONATION, reprinted from *Encyclopedia of Polymer Science and Technology*, Vol. 7, pp. 231–260

MOLECULAR WEIGHT DETERMINATION, reprinted from *Encyclopedia of Industrial Chemical Analysis*, Vol. 2, pp. 611–648

CHEMICAL ANALYSIS, reprinted from *Encyclopedia of Polymer Science and Technology*, Vol. 3, pp. 632–665

INFRARED-ABSORPTION SPECTROSCOPY, reprinted from *Encyclopedia of Polymer Science and Technology*, Vol. 7, pp. 620–643

NUCLEAR MAGNETIC RESONANCE, reprinted from *Encyclopedia of Polymer Science and Technology*, Vol. 9, pp. 356–396

DIFFERENTIAL THERMAL ANALYSIS, reprinted from *Encyclopedia of Polymer Science and Technology*, Vol. 5, pp. 37–65

THERMOGRAVIMETRIC ANALYSIS, reprinted from *Encyclopedia of Polymer Science and Technology*, Vol. 14, pp. 1–41

Library of Congress Catalog Card Number: 78:172950

ISBN 0-471-07231-1

Printed in the United States of America.

10 9 8 7 6 5 4

CONTRIBUTORS

GEORGE L. BEYER, Eastman Kodak Co., Rochester, New York, *Molecular Weight Determination*

NORBERT M. BIKALES, Consulting Chemist, Livingston, New Jersey, *Editor*

F. A. BOVEY, Bell Telephone Laboratories, Murray Hill, New Jersey, *Nuclear Magnetic Resonance*

G. M. BRAUER, National Bureau of Standards, Washington, D.C., *Chemical Analysis*

MANFRED J. R. CANTOW, IBM Corp., Systems Development Division, San Jose, California (formerly at Chevron Research Company), *Fractionation*

JULIAN F. JOHNSON, Department of Chemistry and Institute of Material Science, University of Connecticut, Storrs, Connecticut (formerly at Chevron Research Company), *Fractionation*

BACON KE, Charles F. Kettering Research Laboratory, Yellow Springs, Ohio, *Differential Thermal Analysis*

G. M. KLINE, National Bureau of Standards, Washington, D.C., *Chemical Analysis*

IVO KÖSSLER, Ustav Fysikalni Chemie, Ceskoslovenska Akademie Ved, Praha, Czechoslovakia, *Infrared-Absorption Spectroscopy*

DAVID W. LEVI, Picatinny Arsenal, Dover, New Jersey, *Thermogravimetric Analysis*

ROGER S. PORTER, University of Massachusetts, Amherst, Massachusetts, *Fractionation*

LEO REICH, Picatinny Arsenal, Dover, New Jersey, *Thermogravimetric Analysis*

W. R. SORENSON, Continental Oil Company, Wilton, Connecticut, *Characterization*

PREFACE

Next to polymer preparation itself, characterization is the most important step in the synthesis and development of useful polymers. It provides the means of ascertaining the value, whether scientific or practical, both of the preparative method and of the products therefrom. Characterization techniques range from simple viscometry to elaborate methods of determining chemical structure, molecular weights, and various physical properties.

This book assembles in one volume eight chapters from six different sources contributed by recognized authorities in the field. Because the chapters are reproduced unchanged from their original sources, the reader should ignore the cross references to other pages and to other chapters that appear in the text.

I should like to acknowledge the great help of Jo Conrad in editing the original Encyclopedia articles on which this book is based.

NORBERT M. BIKALES

Livingston, New Jersey
June 1971

CONTENTS

CHARACTERIZATION OF POLYMERS

CHARACTERIZATION OF POLYMERS

Characterization of a new polymer follows closely on the preparation of a polymer in both practice and importance. The synthesis of a polymer is, quite clearly, the sine qua non for polymer characterization, but without the latter the former becomes an exercise of dubious value no matter how sophisticated the synthesis is chemically. It is comparable to the synthesis of an organic compound without a subsequent characterization of its physical properties or structure. Polymer characterization is therefore of very great importance.

The scope of characterization can be relatively narrow or extremely broad, depending on the degree of scientific and commercial interest that develops. The subject can be considered in two wide categories initially: molecular structure and physical properties. The chemical structure of a polymer is most often, though not always, known by the method of synthesis; but some aspects of configuration, conformation, isomerism of various types, or comonomer placement, for example, may require a detailed attack by the chemical and physical methods available.

Some methods for characterizing physical properties are of an elementary type; they are of immediate importance in defining a new material, and are usually applicable to most classes of polymers. A second category of characterization methods then becomes significant; it covers those methods that begin the broadening of the original question of what a polymer is like to what a polymer is good for. The total process of characterization involves the shading of the first question gradually into the second, since it is characterization broadly conceived that guides a previously unknown polymeric chemical to an article of commerce. In both classes of characterization, the results may provide a fundamental property of the polymer or may be entirely empirical. Both kinds of information fill an important role in overall characterization.

There is no definite sequence of characterization methods that can be unfailingly applied to a new polymer, nor can the division of methods mentioned above be considered a clear and distinct breakdown. There will inevitably be overlap, omission, and addition of methods to suit the goals of the investigator and the requirements of the material (5,11,14,18).

Table 1 is an outline of the division of methods as considered above. It should be recognized that in practice there will be interrelation throughout between the various methods of physical-property characterization, and between the latter and molecular-structure characterization. In both, there will be simple and sophisticated techniques available for reaching the same goal; the value of the results in either case will be governed by the needs and objectives of the research worker. Theoretical and practical treatments of some of the areas of polymer characterization have been published. More detailed discussions of the various characterization methods will be found in other articles of the Encyclopedia.

Molecular Structure

The total molecular structure of a polymer can be specified by most of the conventional parameters applied to simpler organic compounds which, in total, define the three-dimensional arrangement of atoms that constitutes a molecule in its most com-

Table 1. Polymer Characterization

Molecular structure	*Physical properties*
molecular weight	crystalline melting point and glass-transition temperature
number-average methods	other softening methods
weight- and *z*-average methods	fabricability and qualitative mechanical properties
solution-viscosity methods	solubility
melt-viscosity methods	density
molecular-weight distribution	mechanical properties—stress–strain, creep, stress relaxation, impact, dynamic tests
crystallinity and orientation	flow properties and processing characteristics
chemical structure, configuration, and conformation	aging tests
elemental analysis	electrical tests
degradation	optical tests
spectroscopy and mass spectrometry	permeability
x-ray diffraction	flammability
optical methods	biological and toxicological properties

plete sense. These would include the chemical structure in terms of atom sequences, functionality, configuration of asymmetric sites, overall molecular conformation, molecular weight of the species, and, when in the solid state, arrangement of molecules in a crystal lattice. The problem is usually enormously more complicated for polymer molecules because of the inhomogeneity of molecular weight, their partial or totally noncrystalline character, and the painful variety of isomeric types (position, geometric, configurational) that may exist in one single molecule. These factors devolve from the large molecular size and the isomeric complexity of the polymers. The latter is a result of the usual origin of the polymers via repetition of a chemical reaction (particularly the propagation step), which is more often than not random in isomer formation. Naturally occurring polymers and those derived from the recently developed stereoregulating catalysts (47) are simpler isomerically, but are never of uniform molecular weight. See also MORPHOLOGY; STRUCTURE OF POLYMERS.

Molecular Weight. There is always some difference of opinion as to what constitutes a polymer in terms of molecular weight. The adjective "high" is a useful, if nonexact, method of distinction in that it implies something very much beyond the reach of conventionally applied and executed chemical reactions. A high polymer is often taken to describe that level at which a considerable change upward in molecular weight ceases to cause a marked change in properties. This is untenable in practice since it must exclude low-molecular-weight (ie, 6000) polyethylene, or styrene–maleic anhydride containing about eight units of each, both of which are commonly regarded as polymers by those who produce them or use them commercially. The need to characterize a material by means of its molecular weight surmounts all semantic barriers, however (23).

The methods of molecular-weight measurement are distinguished by whether the contribution of each molecule to the measured property is proportional to the number of molecules or to a power of the weight of the molecules present. The first category leads to a number-average molecular weight, $\overline{M}_n$, which is defined as

$$\overline{M}_n = \frac{\text{weight of molecules}}{\text{number of molecules}} = \frac{\sum N_i M_i}{\sum N_i}$$

where N_i is the number of molecules of molecular weight M_i for every species i. Or,

$$\bar{M}_n = \sum n_i M_i$$

where n_i is the number fraction of molecules of molecular weight M_i.

The weight-average molecular weight is given by

$$\bar{M}_w = \frac{\sum N_i M_i^2}{\sum N_i M_i} = \sum w_i M_i$$

where w_i is the weight fraction of each species of molecular weight M_i. Light scattering and various types of ultracentrifugation give weight averages.

The z-average molecular weight is defined as

$$\bar{M}_z = \frac{\sum N_i M_i^3}{\sum N_i M_i^2}$$

and is the result of measurements of the radial distribution of the refractive-index gradient in equilibrium ultracentrifugation, a method in which the proportional contribution of the weight of each molecule is substantially greater than in the weight-average methods. Higher averages, as $z + 1$, can be obtained by other ultracentrifugation methods.

For polymers having homogeneous molecular size, all the averages mentioned would be equal. For heterogeneous molecularity, the more usual situation, $M_n < M_w < M_z < M_{z+1}$. See also MOLECULAR WEIGHTS; MOLECULAR-WEIGHT DETERMINATION.

Measurement of Number-Average Molecular Weight. Number-average molecular weights are obtained by measurement of the colligative properties of dilute polymer solutions, ie, freezing-point depression, vapor-pressure lowering, boiling-point elevation, and osmotic pressure (1,2,4). End-group determinations also provide a direct count of the number of molecules when it is known precisely that one or both ends of the molecules are involved.

The choice of the above methods depends on the nature of the material. Cryoscopic and ebullioscopic methods can be used up to a molecular weight of 30,000 at least; molecular weights up to 100,000 have been measured with the latter (29d). The problem encountered is the large amount of solute required to give the necessary molar concentration when the molecular weight is very large. Most vapor-pressure lowering methods are impractical because of long equilibration times; an exception is the thermoelectric differential vapor-pressure method, useful to above mol wt 30,000, though not a thermodynamic method (59).

Osmometry (qv) is inherently more sensitive than the other methods and can consequently be used to obtain number-average molecular weights in the range of 100,000 or higher, with an upper limit of about 1,000,000. A major difficulty in osmometry is the preparation of membranes sufficiently permeable to solvent to allow attainment of equilibrium within practical working times, yet not permeable to solute. Osmometry measures the molecular weight of those molecules larger than the maximum size permitted to diffuse through the membrane. The difficulty is compounded when working at elevated temperature because of possible membrane instability (3).

End-group determination by chemical means can be effective for number-average molecular weights of about 25,000 and lower. Above this value, the concentration of end groups becomes so low as to make most chemical techniques unreliable. Infrared spectroscopy, for example, can establish number-averages of 100,000 or more in certain instances, by identification of end groups. It is essential that the occurrence of end groups in the polymer be known, ie, whether the group under analysis occurs at one or both ends, whether chain branches also bear the group, and whether loss of end groups takes place through side reactions. End-group identification can give insights into the mechanisms of polymerization reactions also, eg, extent of termination by coupling and disproportionation in radical polymerizations, when combined with molecular-weight values determined by other methods (5). See also CHEMICAL ANALYSIS.

Measurement of Weight- and z-Average Molecular Weight. Light scattering (qv) of polymer solutions is the most widely used and generally applied of the methods that yield a weight-average molecular weight. It has the virtue of being an absolute method, independent of other measurements (4,5,19,20). The method is fundamentally simple, depending on the intensity of scattering of monochromatic light by a polymer solution. The intensity of scattering is proportional to the square of the mass of the scattering particle; hence, larger macromolecules contribute to a greater degree than smaller ones, or, to put it another way, the method is more sensitive to the presence of the larger molecules. Under the proper circumstances of materials and operation, information on molecular shape and solvent–polymer interaction can be obtained.

Ultracentrifugation is another absolute method of molecular-weight determination; it can also give information on distribution. The ultracentrifuge has the advantage over light scattering (a) of handling fairly viscous solutions which would be difficult to free of dust, and (b) of requiring smaller sample size. However, although there are many factors to be considered in comparing ultracentrifugation and light scattering, it is generally true that light scattering is usually more convenient. In addition, ultracentrifugation is more difficultly applied to random-coil polymers in terms of quantitative interpretation. Nevertheless, the method has found considerable application (21), much of it in the field of biopolymers (22). Four basic ultracentrifuge experiments are possible: sedimentation velocity, sedimentation equilibrium, equilibrium sedimentation in a mixed-solvent density gradient, and the approach to equilibrium method; the latter may become more useful for random-coil polymers. All the methods depend on a measure of the change in refractive index, and hence concentration, with the distance from the center of rotation. Depending on the particular experiment used, a weight-average or *z*-average molecular weight, or both, may be obtained. See also ULTRACENTRIFUGATION.

Measurement of Viscosity. Solution-viscosity methods offer by far the quickest and simplest diagnostic tool for molecular weight; these advantages are countered by a dependence on some other kind of molecular-weight determination to provide an ultimate basis for actual molecular weight. Nevertheless, solution viscosity is the most widely used method of molecular-weight designation. In operation, the method consists simply of measuring the flow times through a capillary viscometer of a polymer solution and of the pure solvent at specified condition of temperature and concentration, the latter usually expressed in g/dl or g/ml. The resulting ratio is customarily termed relative viscosity, though it and the customary "viscosity" values derived from it do not have the units of viscosity. Other terms have been proposed by the IUPAC (23). The more common terms and those proposed to supplant them are shown in Table 2. The units of intrinsic viscosity are, as can be seen, reciprocal concentration; in most

Table 2. Viscosity Terms

Current term	Defined by	IUPAC term
relative viscosity	$\eta_r = \dfrac{\text{flow time solution}}{\text{flow time solvent}}$	solution/solvent viscosity ratio
specific viscosity	$\eta_{sp} = \eta_r - 1$	
reduced viscosity	$\dfrac{\eta_{sp}}{c}$, c in g/dl	viscosity number
inherent viscosity	$\eta_{inh} = \dfrac{2.3 \log \eta_r}{c}$	logarithmic viscosity number
intrinsic viscosity	$[\eta] = \lim\limits_{c \to 0} \dfrac{\eta_{sp}}{c} = \lim\limits_{c \to 0} \eta_{inh}$	limiting viscosity number

usage, it is dl/g. The units for limiting viscosity number have been taken as ml/g; consequently, the value for the latter will be tenfold greater than what is generally reported for intrinsic viscosity. In either case, the value is obtained as the zero-concentration intercept of the extrapolated curve of η_{sp}/c or η_{inh} vs concentration.

The empirical equation $[\eta] = KM^a$ relates molecular weight to intrinsic viscosity (or limiting viscosity number) through the constants K and a, which will apply only for a particular polymer–solvent system over a specified molecular-weight range and a particular temperature. If the log of intrinsic viscosity is plotted against the log of the molecular weight, K is given by the intercept and a by the slope of the line. The values of K and a are usually, but not always, established for relatively narrow molecular-weight fractions. When applied to polydisperse material, the molecular weight obtained is a viscosity-average value. When $a = 1$, the molecular weight is equal to the weight-average molecular weight. Since a is between 0.5 and 1.0 for most polymers, and very largely in the range 0.6–0.8, the viscosity-average molecular weight is closer to the weight-average molecular weight (4).

Tabulations of K and a values (24) have been published for a variety of polymers; reliability and validity are often debated. Even so, for less critical purposes K and a are sometimes borrowed for use with a new polymer of similar structure and in the same solvent, although the practice cannot be recommended unless the serious limitations are understood. Often it is not necessary to know actual molecular-weight values; rather, it may be enough to know that polymer within a certain range of η_{inh} or η_{sp}/c will be suitable, say, for film extrusion or fiber spinning.

Solution viscosity offers a simple means of noting changes in molecular weight wrought by accelerated aging tests, processing conditions, or end-use exposure, for example; in most of these instances it is not necessary to know actual molecular weights, but much practical information can be had only from the viscosity values and the rate at which they are altered by the condition under scrutiny.

Besides molecular weight, the effective hydrodynamic volume of a macromolecule can be estimated from intrinsic-viscosity data, as can the degree of long-chain branching in certain instances (1). See also VISCOMETRY.

From the kind of dilute-solution measurements that have been discussed, particularly osmotic pressure, light scattering, and solution viscosity, factors other than the molecular weight can be determined. Collectively, these are termed solution properties (qv); they relate to polymer–polymer and polymer–solvent interactions in dilute solution, and to the general shape and dimensions of the polymer molecule at the temperature and in the solvent involved. These parameters derive from statistical considerations of the positions allowable to the segments of a freely bending chain, and further imposing the limitations of bond angle and bond lengths. From the execution of solution measurements, then, and suitable mathematical treatment of the data, such factors may be determined as the root-mean-square distance between chain ends; the radius of gyration (root-mean-square distance of a chain element from the center of the chain, a concept applicable to branched polymers having several ends); the expansion factor (indicative of how "opened up" a coiling macromolecule is in solution relative to its theoretical state); and the coefficients indicative of solute volume and interaction with solvent (eg, second virial coefficient). Closely involved in these concepts is the *theta temperature,* at which conformation is considered to correspond to the statistically ideal model (1).

The viscosity of molten polymers has also been correlated with molecular weight. Methods are discussed under Flow Properties and Processing Characteristics.

More sophisticated methods of determining melt viscosity (qv) involve capillary or rotational viscometers of a precise type, capable of operating over a wide range of shear rates. Quantitative rheological data can be obtained; molecular weight has been related, for example, to the shear modulus (26) using a capillary instrument for polyethylene, and to the melt viscosity using a rotational viscometer for polystyrene (27). Melt viscosity is more closely a function of the weight-average molecular weight.

Molecular-Weight Distribution. Knowledge about the breadth of the spectrum of chain lengths that constitute any polymer sample serves basic and practical purposes. Among the former, for example, would be an understanding of the mechanism of polymer-forming reactions. Practically, the processability, mechanical properties, and use performance characteristics are determined by the shape of the distribution curve around some average. A given average may represent a broad, symmetrically distributed molecular spread or a skewed or very narrow distribution. The extrusion of film, for example, may involve quite different power requirements and produce a product with distinct properties and visual characteristics in each case.

The molecular-weight distribution is experimentally determined by some kind of fractional separation of molecules based on size (3), each fraction being characterized by its molecular weight and its fraction of the total material. Precipitation or extraction of increasingly higher molecular weights from solution by incremental addition of a nonsolvent or improved solvent is a difficult and tedious, but long-standing method. Column extraction has found considerable use (4,29) through its effectiveness and greater convenience of operation. See also FRACTIONATION.

Crystallinity and Orientation. The interpretation of the physical evidence for crystalline and amorphous regions in polymers has undergone a significant change from the late 1950s forward, stimulated by the remarkable achievements in the isolation of polymer single crystals (30). The status of crystallinity theory has been revised and is, no doubt, not completely settled (7). Whatever the actual physical picture of crystallinity in macromolecules, its detection, measurement, alteration, and practical consequences are a matter of moment in polymer characterization. For most forms of polymers, the amount of crystalline material present and the alignment of molecules, either by crystallite orientation or in amorphous regions, together constitute extremely important determinants of mechanical properties and end-use behavior.

Crystallinity is detected and its amount estimated by a number of methods:

x-ray diffraction (qv), infrared-absorption spectroscopy (qv), polarized light microscopy, density, colorimetry, differential thermal analysis (qv), and nuclear magnetic resonance (qv) spectroscopy. Examination of a sample between crossed polarizers in a microscope equipped with a hot stage permits the observation of birefringence (5,11) due to crystallinity and/or orientation; the effects can generally be separated. Quantitative estimation by microscopy is not practical, however, but the temperature of disappearance of crystallinity, ie, the crystalline melting point, T_m, is most easily measured by this method (11). Density changes (31) due to onset or increase of crystallinity provide another simple method of estimating crystallinity changes. The change in density of polyethylene from about 0.91 to about 0.98, for example, is well recognized as a useful index of extent of crystallinity, hence short-chain branching. A density increase can be generally construed as meaning a definite increase in crystalline content; poly-4-methylpentene-1, however, has a lower density in the crystalline regions than in the amorphous (32), and is a rare exception. Solubility differences are often due to crystallinity; amorphous polystyrene is readily soluble in benzene, for example, whereas the crystalline polymer is not. Solubility in hot heptane is a standard method of freeing highly crystalline isotactic polypropylene from its amorphous atactic, and less crystalline stereoblock, counterparts. Solubility difference alone is not, however, a sure indication of crystallinity, since molecular weight can override the influence of crystallinity in some cases (33).

Accurate and undisputed measurement of the volume fraction of crystallinity in a polymer is not easily accomplished because each of the methods of measurement mentioned above is concerned with a different physical aspect of a material. Nevertheless, the great practical consequences of crystallinity and orientation on mechanical properties dictate that at least relative changes in these factors be observed by whatever means are applicable, and that these be correlated with changes in processing and fabrication methods and with end-use behavior. See also CRYSTALLINITY; ORIENTATION.

Chemical Structure, Configuration, and Conformation. In a majority of cases, the chemical structure of a polymer is evident from its method of synthesis, especially with the broad understanding now prevalent of the general mode of operation of polycondensation, addition polymerization of unsaturates, and other polymer-forming reaction types. Amount of head-to-head addition in vinyl polymerization, extent of biuret and allophanate in polyurethans, skeletally rearranged units in the cationic polymerization of branched vinyl monomers, the stereochemical configuration of asymmetric carbon atoms in a polymer backbone, and the random open-chain vs regular helical-coil conformation in proteins and certain types of polyolefins are cases in which one or several aspects of polymer structure are not inferred with certainty by the synthetic method or natural source.

Many, and probably most, of the methods that have been used to resolve conventional organic structural problems have also been used in attacking the structural problems in polymers; special methods of application and additions to the theory of interpretation have often been necessary for polymeric materials, however.

Elemental Analysis (34). Elemental analysis is usually of limited value in homopolymerization, where the method of synthesis leaves few questions as to the gross chemical composition. However, in copolymerization, analysis for an element present in only one of the comonomers is the most direct method of estimating the amount of that comonomer incorporated. The elemental identification of labeling end groups has been mentioned as a method of number-average molecular-weight determination, as was functional-group analysis, which may be accomplished by elemental methods in some cases. See CHEMICAL ANALYSIS.

Polymer Degradation. Polymer degradation (qv) via pyrolysis to identifiable fragments was one of the basic steps in elucidating the structure of natural rubber, for example, when the long-chain, covalently linked nature of polymers was only imperfectly understood. Since then, pyrolytic and chemical degradations have become useful structural tools as well as valuable research topics in their own right (29b,35,36). The hydrolysis of polymers having condensation-type structures to identifiable products is well recognized; for example, the hydrolysis of the polymer from the sodium-catalyzed polymerization of acrylamide gave β-alanine, proving that the material was the polyamide nylon-3 (37).

The latter is an example of a degradative reaction on a polymer. The important problem of backbone stereochemistry has been attacked by reactions that do not operate to cleave the chain. For example, hydrogenation of amorphous poly-*o*-methoxystyrene to the crystalline poly-*o*-methoxycyclohexane allowed the backbone conformation in the unit cell to be established by x ray (38). Reaction-rate differences in hydrolysis of the stereochemically different polyacrylates (39) have aided in their structural assignments.

Gas chromatography has been of great value in identifying the products of degradative processes, and in identifying the polymer itself (29b,45) (see CHROMATOGRAPHY). Thermogravimetric analysis (qv), wherein the weight loss of a polymer in a controlled atmosphere is continuously recorded throughout a programed temperature rise (eg, 150°C/hr), has become a valuable method of characterization of thermal performance of a polymer (46).

Spectroscopy and Mass Spectrometry. All methods of spectroscopy have found extensive application in polymer characterization. The use of infrared for crystallinity estimates has been mentioned. Orientation can also be estimated (40) as well as the customary range of structural features (5,41,42). See INFRARED-ABSORPTION SPECTROSCOPY.

Ultraviolet absorption is less specific for structural assignment than infrared; however, it has been of value in detecting and estimating added materials (stabilizers, accelerators, etc) and degradation products, for example, as well as aiding in structure determination in some cases, eg, lignin (5). See ULTRAVIOLET-ABSORPTION SPECTROSCOPY.

Nuclear magnetic resonance (NMR) has had particular application in two areas: the study of microtacticity, ie, the detailed stereochemical configuration along the polymer chain, where it is in some instances (eg, polymethyl methacrylate) the method of choice (3,29c,43); and the study of molecular motion and chain flexibility (42a,44). NMR has also been used to determine polymer crystallinity. The related phenomenon of electron-spin resonance has been used to detect long-lived free radicals trapped on a polymer backbone or at a chain end as a result of chain immobilization; such radicals usually constitute an extremely small, but chemically very significant, part of the polymer structure (5).

The problem of microtacticity noted above has become more important as more polymers having a controlled stereochemical sequence in the backbone are synthesized. Not only is this a fundamentally important area of polymer structure, but mechanical properties and transition behavior are influenced markedly by microtacticity. Besides

NMR, other methods are also applicable to the problem, such as x-ray diffraction, diffraction based on melting point and extent of crystallinity, dynamic mechanical testing, measurement of solution properties (eg, second virial coefficient and expansion factor differences), dipole-moment studies, and flow birefringence (3). NMR has so far offered the most penetrating analyses of the configuration of asymmetrically substituted carbons from one monomer unit to the next in a way not dependent on crystallinity, as is the x-ray method, for example. See also NUCLEAR MAGNETIC RESONANCE.

X-Ray Diffraction. X-ray diffraction was discussed earlier as a basic method for determining the presence and amount of crystallinity and orientation. A detailed crystallographic study of a number of partially crystalline polymers has established the relative configuration of the asymmetric carbon atoms in the chemical repeat unit and the helical conformation of chains in the crystal lattice. X-ray studies have been generally an extremely important factor in defining the spatial relationships in polymeric systems (3,30,42b). See also X-RAY DIFFRACTION.

Optical Methods. Standard optical methods can be applied to structural characterization also (3,5). Optical rotation has been observed in protein molecules and is attributed mainly to the dissymmetry arising from helical configuration. Certain synthetic polymers have been prepared having distinguishable optical activity (48). The refractive index of a polymer can aid in its identification when compared with samples of known materials, and is an important parameter in the optical application of plastics (49). Birefringence, which is a measure of the difference in refractive index in two directions in an anisotropic substance, can give important information on the dissymmetry induced by strains from drawing operations, molding, or extrusion. Stress analysis in loaded transparent plastics can be carried out by birefringence measurements also (5). A related phenomenon, flow birefringence (qv), depends on a measure of birefringence in polymer solutions under conditions of shear; such dissymmetry arises from orientation of the macromolecules. Information on molecular size and rigidity can be derived. The later section on Optical Tests considers the optical properties of polymers in relation to end uses.

Physical Properties

An arbitrary division of the multitude of physical properties of interest in a polymeric species is possible on the basis of: (a) those that serve to categorize a material broadly as, for example, high or low melting or softening; easily, difficultly, or not soluble; moldable, curable, castable; tough or brittle; and (b) those that define more completely, and often for specific circumstances, the physical performance of a material in terms of mechanical properties, rheological behavior, electrical character, environmental resistance, etc. The two divisions must inevitably overlap. The distinction is intended to point out certain physical parameters that usually must be examined to establish that a new polymer is essentially a plastic- or fiber-like solid or an elastomer, under what conditions and over what temperature range these properties are evident, and, further, how the material might reasonably be handled and worked. This limited level of information may be entirely adequate for many purposes, regardless of what specifically is learned. To many workers, however, the first-mentioned category of physical properties will dictate that further study should be abandoned because of obvious inadequacy for certain purposes (eg, textile fibers or chemically resistant elastomers), or should be pursued in greater detail because of promising initial indications. For some workers, what is considered here a second-category property will be routinely in the first category; others may not be appropriate in either category; and some physical-property tests highly specific to certain end uses will not be considered at all. In summary, the first category properties will be appropriate for characterizing most new polymers; the second category properties represent a more specialized characterization. It should be clear that the two divisions are not intended to represent "basic" or "applied" tests in any sense. One of the real problems in polymer characterization is relating tests of any kind to the end-use performance. Even the most "applied" tests may leave much to be desired in predictive value; usually only combinations of tests and properties, preferably coupled with the insight of experience, can be satisfactorily predictive of ultimate performance.

Crystalline Melting Point and Glass-Transition Temperature. The most important basic parameter of an amorphous polymer is its glass-transition temperature (T_g), for it will determine whether the material will be a hard solid or an elastomer in conventional use temperature ranges, and at what temperature the behavior pattern changes (6,50). T_g is also very important in partially crystalline polymers since it frequently gives an indication of an embrittlement temperature. T_g is related to the onset of segmental motion in the amorphous regions of polymers. Its measurement is often more difficult and the results less clear than is the measurement of crystalline melting point. The measurement is clouded in many instances by a dependency on the time scale of the experiment. The glass-transition temperature, less often called second-order transition temperature, is not generally regarded as a thermodynamic property, though recent studies suggest that it may be a first-order, or thermodynamic, property (51).

A number of methods are known for determining T_g. Many depend on a measurement of a change in some physical parameter, such as specific volume, coefficient of expansion, heat capacity, dielectric constant, or dynamic modulus, as a function of temperature (12). A change in slope of the plot of the property measured as a function of temperature is taken as T_g. Differential thermal analysis is often applicable to T_g measurements (3,52). Among some other methods for T_g is the observation of fiber spiralization under relaxation of stress as the temperature is raised (53), and the measurement of β-absorption as a function of temperature (54). A direct calculation of T_g from a knowledge of chemical structure, the molar cohesion, and numerical values arbitrarily assigned on the basis of the rotatability of groups or atoms has been attempted (55).

The crystalline melting point, T_m, in partially crystalline polymers is the temperature at which the last traces of crystallinity disappear under equilibrium conditions. The simplest laboratory methods for its determination are by means of a polarizing microscope equipped with a heating stage and with the changes noted visually (11); or by passing a light beam through the sample and measuring intensity change (56); or by differential thermal analysis (3,29a). More elaborate dilatometric methods (57) have been used, as well as the disappearance of an x-ray diffraction pattern as the temperature is raised. The time scale of the experiment is important in T_m as well as in T_g determinations; however, practically useful and satisfactory values can be obtained and reproduced at a temperature rise of 1°/min if the sample has been annealed for about 30 min at about 20°C below the expected T_m. Crystallite size can also affect T_m, small crystallites giving lower values than large crystallites. Different methods of measuring T_m give different values because of variations in their sensitivity and because of the T_m range in crystallites differing not only in size but in perfection.

Structural factors that affect crystallinity and T_m, such as chain stiffness, polarity, and symmetry, affect T_g also. Although there is a theoretical basis (58) for expecting the magnitude of the affect on T_m and T_g to be proportionately the same, a useful empirical relation exists for T_m and T_g in partially crystalline polymers: $T_g = \frac{2}{3}T_m$ for nonsymmetrically substituted (60), where temperatures are in degrees absolute.

For most partially crystalline polymers, the practical upper use temperature limit will be well below T_m; the lowest practical processing temperature, however, may exceed T_m only slightly (eg, the nylon molding resins). See also TRANSITION TEMPERATURES.

Other Softening Methods. A number of purely pragmatic and somewhat subjective methods exist for determining the softening temperatures of crystalline and amorphous polymers. For the former, some of the methods can give a value which is a rough approximation of the crystalline melting point, and for the latter, the glass-transition temperature. The first methods considered here are for bulk polymer as a powder, film, or plug; the Vicat softening and deflection temperature require molded bars. The following methods have particular value where polymers of a generally similar kind are compared routinely.

Polymer melt temperature is defined as the temperature at which a polymer sample advanced across a temperature gradient bar begins to leave a molten trail (11). The method is rapid and reasonably reproducible, and can be done acceptably, though with greater consumption of time, on an ordinary melting-point block. Capillary melting points in the fashion of simple organic compounds are generally not satisfactory because of the high viscosity of molten polymers and the difficulty therefrom in clearly detecting the often not very abrupt changes that take place as softening or melting occur.

The *Vicat softening-point test* (42c,61) determines the temperature at which a 1-mm^2 flat-ended needle under a load of 1000 g penetrates a sample bar 1 mm as the temperature is raised 50°C/hr. It is similar in form and apparatus to the deflection-temperature measurement (62) which is taken as the temperature at which a standard size bar is caused to deflect 1 mm as the temperature is raised 2°C/min at a fiber stress on the sample of 264 or 66 psi. Both methods are applicable primarily to rigid plastic materials and are most suited for control and developmental work.

Softening points corresponding closely to glass-transition temperatures can be determined by a *light-intensity method* similar to that noted for the determination of crystalline melting point (56).

The *Durrans mercury method* (11) is most often encountered in characterizing thermoplastic epoxy resins. It determines the temperature at which a plug of the sample will no longer support the weight of a layer of mercury and the latter penetrates the resin.

The softening point of thin films (0.001–0.060 in.) can be determined by the tensile heat-distortion temperature (63,64); in this test a sample of film is loaded with a tensile stress of 50 psi and is at the same time subjected to a temperature increase of 2°C/min. The temperature at which an elongation or shrinkage of 2% has occurred is taken as the tensile heat-distortion temperature. Shrinkage occurs when the film is oriented and subsequently relaxes on heating. The test is limited to materials with a tensile modulus of at least 10,000 psi.

Fabricability and Qualitative Mechanical Properties. Conversion of a gummy, powdery, or solid chunk polymer to a shaped structure may often be done very simply in the laboratory, giving much qualitative information about its general character and value for future investigation. The simplest method is to dissolve the polymer in a solvent to make a fairly viscous dope (usually 10–30% by wt) and then to cast it as a thin layer on a glass plate, merely by means of a glass rod in the crudest case. Thorough removal of the solvent by slow evaporation to avoid bubble formation will usually give a film sample from which such factors as toughness, stiffness, orientability, crystallizability, clarity, and others can be assessed qualitatively; the effect of post-treating operations and environmental exposure can also be judged. By operating in a controlled and refined manner, small samples suitable for quantitative measurement can also be made (11). Proper selection of a solvent (providing more than one can be found that dissolves the polymer) can affect film properties. For high-boiling solvents or heat-sensitive polymers where evaporation under heat and vacuum are not practical, it may be necessary to wet-cast a film by immersing the freshly cast solution in a liquid which is a nonsolvent for the polymer but is miscible with the polymer solvent. Such a technique is more difficult to execute. For non-self-supporting films, the polymer can be cast onto a supporting surface simulating an end-use substrate, such as a metal plate, and subsequently baked dry; such a process may also involve curing when the test polymer is ultimately to be crosslinked. Flexibility, hardness, color, etc can then be evaluated for the coating. An elastomeric material can be cast to a self-supporting film if provision is made to bring about crosslinking; for example, a urethan prepolymer, rich in isocyanate groups, can be cast from methylene chloride and exposed to a moist atmosphere to give a nontacky, strong, elastomeric film in which crosslinking has occurred through water–isocyanate reaction.

Essentially the same solution techniques described above are applicable to spinning fibers (11). Here, the polymer solution is extruded from a hand or motor-driven syringe into heated air or a nonsolvent for the polymer. The method is more difficult for fibers than films and may require at least modest equipment for the extrusion, solvent removal, and takeup of the fiber.

For polymers that are stable even for a short time in the molten state, and when a simple laboratory press is available, it is very convenient to press films, mold chips or bars, and extrude fibers (11). A polymer may simply be pressed above its melt temperature between sheets of aluminum foil to give a circular film of uneven thickness; by using metal plates and shims to shape the film and control the thickness to a uniform and desired level, it is possible to prepare a film or sheet from which satisfactory tensile-strength test specimens may be cut, for example. Fibers can be melt-extruded from relatively simple glass apparatus in the laboratory, but for a hand evaluation it may suffice to draw out filaments manually from molten polymer in a test tube by means of a glass rod (11). Elastomers are most often compounded with a variety of fillers, curing agents, etc, by blending the hot mass on a two-roll mill to a rough sheet which can then be "polished" by hot pressing and curing as described above for films. Such polished sheets of cured elastomer or plasticized poly(vinyl chloride) can be used for test specimens.

Solubility. The solubility of a polymer can be a deciding factor in the kinds of end uses to which it can be put, and, more directly, finding a solvent may be essential to its fabrication. As was discussed under Molecular Weight, most molecular-weight methods necessitate a solvent for the polymer. Finding a suitable solvent can be accomplished to some extent on a systematic basis through the solubility parameter, by which an estimate of this value for the structure of the polymer permits solvents with similar values to be tested (6). Recourse to trial and error is usually necessary, however, although there are some well-recognized polymer class–solvent class relationships (eg, cresols and formic acid for polyamides, and ketones for vinyl chloride resins). Solubility categories have been useful for comparing the interactions of different polymers and solvents at different temperatures in a qualitative manner (11). Chemical and solvent sensitivity are important in polymer usage and can be considered along with outright solubility. A common method is to simply immerse strips or bars of polymer in a solvent or chemical solution at room and/or elevated temperature and note the gain or loss in weight as solvent absorption, polymer dissolution, or chemical attack occurs. The solvents or chemicals tested will depend on anticipated uses (eg, gasoline and oil for automotive rubber uses, dry cleaning solvents and laundry soap solutions for fibers, animal fats and vegetable oils for plastics in food-packing uses). Mechanical property changes which occur as a result of such exposure are also often noted.

Density. The density of a polymer may in some cases be critical to the ultimate end-use potential since a polymer with a low density will, obviously, occupy more volume or cover more surface per pound than one of higher density. In any event, density can add to the basic understanding of the material in every instance. For most materials, it is convenient to prepare a solid piece of the polymer by either melt or solution casting methods and to place it in a density-gradient tube (64). The density of molten polymers can be established by dilatometric and other means (65).

The physical properties discussed so far would mainly be regarded in the first category of properties outlined in the introduction to this section (on Physical Properties); those that follow are in the second category, for the most part. They are potentially very great in number; in principle, they extend to the final end-use evaluation, such as observing the performance of plastic milk bottles at the consumer level. The following treatment cannot extend over the entire ground. It will endeavor to bring out some of the more important, broad categories of testing that are beyond the first category, as it was defined, but necessary before the extreme hypothetical example cited above can come about.

Mechanical Properties (12,13,16,17). The mechanical properties of a polymer are the facets of behavior that are evident when the polymer is subjected in some form to a mechanical stress. The latter, for example, may be stresses of the classically defined types, (tension, shear, torsion, and compression) or the less well understood impact stress. The rate of application of stress may vary over a broad range, and may be applied in a continuous or periodic manner. It is from the mechanical properties that end uses become apparent and limitations recognized.

The precise determination of mechanical properties requires that samples used be free of defects or flaws, such as voids, cracks, nicks, and other abnormalities that could act, for example, as stress concentrators. Because these factors are hard to eliminate under the best of circumstances, a number of samples are usually tested and averaged, the number depending on the kind of sample, the test involved, and the confidence level desired. Variations in temperature and humidity can also affect results; samples are therefore customarily conditioned for 24 hr and tested at a standardized temperature/humidity; this is especially important for forms with a large surface area, such as fibers and films. Sample size can also influence results considerably in some tests, and standards have been set for most.

One of the most important mechanical characterizations of a polymer is its tensile stress–strain properties. Here a specimen of some standard shape and dimensions is subjected to a stress applied at some determined rate. It is desirable that the testing instrument be connected to a recorder so that a plot of the force required to produce a given elongation is produced; the stress and strain only at specimen failure are determined otherwise. The stress–strain plot shows at once the presence or absence of a yield point, for example, and the slope of the initial portion of the curve can be used to calculate an elastic modulus in tension, which is related to the stiffness and resistance to deformation. The overall shape of the curve reveals whether the material is basically *brittle* in tension (the curve is characterized by absence of a yield point, relatively low elongation at break, and often a steep initial portion), or *ductile* (the curve is characterized by a yield point and relatively high elongation at break). The yield point refers to that event in the application of stress where the latter remains constant or decreases while elongation (strain) continues to increase. Beyond the yield point, leveling off of the stress will often occur, followed by an increase in stress until the sample fails. In metals, this is sometimes called strain hardening, and in polymers represents the onset of orientation in the specimen.

Because of the fundamentally viscoelastic nature of polymeric materials (6), the *stress–strain properties* and other aspects of mechanical behavior are often very strongly influenced by the rate of application of stress and the temperature of the test. Thus, a polymer having a tough, ductile character as reflected in its stress–strain curve at a slow rate of loading, say, 0.02 in./min, may show a considerably more brittle type of failure at a high rate of loading, possibly 50 in./min. Similarly, the brittle material at 25°C may display typically ductile behavior at 75°C. These interrelationships are unified in the time–temperature superposition principle (66).

An underlying molecular conception of the principles of mechanical behavior is fairly well in hand for amorphous polymers; the theories encounter considerably more difficulty in developing a coherent picture for partially crystalline materials. Nevertheless, certain generalizations can be made on mechanical properties as a function of the crystalline-amorphous character. For example, amorphous polymers below the glass-transition temperature T_g are generally hard and brittle, and have a low elongation, high modulus of elasticity high tensile strength, and low impact strength. Examples at room temperature are polystyrene and poly(methyl methacrylate). Amorphous polymers above T_g often possess the reverse of these properties, as, for example, all commonly used elastomers at room temperature, and the previously mentioned polymers at around 80°C. Crystalline polymers show mechanical properties which are a function both of the amount of crystallinity present and of the location of T_g relative to the use temperature. A crystalline polymer well above its T_g generally has better impact strength than when at or below T_g. However, the greater the amount of crystalline content, the lower is the impact strength. Increased crystallinity (among samples of the same polymer) tends generally to give a higher modulus, lower elongation, higher tensile strength, and greater resistance to creep. Any added material which acts to depress the glass transition or disrupt crystallinity will shift mechanical properties as indicated by these effects, if other factors such as compatibility are satisfied; plasticizers in poly(vinyl chloride), for example, convert this hard, stiff material to one with elastomeric character by depression of T_g.

Superimposed on considerations of crystallinity or its absence is the question of molecular orientation. Fibers and films can be highly oriented (in the case of films, in one or both directions) to vastly alter strength, elongation, modulus, impact, and other values. Fibers are almost always used in an oriented state; some films must be oriented to develop useful properties (eg, from poly(ethylene terephthalate)), others find separate and important uses in their unoriented and oriented states (eg, polypropylene), and some are rarely used oriented (eg, polyethylene). Molded thermoplastics may undergo purposeful or accidental orientation under the high shear conditions which may be encountered; however, this involves orientation only at the surface of the finished structure usually, and the effects are less felt the thicker the molded piece. Elastomers are incapable of sustaining molecular orientation, except as stress is applied, as long as they are above their T_g. Nonuniformly oriented polymeric materials have disparate directional properties; generally, they show greater tensile strength, lower elongation, and higher modulus in the direction of orientation than transverse to it; impact strength is higher normal to the direction of orientation than parallel to it.

Two closely related mechanical properties are *creep* and *stress relaxation*. The former generally has more practical importance and is the more widely studied of the two; the latter is not without its practical aspects and is usually more valuable in interpreting the basic viscoelastic behavior of polymers than creep. Creep involves the elongation or strain induced in a material under the application of a constant stress (changing strain at constant stress); stress-relaxation studies measure the decrease in stress occurring when a specimen is held at a constant, essentially instantaneously induced, strain (changing stress at constant strain). The changing strain or stress is measured as a function of time. The practical importance of creep is readily evident. Polymeric materials that readily elongate (creep) under mild but steady stresses would be quite useless for certain applications. Creep becomes more important at increasing stresses and temperatures.

Associated with creep testing is the general area of deformation under nondestructive loads for varying periods; the amount of permanent deformation in compression is an important part of elastomer behavior, where nearly complete recovery from deformation is desired in many uses. Stress relaxation is often used in determining what is, in effect, the tendency for unrecoverable deformation in elastic fibers where compression deformation testing is not so practical as in bulkier forms of rubber such as tire tread. Creep can be measured in tension, shear, or compression; the first is far more commonly performed, and can be accomplished by simply attaching weights to a suspended sample and measuring the distance between gage marks by a cathetometer or possibly a ruler. For very small deformations, strain gages are necessary. Stress relaxation is difficult for rigid materials because the apparatus used to maintain a constant deformation must not itself deform, or corrections for this must be possible. For this reason, stress relaxation is more commonly used for low-modulus polymers, such as elastomers, where the apparatus can be relatively simple.

Impact testing is widely performed and little understood both in the practical merit of the tests in predicting behavior and in theory of impact failure. Nevertheless, impact-testing results have value as an empirical basis for comparison and control. Two common test methods are the Izod and Charpy (67); both employ the contact of a weighted pendulum with a bar of sample. In the Izod method, the bar is supported vertically as a cantilever beam; in the Charpy method it is clamped at both ends as a simple beam. The specimen is notched in the Izod method to amplify the effect of a point of stress concentration; notching is optional in the Charpy. A dart-drop test has been devised for polyethylene film (68), and a falling-weight test for thick sheet (17). The latter is said to give more meaningful and predictive impact information. The high-speed application of tensile stress is also used to study the effect of impact, as is repeated impacting at high speeds (69).

Dynamic mechanical testing of polymers has grown in importance as a method for fundamental characterization and information related to practical properties. The method depends basically on the periodic application of a stress and the measurement of the response of the material. In a simple application of the concept, a fairly thin, rectangular sample is fixed at one end while the other is attached to a bar of known moment of inertia; the latter is set in oscillation and the sample is subject to a twisting action, restored to its original position, then twisted in the opposite direction. The shear modulus and mechanical damping of the sample can be calculated from the frequency of oscillation and a determination of the amplitude of each cycle in the kind of test described. Frequency and temperature can both be varied. Dynamic mechanical properties can give valuable information on glass-transition temperature, crystalline character, crosslinking, and components of copolymers and blends. Damping has practical significance as an indication of the amount of energy converted to heat in viscoelastic materials.

A great many specific mechanical property tests have been devised but have not been treated here, even briefly; among these are hardness, friction, abrasion resistance, and fatigue (12). See also MECHANICAL PROPERTIES.

Flow Properties and Processing Characteristics (9,10,70). Virtually all polymer materials are ultimately brought to final, useful form through some technique that depends on molten or solution behavior. It becomes necessary, therefore, to characterize the flow properties of polymers.

Since the greatest number of specific polymers, and the greatest poundage volume of them also, are processed as melts, characterization of the melt flow is most often required. The equivalent of Hooke's law for deformation of solids is paralleled in liquids by Newtonian treatment of deformation, in which shear stress, τ, is proportional to shear rate, γ, and the proportionality constant is the viscosity, η. It has been found that most polymeric melts and solutions are non-Newtonian over a shear-rate range that covers many practical operating situations. In this realm, shear stress is no longer proportional to shear rate and the viscosity at any combination of shear stress and rate is therefore only an "apparent" viscosity, valid at that combination only. Because the processability of melts (in terms of speed of operation, appearance of the final product, and power requirements) is to so great an extent a function of the viscosity, it is often essential to determine apparent viscosities and shear stresses at a variety of shear rates. This is conveniently done in a capillary viscometer in which the shear rate can be varied (28). The quantities mentioned and other parameters of polymer melts can be readily measured; under certain conditions the molecular weight of a polymer can also be derived (26). A simpler instrument lacking this ability is the melt index or extrusion plastometer (25), which customarily measures only one combination of shear stress and shear rate conditions for a given polymer type. The latter is very useful in control and comparison, and is therefore widely used. Other types of viscometers, such as the rotational type, often involving a cone and plate or concentric cylinders, are used (27). The capillary methods, employing a range of shear rates and temperatures, permit direct conclusions on molding and extrusion behavior by defining conditions where, for example, nonuniformities in extrudate surfaces occur.

In elastomer technology, the Mooney viscosity is the most used standard plasticity test, and involves the measurement of torque developed by a rotor turning in a mass of unvulcanized rubber (15). The method can be employed to determine cure behavior also. A related procedure and instrument, the Brabender Plasticorder (10), is often used for determining processing characteristics of elastomers and plastics, eg, the fluxing behavior of plasticized poly(vinyl chloride). For gaging the flow properties of thermoplastics under actual injection-molding conditions, a spiral mold cavity has been used; the depth of flow into the spiral as a function of cylinder and mold temperature and injection pressure can be easily determined (71).

Aging Tests (13,15,16). Most polymeric materials will be exposed either continuously or in part to environmental circumstances that will lead to degradation and failure (eg, sunlight, atmospheric oxygen or ozone, moisture, temperature extremes, solvents, high-energy radiation, and corrosive chemicals). Many of these will occur while the polymer is also under stress, and the combination may be decidedly more destructive than either alone; the stress-cracking propensity of polyethylene is an example. Environmental failure may occur very slowly (many years) or quite rapidly (days or weeks). Accelerated aging tests are therefore necessary to predict the long-term behavior of a new material and to facilitate the development of chemical modifications in the polymer, added stabilizers, or processing conditions that can retard aging. Much difficulty is encountered in developing accelerated tests that adequately duplicate the environment under test, but for most commercially established polymers acceptable test procedures have been developed, and these can serve in many cases either directly or as a starting point for aging new materials. The single most common environment leading to degradation involves oxygen and ordinary sunlight; in the latter, the ultraviolet content is the most harmful part of the spectrum. Accelerated tests in the laboratory generally use a carbon arc, xenon lamp, or fluorescent light as a source, often in combination with a water spray. Although these are worthwhile comparison methods, they are usually followed by actual outdoor exposure in Florida or Arizona for maximum sunlight in the shortest time, though such tests may take a year or more. Use at high temperature accentuates the oxidation problem and polymers destined for such use are naturally screened under this added condition.

Electrical Tests (13,15,16,72,73). Many polymeric materials have been very successful in electrical applications because of one or more outstanding properties, such as dielectric strength, volume and surface resistivity, dielectric constant, arc resistance, loss factor, and dissipation factor. Electrical properties are not a function of molecular weight in high polymers. Crystallinity can affect the dielectric constant, however, through the differences in density between crystalline and amorphous regions. Electrical properties are influenced more by dipole asymmetry than by the presence of polar groups as such. Polytetrafluoroethylene, for example, has a lower dissipation factor than poly(vinyl fluoride); the latter possesses a greater dipole asymmetry than the former. When the dipoles are able to respond readily to changes in the electric field (low frequency or high temperature) the dielectric constant is high. At high and low frequencies, the dipoles can either respond completely or not at all to the change in field, and the loss factor, the product of dielectric constant and power factor, is low. The power factor is the sine of the angle of phase difference due to the delay in movement of the dipole with the change in field. It is possible to draw some analogies between electrical and mechanical properties (12). See also ELECTRICAL PROPERTIES.

Optical Tests (13,16). The use of optical methods to aid in identifying a polymer or its molecular structure was considered in the section on Optical Methods. Here, testing of optical properties for practical end-use application is involved. The clarity and color of plastics become important in some film and sheeting uses, for esthetic as well as practical reasons (74); but in critical optical uses, such as lenses in eyeglasses or precision instruments, their entire ability to perform satisfactorily may hinge on small differences in inherent optical properties or those due to foreign contaminants (eg, catalyst residues, stabilizers), degradation products from processing, or the like. Index of refraction was mentioned under Molecular Structure (49) and is important in determining optical applications. The appearance and covering power of fibers are practical manifestations of their optical properties.

Permeability. Permeability (qv) to moisture, oxygen, and carbon dioxide are usually of importance in categorizing a polymeric film for certain broad end uses, where the contents to be protected suffer from contact with these agents, or a failure to "breathe" moisture, for example, which is originally held by the contents. This can be especially important in the packaging of food. Permeability to other gases may be important for special purposes, but these are less often determined. The chemical nature of a polymer, its density, and its crystallinity, all of which are interrelated, are reflected in the extent and type of gas permeability (75).

Flammability. Flammability as a limitation depends on the end uses envisioned. Synthetic apparel and carpet fibers clearly must not be notably more easily ignited or rapid-burning than their natural counterparts (16), nor can polymers used in basic construction be substantially more combustible than wood. In the latter instance, the products of combustion must also be considered. For many purposes, polymeric materials are used in lieu of normally combustible materials because of their inherently superior mechanical properties, and where flammability is relatively the same as the prior material the issue is not important; for example, plastic films are widely used in place of paper for consumer and industrial packaging. Flammability is often categorized in terms of (a) flammable, ie, the sample will support combustion when an igniting flame has been removed, (b) self-extinguishing, ie, the sample will not continue to burn on removal of the flame, and (c) nonflammable, ie, the sample does not ignite in a flame, although it may char and decompose (76). Flammability tests take cognizance not only of these categories, but also rate of burning, nature of the residue, if any, and burning character, eg, some polymers may be self-extinguishing by a melting and dripping away of the burning portion.

Biological and Toxicological Properties. Most polymers are inert and nontoxic, but residual monomers, degradation products, stabilizers, or plasticizers may not be; thus, the polymer product as it is used must be tested for its safety in certain applications, particularly in food contact uses (77). Polymers are sometimes subject to attack by fungus and mold, for example, on painted surfaces, and, again, additives such as plasticizers may be attacked, if not the polymer itself.

Bibliography

1. F. W. Billmeyer, *Textbook of Polymer Science*, Interscience Publishers, a division of John Wiley & Sons, Inc., New York, 1962.
   Virtually all phases of polymer science treated in at least their essentials; a basic source for the fundamentals of most subjects before recourse to specific literature sources.
2. R. V. Bonnar, M. Dimbat, and F. H. Stross, *Number Average Molecular Weights*, Interscience Publishers, Inc., New York, 1958.
   Detailed discussion of theory, methods, limitations, and interpretation of cryoscopy, ebullioscopy, osmometry, end-group determination, and vapor-pressure lowering methods as applied to molecular-weight determination.
3. B. Ke, ed., *Newer Methods of Polymer Characterization*, Vol. 6 in *Polymer Reviews* Series, Interscience Publishers, a division of John Wiley & Sons, Inc., New York, 1964.
   Provides brief theoretical, historical, and experimental background necessary for a description of latest advances in, and current status of, the characterization methods covered.
4. P. W. Allen, ed., *Techniques of Polymer Characterization*, Butterworths Publications Ltd., London, 1959.
   Devoted mainly to molecular weight determinations; one chapter concerned with characterization of block and graft copolymers.
5. G. M. Kline, ed., *Analytical Chemistry of Polymers*, Vol. 12 of *High Polymers* Series, Interscience Publishers, a division of John Wiley & Sons, Inc., New York. In 3 parts: I: *Analysis of Monomers and Polymeric Materials. Plastics—Resins—Rubbers—Fibers*, 1959; II: *Analysis of Molecular Structure and Chemical Groups*, 1962; III: *Identification Procedures and Chemical Analysis*, 1962.
   Invaluable information on chemical- and physical-property characterization.
6. A. Tobolsky, *Properties and Structure in Polymers*, John Wiley & Sons, Inc., New York, 1960.
   Delineates principles underlying mechanical properties of polymers in terms of molecular structure and the familiar concepts of polymer morphology; highly sophisticated in its mathematics.
7. L. Mandelkern, *Crystallization of Polymers*, McGraw-Hill Book Co., Inc., New York, 1964.
   Review of modern thought in polymer crystallinity. Ref. 30 (P. H. Geil) should also be consulted for a treatment of polymer single crystals.
8. A. E. Lever and J. Rhys, *The Properties and Testing of Plastics Materials*, Chemical Publishing Co., Inc., New York, 1958.
   Source of references to methods of plastics testing, with brief descriptions of methods provided; a section on tests for specific commercial polymers.
9. E. T. Severs, *Rheology of Polymers*, Reinhold Publishing Corp., New York, 1962.
   Principles of flow in polymers presented in nonmathematical, completely descriptive terms; considers melts, emulsions, plastisols, etc. Generally informative, but not for basic rheological practitioner.
10. J. R. Van Wazer, J. W. Lyons, K. Y. Kim, and R. E. Colwell, *Viscosity and Flow Measurement—A Laboratory Handbook of Rheology*, Interscience Publishers, a division of John Wiley & Sons, Inc., New York, 1963.
    Practice of rheology in great detail, with descriptions of most of the instruments used in all phases of the subject; also contains theory.
11. W. R. Sorenson and T. W. Campbell, *Preparative Methods of Polymer Chemistry*, Interscience Publishers, Inc., 1961.
    Chap. 2 ("Preparation, Fabrication, and Characterization of Polymers") is valuable for the laboratory-level operations.
12. L. E. Nielsen, *Mechanical Properties of Polymers*, Reinhold Publishing Corp., New York, 1962.
    Excellent practical and experimental insights into mechanical properties; high level of theoretical treatment; coverage broad; prime source of direct information and leading references for any phase of mechanical behavior.
13. *Standards on Specifications, Methods of Testing, Nomenclature, Definitions*, American Society for Testing and Materials, Philadelphia. 12th ed., March 1961: Part 9, *Plastics;* Part 10, *Fibers and Textiles;* Part 11, *Rubber.*
14. E. Müller, ed., *Methoden der organischen Chemie* (*Houben-Weyl*), Part II, Vol. 14: *Makromoleculare Stoffe*, Georg Thieme Verlag, Stuttgart, 1963.

Contains a 110-page listing of publications on determination of polymer properties according to type of analysis used.

15. A. E. Juve, in M. Morton, ed., *Introduction to Rubber Technology*, Reinhold Publishing Corp., New York, 1959.
    Survey of types of physical tests carried out on rubbers, experimental methods used, and significance of results.
16. W. E. Morton and J. W. S. Hearle, *Physical Properties of Textile Fibers*, The Textile Institute and Butterworths & Co. Ltd., 1962.
    Covers virtually all phases of fiber properties, theory, and experimental methods.
17. P. I. Vincent, *Plastics* **26** (1961): p. 121, Oct.; p. 141, Nov.; *Ibid.* **27** (1962): p. 115, Jan.; p. 117, Feb.; p. 116, April; p. 133, May; p. 136, June; p. 110, July; p. 105, Aug.; *Ibid.* **28** (1963): p. 109, Feb.; p. 107, March; p. 120, April; p. 95, May; p. 109, Sept.; p. 104, Dec.
    A series, "Strength of Plastics"; a short course in the basic factors affecting mechanical properties of plastics, methods of testing for them, and behavior of specific commercial plastics.
18. C. H. Adams, R. J. Bourke, G. B. Jackson, and J. R. Taylor, *Polymer Evaluation Handbook*, Armed Service Technical Information Agency (ASTIA) Document No. AD 110557; Wright Air Development Center Report 56-399.
    Descriptions of laboratory characterizations.
19. K. Stacey, *Light Scattering in Physical Chemistry*, Academic Press, Inc., New York, 1956.
20. M. M. Fishman, *Light Scattering by Colloidal Systems*, Technical Service Laboratories, River Edge, N.J., 1957.
21. R. L. Baldwin and K. E. Van Holde, *Fortschr. Hochpolymer.-Forsch.* **1**, 451 (1960).
22. H. K. Schachman, *Ultracentrifugation in Biochemistry*, Academic Press, Inc., New York, 1959.
23. "IUPAC Report," *J. Polymer Sci.* **8**, 257 (1952).
24. G. Meyerhoff, *Fortschr. Hochpolymer.-Forsch.* **3**, 59 (1961).
25. ASTM D 1238-57T; A. Rudin and H. P. Schreiber, *SPE J.* **20**, 533 (1964).
26. E. B. Bagley, *J. Appl. Phys.* **31**, 1126 (1960).
27. H. J. Cantow, *Plastics Inst. (London), Trans. J.* **31**, 141 (1963).
28. G. Pezzin, *Atti Congr. Intern. Materie Plastiche* **14**, 80 (1962); E. H. Merz and R. E. Colwell, *ASTM (Am. Soc. Testing Mater.) Bull.* **232**, 63 (1958).
29. M. F. Vaughan and J. H. S. Green, in *Techniques of Polymer Science*, Society of Chemical Industry Monograph No. 17, Gordon & Breach Science Publishers, Inc., New York, 1963, p. 81.

29a. T. R. Manley, p. 175 in Ref. 29.
29b. A. Barlow, R. S. Lehrle, and J. C. Robb, p. 267 in Ref. 29.
29c. C. E. H. Bawn, p. 251 in Ref. 29.
29d. E. I. Zichy, p. 122 in Ref. 29.

30. P. H. Geil, *Polymer Single Crystals*, Vol. 5 in *Polymer Reviews* Series, Interscience Publishers, a division of John Wiley & Sons, Inc., New York, 1963.
31. R. P. Sheldon, *J. Polymer Sci.* [B] **1**, 655 (1963).
32. J. H. Griffith and B. G. Ranby, *J. Polymer Sci.* **44**, 369 (1960).
33. G. Natta, I. Pasquon, A. Zambelli, and G. Gatti, *Makromol. Chem.* **70**, 191 (1964).
34. A. Gruben and H. H. Leiner, *Mod. Plastics* **37**, 88 (July 1960).
35. N. Grassie, *Chemistry of High Polymer Degradation Processes*, Butterworths Publications Ltd., London, 1956.
36. *High Temperature Resistance and Thermal Degradation of Polymers*, Society of Chemical Industry Monograph No. 13, Page Bros. (Norwich) Ltd., Norwich, 1961.
37. A. S. Matlack (to Hercules Powder Co.), U.S. Pat. 2,672,480 (March 16, 1954).
38. G. Natta, G. Dall'Asta, G. Mazzanti, and A. Casale, *Makromol. Chem.* **58**, 217 (1962).
39. E. Gaetjens and H. Morawetz, *J. Am. Chem. Soc.* **83**, 1738 (1961).
40. J. A. Gailey, *Anal. Chem.* **33**, 1831 (1961).
41. A. Davis in H. A. Szymanski, ed., *Progress in Infrared Spectroscopy*, Vol. I, Plenum Press, New York, 1962.
42. W. J. Potts, in "International Symposium on Plastics Testing and Standardization," *Am. Soc. Testing Mater. Spec. Techn. Publ.* **247**, p. 225.

42a. W. P. Schlichter, p. 257 in Ref. 42.
42b. W. O. Statton, p. 242 in Ref. 42.
42c. C. E. Stephenson and A. H. Willbourn, p. 169 in Ref. 42.
43. F. A. Bovey and G. V. D. Tiers, *Fortsch. Hochpolymer.-Forsch.* **3,** 139 (1963).
44. W. P. Schlichter, *Fortsch. Hochpolymer.-Forsch.* **1,** 35 (1958).
45. K. Ettre and P. F. Varadi, *Anal. Chem.* **34,** 752 (1962).
46. D. A. Anderson and E. S. Freeman, *J. Polymer Sci.* **54,** 253 (1961).
47. F. Danusso, *Polymer* **3,** 423 (1962).
48. P. Pino, F. Ciardelli, and G. P. Lorenzi, *J. Am. Chem. Soc.* **85,** 3888 (1963).
49. ASTM D 542–50.
50. A. H. Willbourn, *Trans. Faraday Soc.* **54,** 717 (1958).
51. E. A. Dimarzio and J. H. Gibbs, *J. Polymer Sci.* [A] **1,** 1417 (1963).
52. S. Strella, *J. Appl. Polymer Sci.* **7,** 569 (1963).
53. R. P. Sheldon, *J. Appl. Polymer Sci.* **6,** S43 (1962).
54. R. Zanneth, P. Manaresi, and L. Baldi, *J. Polymer Sci.* **62,** S33 (1962).
55. R. A. Hayes, *J. Polymer Sci.* **5,** 318 (1961).
56. K. R. Dunham, J. Vandenberghe, J. W. H. Faber, and L. E. Contois, *J. Polymer Sci.* [A] **1,** 752 (1963).
57. P. J. Flory, L. Mandelkern, and H. K. Hall, Jr., *J. Am. Chem. Soc.* **73,** 2532 (1951).
58. J. R. Knox, *Am. Chem. Soc., Polymer Div. Preprints* **3** (1), 234 (1962).
59. M. Ezrin and G. C. Clever, *ibid.*, p. 308.
60. R. G. Beaman, *J. Polymer Sci.* **9,** 470 (1952).
61. ASTM D 1525-58T.
62. ASTM D 648-56.
63. D. J. H. Sandiford and K. A. Buckingham, *Brit. Plastics* **34** (11), 594 (1961); ASTM D 1637-59T.
64. C. H. Lindsley, *J. Polymer Sci.* **46,** 543 (1960); ASTM D 1505-60T.
65. P. R. Swan, *J. Polymer Sci.* **42,** 525 (1960); B. W. Terry and K. Yang, *SPE J.* **20,** 540 (1964).
66. J. D. Ferry, *Viscoelastic Properties of Polymers,* John Wiley & Sons, Inc., New York, 1961.
67. ASTM D 256-56.
68. ASTM D 1709-62T.
69. A. G. H. Dietz and F. R. Eirich, *High Speed Testing,* Vol. IV, Interscience Publishers, a division of John Wiley & Sons, Inc., New York, 1964; ASTM D 1822–61T; J. R. Heater and E. M. Lacey, *Mod. Plastics* **41** (9), 123 (1964).
70. A. B. Metzner in E. C. Bernhardt, ed., *Processing of Thermoplastic Materials,* Reinhold Publishing Corp., New York, 1959.
71. J. J. Gouza and G. G. Freygang, *SPE J.* **17,** 1211 (1961).
72. G. F. Kinney, *Engineering Properties and Applications of Plastics,* John Wiley & Sons, Inc., New York, 1957, p. 221.
73. A. H. Scott, *SPE J.* **18,** 1375 (1962).
74. J. Miles and A. E. Thornton, *Brit. Plastics* **35,** 26 (Jan. 1962).
75. ASTM D 1434-58; E-96-53T; see also C. J. Major and K. Kammermeyer, *Mod. Plastics* **39,** 135 (July 1962).
76. ASTM D 568-56T; D 635-56T; D 757-49; D 1433-58; D 1692-59T.
77. A. E. Molzon, *AD-276001,* Plastics Technical Evaluation Center, Picatinny Arsenal, Dover, N.J.; R. Burgess and A. E. Darby, *Brit. Plastics* **37** (1), 32 (1964).

W. R. Sorenson
Continental Oil Company

FRACTIONATION

Fractionation entails the orderly separation of a polymer into a number of fractions of different molecular weights. Thus, polymer samples with narrow molecular-weight distributions may be prepared for subsequent testing. In addition, determination of the amount and molecular weight of each provides the information necessary to compute the molecular-weight distribution of the polymer. The molecular-weight distribution affects many physical properties of polymers (see, for example, FRACTURE; MECHANICAL PROPERTIES; see also articles on individual polymers). Because of the importance of methods of determination of molecular-weight distribution, such methods, based on fractionation, are included in this article although fractions as such are not isolated. Thus, turbidimetric titrations, ultracentrifugation, and rheological techniques are discussed.

The principles of the classical techniques of fractional precipitation are discussed in this article, but no attempt is made to review exhaustively the several hundred applications. More detail is given to the newer methods of fractionation, column chromatography and gel-permeation chromatography. Less widely used methods, such as thermal diffusion, are discussed briefly.

Fractional Precipitation

Polymer fractionation is often carried out on the basis of the decreasing solubility of a polydisperse sample with increasing molecular weight. Recent reviews of the pertinent theory have been given by Tompa (1), by Voorn (2), and by Huggins and Okamoto (3). The Flory-Huggins treatment of polymer solutions (4–7) arrives at an expression for the partial molal free energy of mixing of a solvent with a polymer, ΔG_1. This quantity can be approximated by

$$\Delta G_1/RT = \ln (1 - v_2) + (1 - \overline{\mathrm{DP}}_n^{-1})v_2 + \chi_1 v_2^2 \quad (1)$$

where R is the gas constant, T the absolute temperature, v_2 the volume fraction of the polymer, and $\overline{\mathrm{DP}}_n$ its number-average degree of polymerization. The interaction between polymer and solvent molecules is represented by the Huggins interaction coefficient χ_1. This coefficient can be obtained experimentally by determining the activity a_1 of the solvent by vapor pressure or osmotic measurements according to

$$(\ln a_1 - \ln v_1)v_2 - 1 = \overline{\mathrm{DP}}_n^{-1} + \chi_1 v_2 \quad (2)$$

Here v_1 is the volume fraction of the solvent. It is found that the interaction coefficient is a measure for the solvent power, with the poorer solvents having the larger values for χ_1. The Huggins coefficient also depends upon temperature, approximately according to the relation

$$\chi_1 = A + B/T \tag{3}$$

Here A and B are constants.

The variation of χ_1 with solvent power and temperature forms a basis for several methods of polymer fractionation, employing solvent and/or temperature gradients. The conditions for phase separation in a polymer solution at the critical point can be derived from equation 1. The critical values for χ_1 and v_2 are found to be

$$\chi_{1,\,\mathrm{crit}} = {}^1/_2(1 + \overline{\mathrm{DP}}_n{}^{-1/2})^2 \tag{4}$$

and

$$v_{2,\,\mathrm{crit}} = (1 + \overline{\mathrm{DP}}_n{}^{1/2})^{-1} \tag{5}$$

Introducing equation 3, the critical temperature is obtained

$$T_{\mathrm{crit}} = B/(\chi_{1,\,\mathrm{crit}} - A) \tag{6}$$

Equations 4–6 show the expected molecular-weight dependence of the critical variables. Upon phase separation, a solvent-rich phase I and a polymer-rich phase II are formed. The distribution of a polymer of degree of polymerization DP between these two phases is calculated to be

$$v_{\mathrm{P}}{}^{\mathrm{I}}/v_{\mathrm{P}}{}^{\mathrm{II}} = \exp\,(-\delta\mathrm{DP}) \tag{7}$$

The quantity δ is a function of the total polymer concentration in both phases and of the average degree of polymerization; it is, however, independent of the particular degree of polymerization, DP, and is always positive. The numerical value of δ increases with increasing χ_1, ie, with decreasing solvent power. See also COHESIVE-ENERGY DENSITY.

Inspection of the above equations suggests the following conclusions: any polymer of a given degree of polymerization is always present in phases I and II; the difference in concentration in the two phases for any species increases with increasing molecular weight; ie, the higher molecular weight is enriched in phase II. It can also be shown that fractionation becomes more efficient in highly dilute solutions.

Experimental. In the fractional precipitation method polymer is dissolved in a solvent, and a nonsolvent is added gradually to the solution at constant temperature. The solution is continuously agitated. After addition of nonsolvent until the solution becomes turbid, the solution is aged for 2 to 4 hr or longer. The slow dropwise addition of further nonsolvent is continued until the solution turns milky. The solution is then warmed until it becomes transparent (8) indicating that the polymer has redissolved. The solution is now cooled gradually to the original temperature with agitation, and kept at this temperature for several hours to let the precipitate settle. The precipitate is separated by decanting or by siphoning, and is filtered and dried to constant weight in vacuum. The supernatant liquid is further titrated with nonsolvent using the previous procedure to obtain subsequent fractions. In choosing a solvent for preparing the original solution, it is desirable that the solvent be thermodynamically poor, ie, have a large value for χ_1, so that only small amounts of nonsolvent will be required to produce insoluble fractions.

The precipitation is carried out by steps in order of decreasing solvent power. The separations may be obtained either by addition of nonsolvent or elimination of solvent by evaporation, or by lowering the solution temperature to achieve separation of fractions. These procedures have been employed in most fractionations carried out on polymer systems. The method can be performed in either small or large scale (9).

Molecular-weight distributions can be obtained from fractional precipitation of polymer from solution. The first fraction generally has the highest molecular weight, and succeeding fractions have progressively lower molecular weights.

Some solvent–nonsolvent pairs have been found to be unsatisfactory because they produce fractionation in reverse order of molecular weight. This can, however, be caused by too high an initial polymer concentration (10,11).

An inherent problem with the fractional precipitation method is that each fraction contains appreciable concentrations of lower molecular-weight species. Because of the nature of this method, fractionation of higher molecular-weight species is carried out at concentrations higher than those of lower molecular weight, as sample concentration is decreased by nonsolvent addition and separation of earlier fractions. Thus separation efficiency increases with decreasing molecular weight.

Polymer sample size is determined by the amount required for examination of the molecular-weight distribution or for obtaining fractions of narrow molecular-weight distribution for subsequent study. For determination of molecular-weight distribution, it is possible to start with less than 1 g and to take twenty or more fractions. The molecular weight of fractions can be determined by various methods, but perhaps the most widely used and easiest is by solution viscosity measurements (see also MOLECULAR-WEIGHT DETERMINATION; VISCOMETRY). The largest sample size practical in laboratory experiments is about 25 g; this involves a volume of about 5 liters of 1% solution.

It is imperative that experiments be carried out at constant temperature with deviations not larger than $\pm$ 0.05°C.

Fractionation by solvent evaporation can be carried out when the nonsolvent is less volatile than the solvent. The concentration of nonsolvent is increased by gradually evaporating the more volatile solvent. This technique depends on a decrease in solvent power caused by evaporation of the solvent component, and differs from the nonsolvent-addition technique only in the manner in which the composition is altered. This approach has several advantages. For example, the scale of the experiment can be increased because the volume of the system decreases as the fractionation proceeds. Also, a continuous change of the solvent–nonsolvent composition is possible and local concentration of the nonsolvent can be avoided. This method has been used to advantage in preference to the nonsolvent-addition method for polymers which are soluble only at high temperatures.

Fractionation by cooling depends on a decrease in solvent power caused by a decrease in the temperature of the system (12). Advantages of this method of fractional precipitation include the need for only one solvent, constancy of solution volume during fractionation, and the possibility of close adjustment of fraction size. Fractionation by cooling has disadvantages in that degradation must be considered in high-temperature experiments and that it is sometimes difficult to find suitable solvents which will completely precipitate polymer on cooling. This fractionation method has not been widely used although it forms a part of experiments using the nonsolvent-addition method (13–15). The procedure is simple and requires fewer steps than precipitation either by nonsolvent addition or by solvent evaporation. The cooling process should be sufficiently slow that thermodynamic equilibrium is approached. In general, for the cooling technique, the solvent must be as poor as possible and the concentration of polymer must be above some lower critical value dependent on the nature of the system.

Everything that has been said so far applies to fractional precipitation in which the solvent-rich phase is amorphous, not crystalline. None of the theoretical considerations in the introductory section apply when the precipitated phase is crystalline, and, in general, separation according to molecular-weight differences does not take place under these circumstances. An exception is the work of Koningsveld and Pennings who have shown that, for sufficiently low-mw samples, large amounts of fairly sharp fractions can be obtained by crystallization from solution (16).

Column Chromatography

Of the several techniques embraced by the term chromatography, two, simple column elution, or adsorption chromatography, and column elution with a superimposed temperature gradient, or precipitation chromatography, will be discussed in this section. The former method is a one-stage extraction process, whereas the latter is a multiple extraction in which fractional precipitation and solution occur repeatedly over the length of the column. See also CHROMATOGRAPHY.

The basic apparatus is a mixing device which provides a solvent gradient, a packed chromatographic column, a temperature-control system, and equipment for sample collection. Polymer samples are generally coated on a small amount of solid support and introduced at the start of the column. The solvent gradient usually consists of a change in composition of a binary solvent with time. For precipitation chromatography, a temperature gradient may be imposed on the column by heating the upper end and cooling the lower end, or, for elution analysis, the column may be operated isothermally. Molecular-weight distributions are determined by recovery of polymer fractions and measurement of the amount and molecular weight of each fraction. Integral and differential molecular-weight distributions can then be constructed by simple computation (see Fig. 1).

Applications of precipitation chromatography, first used by Baker and Williams (17), include the separation by molecular weight of the polymers listed in Table 1.

Table 1. Polymers Fractionated by the Baker-Williams Method

Polymer	Reference
polystyrene	17–20
polyisobutylene	21, 22
poly(methyl methacrylate)	23, 24
polyethylene	25
polypropylene	26
polyester (hexanediol-succinic acid)	27
cis-1,4-polybutadiene	28
polyalkylene (C_{10}–C_{18} olefin)	18
polysarcosine	29
polypeptide	30
poly(isobutylene-*co*-styrene)	31
copolymer of natural rubber and methyl methacrylate	32

Other fractionation methods employing the principle of fractional solution are coacervation, ie, the preferential extraction of low-mw species from a liquid coacervate, and film extraction. In the latter method a thin coating of polymer on a metal foil is extracted with progressively richer solvent–nonsolvent mixtures.

Sample Preparation. Numerous methods have been used to prepare polymer for introduction into columns (17,18,33,34). Polymer is commonly deposited on the column support, which is usually glass beads, diatomaceous earth (35), or sand. The support with polymer is deposited at the head of the column. The polymer coating on the support may be prepared by solvent evaporation or by cooling a solution in the presence of support until the polymer is all precipitated (26,33). Care in preparing the sample on the support can improve fractionation as the highest molecular weight should be deposited first onto the support and should be the last eluted from the column. The size of polymer samples used in analytical fractionations ranges from milligrams to several grams. Efficiency of fractionation decreases with increasing sample size. The amount of polymer that can be successfully chromatographed decreases with increasing molecular weight. Efficiency of fractionation theoretically should be inversely proportional to the square root of molecular weight (23,25).

Column Operation. Dimensions of chromatographic columns range from 0.5 to 5 cm for the internal diameter and from 10 to 150 cm in length. No systematic study of the effect of column dimensions has been made. Heat transfer can be facilitated by using metal tubing with heavy walls. Catalytic reactions on metal columns can, however, be a problem, and glass tubing may be used. Temperature regulators are helpful because total times for fractionation are relatively long. Column packings are chosen to provide sufficient surface area to support the swollen polymer as a thin film. Small glass beads, 40–70 μ in diameter, are commonly used (17,18,36,37). Metal particles and highly crosslinked polymers, such as crosslinked polystyrene, have also been used as packings (38,39).

The solvent gradient chosen for chromatographic fractionation can have a marked influence on separation efficiency. Probably no one gradient is optimum for separation. The logarithmic gradient (sometimes called the exponential gradient) is widely used and the equipment required to produce this gradient is very simple (40–43). A closed and, therefore, constant-volume mixing vessel is used, producing an eluant gradient in which the solvent concentration increases rapidly at first, then approaches the reservoir concentration asymptotically. With an open mixing vessel connected by a syphon to a reservoir of the same volume, a linear gradient is produced (44).

Columns are usually operated in a vertical position. Descriptions have been given of several methods of control of flow (17,18,25,37,38,45). Upward flow of solvent has been used, which has advantages for packings of polymer gel (23). Upward flow, however, requires solvent degassing.

Fraction Collection and Data Interpretation. For calculation of molecular-weight distribution, a number of fractions must be recovered and the amount and molecular weight of each fraction determined. Integral and differential molecular-weight distribution curves can be constructed from such data. Distribution definition increases as the number of defined fractions. Fraction evaluation, however, accounts for the greatest part of effort required for a fractionation. Usually a minimum of ten fractions is necessary. Figure 1 shows typical integral and differential curves for a polyisobutene.

Commercial and custom-made fraction collectors are widely available (17,23,28).

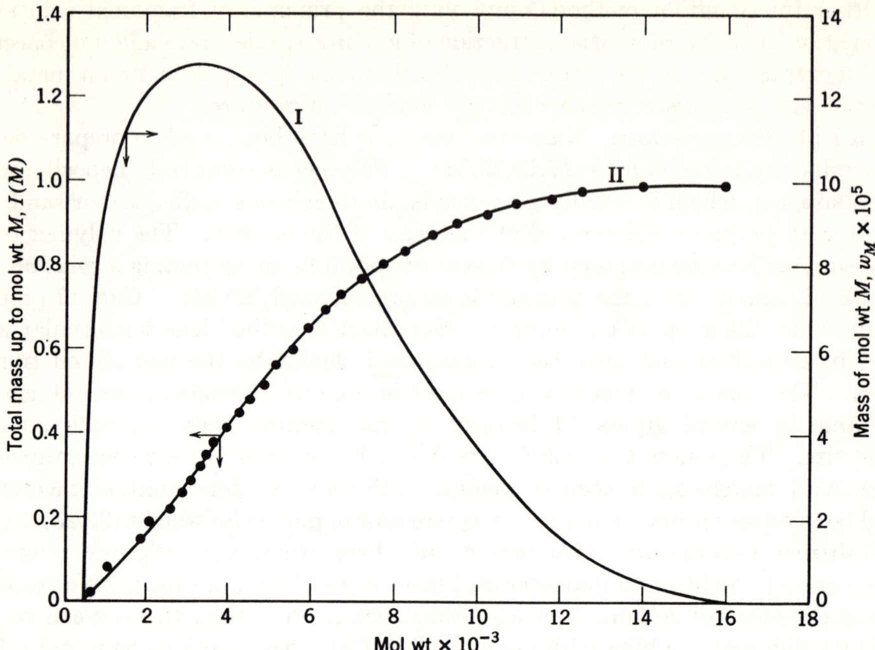

Fig. 1. Molecular-weight distributions for polyisobutylene (21). I, differential distribution; II, integral distribution.

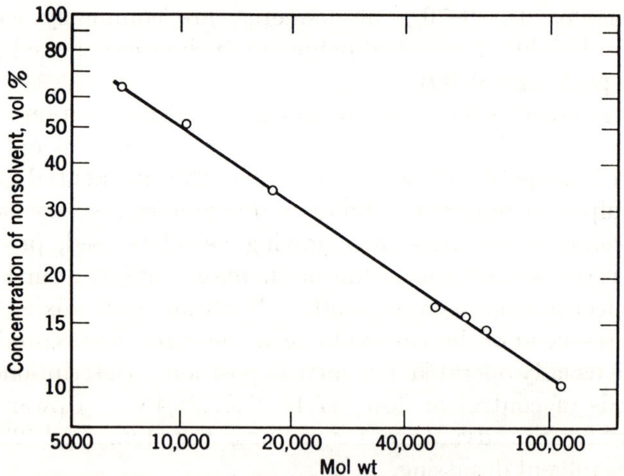

Fig. 2. Variation of molecular weight of polyisobutylene fraction with composition of eluant. Acetone is the nonsolvent and benzene the solvent (37).

Fractions may be collected on the basis of either time or volume (23). If fraction receivers are changed on a time signal, variations in elution flow rate will cause complications. If fractions are collected on a volume basis, care must be taken to avoid solvent evaporation. Fractions have also been collected with a paper-strip collector or by a colorimetric tracer method. In general, fraction recovery methods vary widely with the type of polymer. Addition of cold nonsolvent to precipitate fractions makes

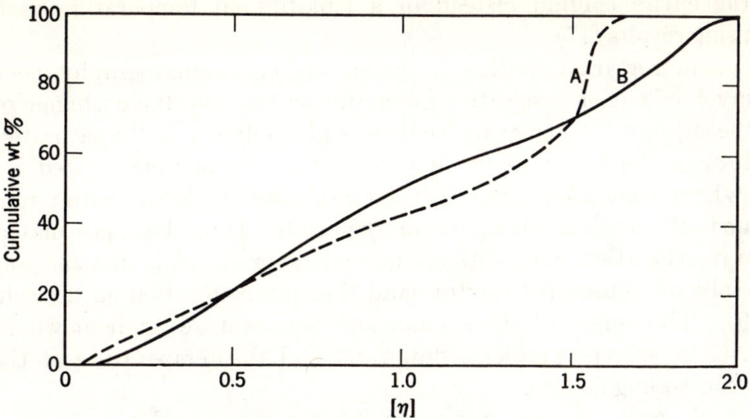

Fig. 3. Fractionation of polyethylene showing the effect of a temperature gradient on efficiency (25). A, column temperature without temperature gradient: top and bottom, 133°C. B, column temperature with temperature gradient: top, 152°C; bottom, 100°C.

possible recovery by filtration (25). Fractions are also commonly recovered by solvent evaporation.

Fraction evaluation generally involves determination of molecular weight by solution viscometry (see VISCOMETRY), ultracentrifugation (qv), light scattering (see SCATTERING), or osmometry (qv). Spectroscopic tests can be used to test for fractionation by chemical type. For repetitive chromatographic separations of polymers of the same chemical composition, a relationship can be experimentally determined between molecular weight of eluted polymer and its elution volume. In this case, fraction weight only need be determined for construction of distribution curves (29). Figure 2 shows a typical double logarithmic plot of solvent composition versus molecular weight which has been established for polyisobutene in benzene–acetone (37).

Solvent Gradient. A variety of nonsolvent–solvent systems has been used for column chromatographic fractionations. A method for selecting nonsolvent–solvent combinations for fractionation has been given by Hulme and McLeod (28). The polymer, dissolved in a "good" solvent, is titrated to a cloud point with different nonsolvents; the nonsolvent–solvent system giving the widest difference in composition is used. However, it has been argued that any two compositional limits may be approached experimentally as slowly as desired (38). It has been suggested that linear gradients can give better resolution for comparatively narrow distributions than logarithmic gradients (29), although experimental tests are somewhat ambiguous (25).

Temperature Gradient. Fractionation by precipitation chromatography and by simple elution analysis without a temperature gradient have been compared by a number of investigators. The use of a thermal gradient was found to give somewhat better resolution and reproducibility, although the differences between the two methods were not appreciable (33,34,46). A comparison of the fractionation of polyethylene with and without a temperature gradient is shown in Figure 3 (25); the results with a temperature gradient are superior, particularly in the region of high-mw. In certain cases better fraction may be obtained without a thermal gradient (36). A modified method of precipitation chromatography using periodic temperature changes of the entire column instead of a longitudinal temperature gradient has yielded promising results (47).

A mechanism for fractionation in precipitation chromatography has been suggested that involves not only selective precipitation but also the exchange of low-mw polymer on the support for higher molecular-weight polymer in the saturated solution as it flows through the lower, as yet uneluted, zones of polymer-coated beads (36). Other workers have concluded that enhanced resolution in elution with a temperature gradient is due to the original adsorption on the beads. It has been postulated that the mechanism of fractionation, even with a temperature gradient, is that the polymer dissolves in the solvent–nonsolvent mixture and that reprecipitation on the column does not occur (39). The length of the column and temperature gradient would thus be unimportant. The solvent gradient, flow rate, and the temperature at the column exit would be controlling factors.

The effect, if any, of a temperature gradient in classic column-chromatographic fractionations of polymers is still not established. Extremely well-controlled conditions are required to conclusively determine the effect of thermal gradient.

Flow Rate. The ratio of volume of eluant to weight of polymer is a critical factor. Too small a volume of eluant results in poor fractionation. On the other hand, recovery of polymer fractions is difficult if the eluted solutions are highly dilute. From experimental and theoretical considerations, it has been concluded that the volume of the eluant should be 200–500 times the volume of the polymer fractionated (34,38,48). For a logarithmic gradient, this requires that the volume of the mixing vessel be at least a hundred times that of the polymer volume.

Maximum column flow rate must be sufficiently slow that equilibrium is established between solution and precipitated polymer. However, at very low flow rates, back diffusion causes a loss of resolution. For an average analytical column with a polymer sample size of about 1 g, optimum flow rates are in the range of 2–6 ml/min (25).

Theory. Two basic models have been evolved for predicting the effect of solvent and temperature gradient on fractionation. Both predict a pronounced temperature effect on fractionation. For amorphous polymers, Caplan has suggested that the polymer phase diagram is an asymmetric binodal with a critical point close to the solvent ordinate (7,29). A dilute polymer solution upon cooling crosses the binodal and a gel phase is precipitated in equilibrium with a larger volume of essentially pure solvent. As the fractionation proceeds, the gel is dissolved by a better solvent combination; this is followed by reprecipitation, at lower temperature, farther down the column. These and other considerations have led to equation 8 for the molecular weight–eluant volume relationship for the case of a single constant-volume solvent-precipitant mixing chamber.

$$M = \left[\frac{KT_2}{\sigma e^{-V/G} + \tau - T_2}\right]^2 \qquad (8)$$

where M is the molecular weight of the polymer species being eluted; T_2 is temperature of the bottom of the column; σ and τ are parameters that may be equated with the Θ temperatures of the solvent and the precipitant, respectively; G is volume of mixing chamber; and V is volume of the solvent which has passed any given point since the start of fractionation.

Estimation of parameters has led to good agreement of equation 8 with fractionation results obtained with a polystyrene. An exact evaluation is not possible because of the inability to precisely define the constants τ and σ.

A different model, based on transport equations for the exchange process between a gel layer on a support, and a sol layer in motion, has been proposed by Schulz and co-workers (49). The resultant equations were applied to both the solvent and temperature gradient. Fractionation was considered in two steps: a process during which a continuous gel phase is formed on the support and, the principal process, exchange distribution between sol phase and gel phase. In the principal process, every component of the polymer exhibits a characteristic equilibrium between the two phases, depending only on the molecular weight of the component. The combination of solvent and temperature gradients has two effects. First, each species moves through the column at a characteristic velocity. Second, because of the temperature gradient, the leading edge of the component moves more slowly than the trailing edge. Considerations based on this theoretical model indicate that extremely high resolution of polymers into fractions is potentially possible. In practice, it is apparently limited principally by the finite time required to establish phase equilibrium. Degradation can be checked for by means of equation 9

$$[\eta]_s = \Sigma w_i [\eta]_i \tag{9}$$

where $[\eta]_s$ is the intrinsic viscosity of the original sample, and $[\eta]_i$ is the intrinsic viscosity of the ith fraction of weight fraction w_i. If $[\eta]_s$ is greater than the summation, degradation has occurred.

Efficiency. Fractionation efficiency can be evaluated in several ways. One criterion is that the recovery of polymer from fractions should equal the amount of polymer introduced. A regular progression to higher molecular weights of fractions with increase in total elution volume should also be observed, that is, there should be no "backlash."

The distribution of molecular weights within each fraction is also a measure of fractionation efficiency. This distribution may be obtained by refractionation of fractions (17,36). However, this is a laborious process not necessarily leading to further resolution of fractions. Ratios of separately determined weight- and number-average molecular weights also provide a measure of fraction homogeneity. This method has the limitation that individual weight- and number-average determinations are not sufficiently precise to obtain an accurate ratio below about 1.03. Use of other methods for distribution determinations on fractions, such as ultracentrifugation (50) and turbidimetric titration, may be preferable.

Large-Scale Fractionation. Although it is often desirable to prepare large amounts of polymer fractions with narrow molecular-weight distribution for study of bulk properties, the techniques of preparation of large fractions are relatively unexplored. Up to 500 g of polyethylene has been fractionated on 4-in. diameter columns (51). An increase in sample and fraction size can be obtained by operating multiple columns in parallel (Fig. 4), a method suggested by the difficulty of scale-up by increasing column diameter (37,52,53). Experience in distillation and in other types of chromatography indicates that above a maximum column diameter a sharp decrease in efficiency of separation usually results, due to channeling and/or lateral temperature gradients. Figures 5 and 6 illustrate the molecular-weight distributions determined by large-scale fractionation.

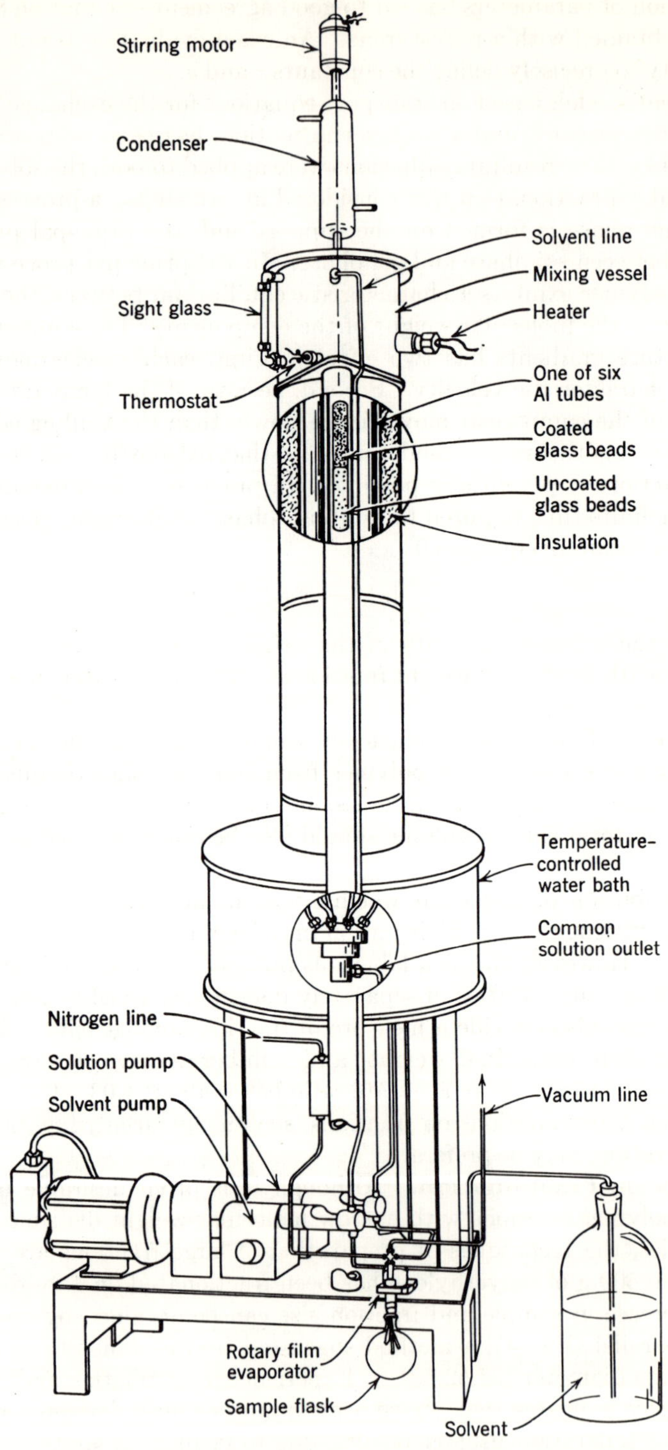

Fig. 4. Multicolumn polymer fractionation apparatus (37).

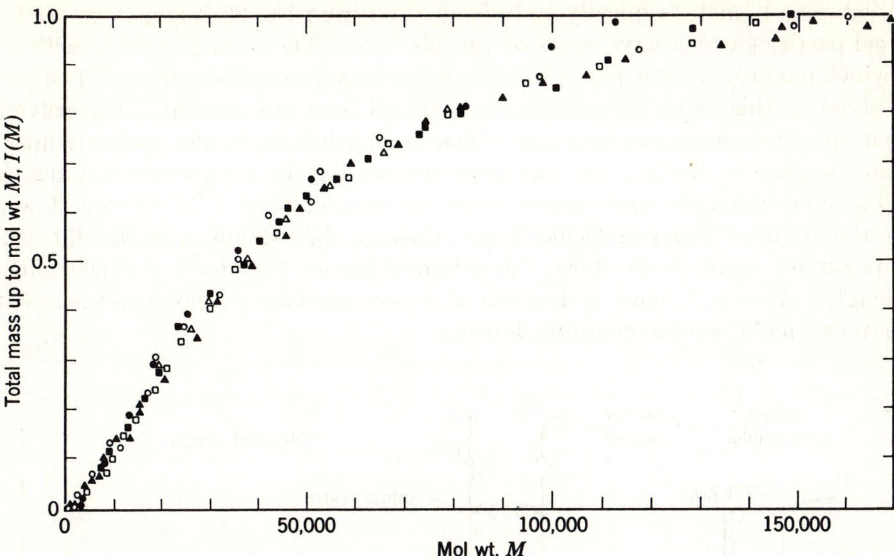

Fig. 5. Integral molecular-weight distribution of polyisobutylene using multicolumn fractionation and initial samples ranging from 35 to 50 g (37). ○, Run A; ●, run B; □, run E; ■, run F; △, run G; ▲, run H.

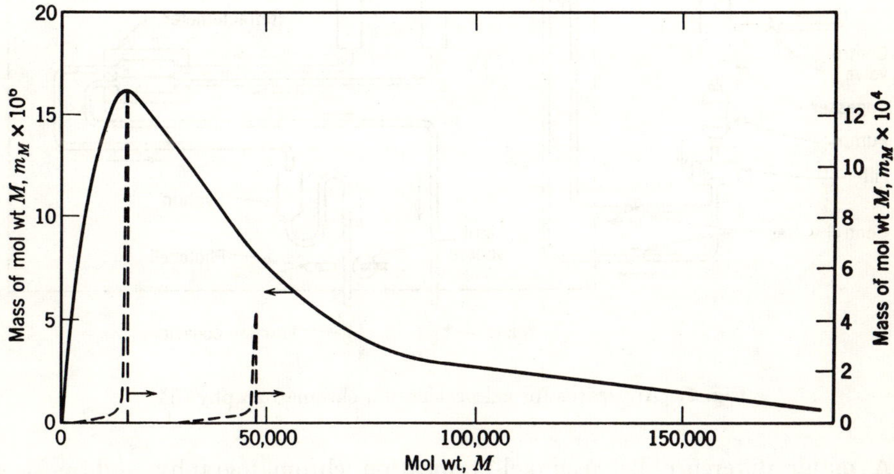

Fig. 6. Differential molecular-weight distribution of unfractionated polyisobutylene and of two of the fractions obtained. The values of $\bar{M}_w/\bar{M}_n$ for the two fractions are 1.013 and 1.020, as determined by refractionation (37).

Gel-Permeation Chromatography

Gel-permeation chromatography, or gel filtration, is a valuable technique for fractionating a wide variety of both natural and synthetic polymers. It is susceptible to automation for readily determining molecular-weight distributions with a minimum of effort and elapsed time. See also CHROMATOGRAPHY.

A column, diameter typically ⅛ to ¾ in., and from 4 to 20 ft long, is packed with small gel particles which have pores of various sizes. The column is filled with a solvent, which not only fills the pores, but fills the void spaces in the column. The sample is dissolved in this same solvent and introduced into the column. The solvent is pumped into the column continuously. The small solute molecules go freely into and through the pores of the gel, whereas large species may be completely excluded from the gel and intermediate sizes cannot enter the smaller pores. As solvent flow continues, the excluded larger molecules travel through the column more rapidly. Small molecules move more slowly down the column because they have a longer effective path length. Figure 7 shows a diagram of a commercially available gel-permeation chromatograph (54) with a recording detector.

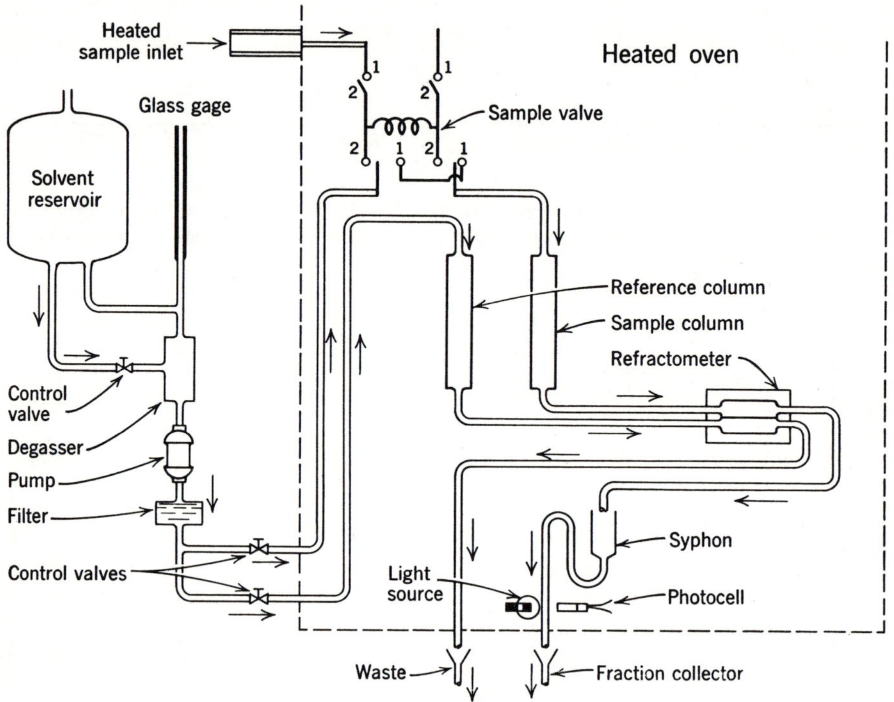

Fig. 7. Apparatus for gel-permeation chromatography (54).

A major difference between gel-permeation chromatography and most other methods of fractionation is that separation is primarily based on molecular size, not on chemical constitution. There are, to be sure, minor effects due to the composition of the molecules. However, in molecular weights above about 1000, polarity of the molecule becomes relatively unimportant, and separation is by size alone. The larger molecules emerge first, the smaller ones at the end of the separation.

Two different techniques are employed; in one, fractionation and isolation of fractions are the objectives; in the other, separation is used only to determine molecular-weight distributions without the collection and isolation of samples.

History. As early as 1925 it was known that ions of different sizes were separated during absorption on clay (55). Later natural polymers were separated from low-mw

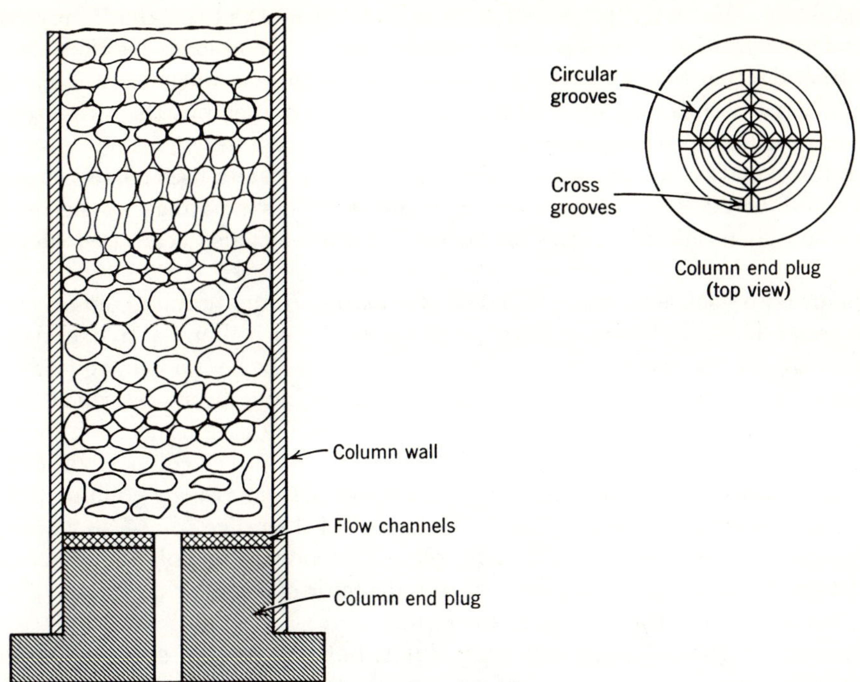

Fig. 8. Cross section of gel-permeation column and detail of end plug (54).

substances on crosslinked gels (56). In 1953 Wheaton and Bauman described uses and theories of gels permeable only to relatively small molecules (57).

A major expansion of gel-permeation chromatography came with the introduction of dextran gels (58). A number of papers occurred in rapid succession describing the use of this technique to fractionate amino acids (59), peptides (60), and proteins (61,62). Shortly, a number of additional systems appeared, principally in the field of biological materials. Very few reports of the use of gel-permeation chromatography with organic solvents appeared because of the lack of suitable gels. Much early work on gels for use with organic solvents dealt with separations in molecular weights below 20,000, a range of limited interest with respect to synthetic polymers (63–67). However, in 1962 the successful fractionation of synthetic polymers with molecular weights up to one million was reported (68). Subsequently, descriptions of the automation of gel-permeation chromatography to rapidly give molecular-weight distribution appeared (54).

Experimental. A wide variety of gels have been used to effect fractionation. Usually these are polymers of various degrees of crosslinking. Aside from the varying pore sizes, ie, permeability, an important parameter of the gel is its rigidity. Very soft particles tend to plug columns. It is thus necessary to produce a gel with both the desired permeability and degree of rigidity.

Gels used in aqueous systems fall into two main classes. A number of crosslinked dextrans, trademarked Sephadex, are produced by A. B. Pharmacia, Uppsala, Sweden. The Sephadex gels contain different degrees of crosslinking and, therefore, of permeability. Standardization is reported as the molecular weights of soluble dextran fractions which are excluded from the gel particles. Reference 69 describes the preparation of gels in detail. Recently, polyacrylamide gels, trade marked Bio-Gel P, have become commercially available from Bio-Rad Laboratories in Richmond, California. These permit fractionation to relatively high molecular weights in aqueous systems. Many other gels have been found suitable for specialized purposes, but the two cited are in widespread use for aqueous systems.

For fractionation using organic solvents, the most widely used gels are crosslinked polystyrenes (68,70,71). These can be made with a wide range of permeability, and also have good rigidity. An expanded silicate gel has also been used successfully (64).

The usual precautions in any kind of chromatography apply to column construction for gel-permeation chromatography. Columns must be nonreactive, and stainless steel is usually a satisfactory choice. Dead volume must be kept down, and this implies good packing arrangements. Packing techniques have been described in detail (69,72). Pressure filtration is often an effective packing technique. Figure 8 shows a column cross section and an end plug to retain the gel in the column without creating undue pressure drop (54). Resolution improves with increasing column length but so also do fractionation times. Usually, column lengths range from 1 to 25 ft.

The solvent must, of course, be suitable for dissolving the polymer. In addition, it should be stable and compatible with the detector system, or allow sample to be readily recovered from the separated fractions.

For aqueous solvents, not only pure water, but a variety of aqueous buffer solutions and mixtures of water with organic solvents have been employed (73). For the crosslinked polystyrene gels a number of solvents have been used, including methylene chloride, *o*-dichlorobenzene, perchloroethylene, *m*-cresol, chloroform, tetrahydrofuran, toluene, benzene, *o*-chlorophenol, carbon tetrachloride, dimethylformamide, 1,2,4-trichlorobenzene, and others. Frequently, antioxidants are added to protect either the solvent or the sample from oxidative or temperature degradation.

Separations have been achieved at temperatures from 20 to 150°C. Higher temperatures lower the viscosity of the solvent, which can lead to more rapid analysis and more resolution in the column. Temperature control of the columns is usually achieved in thermostatted, stirred air baths.

Sample size is chosen to be as small as possible to give the desired amount of polymer. The smaller the sample size, the better the resolution.

The solvent-handling system is required to store the solvent, degas the solvent if dissolved gases are present, and force the solvent through the column at a constant flow rate. Resolution can be lost if the flow rate is not constant. Further, if an automatic detection device is used, a constant flow rate facilitates comparison of measurements. A common approach is to use a variable stroke liquid pump. A buffering system is required to damp out the pulses. Recently, a nonpulsing liquid pump has been described that may be satisfactory without a damping system (74).

Samples are usually prepared as a dilute solution, 0.1–2 wt %, in an identical batch of the solvent that is used as the eluant. It is desirable to introduce the sample as near the head of the column as possible to prevent loss of resolution. One technique is to inject the liquid sample into the flowing solvent from a hypodermic needle through a rubber system, a method widely used in gas chromatography. A more sophisticated approach is to use a four-port valve with a sample loop. Initially, the sample loop of known and easily variable volume is filled at atmospheric pressure with the valve in the "fill" position. The sample may be brought to the temperature of the column, and the

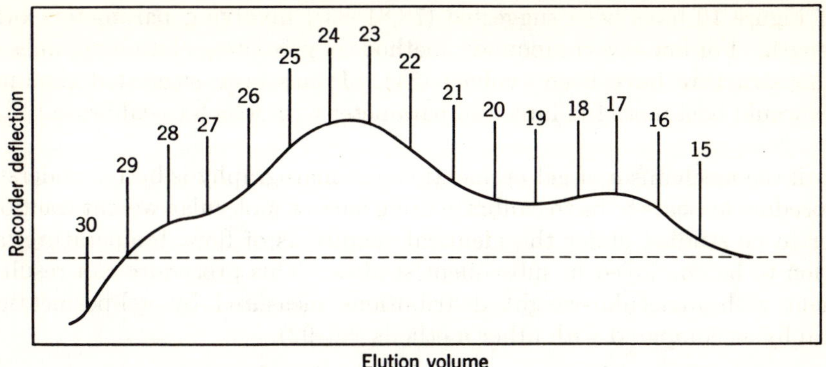

Fig. 9. Gel-permeation chromatogram. The vertical lines measure the elution volume.

valve then turned to the "run" position, which places the sample loop directly into the flowing solvent stream where it is carried onto the column.

Determination of Molecular-Weight Distribution. In many cases, fractionation is carried out to determine the molecular-weight distribution. This is conventionally done by collecting fractions, drying them, weighing to determine the amount of the polymer, and measuring its molecular weight by some independent technique—a time-consuming procedure. A major advantage of gel-permeation chromatography is that these steps can be eliminated by use of a detector that determines the concentration of the polymer in the effluent, since neither a solvent nor a temperature gradient is present. A differential refractometer is usually used for this purpose because the change in refractive index is directly proportional to concentration in dilute solutions of polymer. Moreover, it is usually independent of molecular weight in the range of molecular weights above a few hundred. Figure 9 shows an example of the chart record produced by a differential refractometer. The displacement from the base line is proportional to the difference in refractive index and, hence, the concentration. The vertical lines measure the elution volume. A siphon at the end of the column is dumped when 5 cc of eluant has been collected. This dumping action sends an impulse to the recorder to automatically produce the spike on the curve. Thus, the chart also contains a record of the volume eluted. Other types of detectors for example, an infrared spectrometer (75) and an ultraviolet spectrometer (76), have been employed. Spectroscopic detectors have the advantages of higher sensitivity in some cases and the ability to follow specific chemical groups. They are much more limited in the number of polymers to which they are applicable than is the differential refractometer. See also INFRARED-ABSORPTION SPECTROSCOPY; ULTRAVIOLET-ABSORPTION SPECTROSCOPY.

Thus, if the concentration is recorded directly, then the molecular weight remains to be determined. This may be done by calibrating a specific set of columns at a given temperature by use of fractions of known molecular weight and plotting elution volume as a function of the logarithm of chain length (see, for example, Figure 10).

Some precautions on calibration must be noted. Earlier work tacitly assumed that polymer elution volumes were dependent only on polymer chain length and were independent of polymer type, concentration, temperature, etc. A considerable body of evidence has since been accumulated, showing that this is not strictly true. Concentration effects on elution volume have been demonstrated (75,77–80). Alternative plots of Figure 10 have been suggested (78,81–83), involving parameters other than chain length. For low-mw compounds, methods of predicting elution volumes based on molecular structure have been evolved (84). It has been suggested that molecular volumes would be a more fundamental parameter with which to calibrate the columns (85).

Until the mechanism of gel-permeation chromatography is better understood, the best procedure appears to be to calibrate using narrow molecular-weight fractions of the polymer to be studied under the identical conditions of flow, temperature, and concentration to be employed in subsequent studies. This procedure can result in good agreement with molecular-weight distributions measured by gel-permeation chromatography as compared with other methods (86–92).

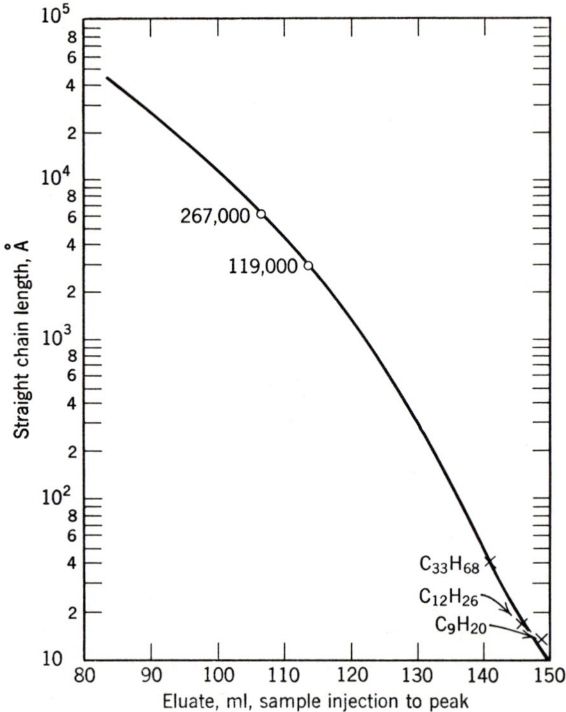

Fig. 10. Calibration curve for gel-permeation chromatograph, 1,2,4-trichlorobenzene at 130°C 1 ml/min (54).

Once a relationship between molecular weight and elution volume has been established, the computation of molecular-weight distributions is straightforward although laborious. A computer program to readily perform these operations has been described (93). In this program, input data from the recorded chromatogram, ie, the elution volume, recorder deflection, recorder sensitivity, and sample concentration, are converted to unit sensitivity; if required, a correction for a constant base-line drift is applied. Numerical integration of the curve follows. A calibration curve as previously described is used to obtain molecular weights from elution volumes. The output lists cumulative and differential molecular-weight distributions, plus number-average, viscosity-average, weight-average, Z-average, and $(Z + 1)$-average molecular weights. An additional saving of time is effected by printing out differential and cumulative molecular-weight-distribution curves and plotting a differential histogram.

Computation methods based on a single calibration curve as shown in Figure 10 are not exact. The resolution of gel-permeation chromatography is not infinite. A monodisperse material produces a chromatogram that is not a straight line but a Gaussian bell-shaped curve. The position of the center of the peak depends on the molecular weight, and the width of the curve depends on the resolution of the column. The chromatogram of a polydisperse polymer then becomes a composite of the Gaussian curves of all the components. The height of the curve is not proportional to the amount of polymer with molecular weight determined by the elution volume because the height depends also on the amount of neighboring components. At each end of the chromatogram, there are areas of the curve that correspond to molecular weights of polymer not present in the sample. This effect must be accounted for before a true molecular-weight distribution can be obtained. Various procedures to accomplish this have been suggested (94–97). Branched polymers may be treated on the basis of the hydrodynamic volume (94a).

Thermal Diffusion

The use of thermal diffusion for separating organic liquids is well known. The same technique has been used to fractionate some polymers on a molecular-weight basis. A typical plate-type thermal diffusion apparatus is shown in Figure 11 (98). The apparatus is filled with polymer solution using the injection syringe. One wall is

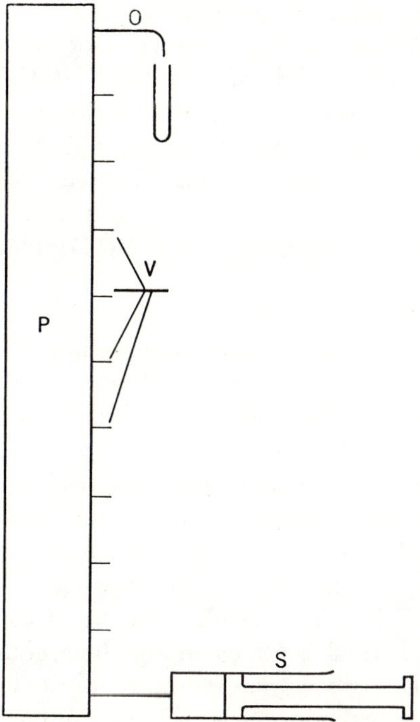

Fig. 11. Schematic diagram of thermal diffusion apparatus. P, working space; S, syringe; O, hole for sampling of fractions; V, sampling holes (98).

heated, the other cooled. The width between plates is 0.5 mm, and the average temperature differential is 35°C. After the time necessary to reach equilibrium, solvent alone is added using the syringe and an equivalent volume of solution removed from "O" simultaneously. Further additions of solvent may be made subsequently. The sampling program may be automated.

There are many variations on the design of thermal diffusion apparatus. Concentric cylinders are often used. Continuous separations are possible. Cascade arrangements of multiple apparatus can give increased efficiency.

The first actual fractionation of a polymer was reported in 1948 by Debye and Bueche, who, using a concentric tube column and a solution of polystyrene, found that the molecular weight of the polymer at the bottom of the column was higher than that at the top (99). To illustrate the confusing nature of results from thermal diffusion, Fritzmeier and Hermans found, also in 1948, that poly(methyl methacrylate) in a parallel plate-type column did not fractionate with respect to molecular weight (100). A cascade series of columns was first used in 1955 (101); somewhat later, a similar arrangement was used for fractionating poly(vinyl chloride) in cyclohexanone (102). Application of theory to predict results from thermal diffusion has been very difficult. Essentially, the only way to determine if fractionation will occur is to try it experimentally.

Molecular-weight distributions of only a small number of polymers have been determined by thermal diffusion. The molecular-weight distribution of poly(methyl methacrylate) in benzene, determined by collecting samples from various points in a column and determining the molecular weights, is in good agreement with that measured by fractional precipitation (103). Similarly, good agreement was found for polystyrene fractionated by thermal diffusion and by other methods (104,105). Other results, for example, Refs. 106 and 107, have shown poor agreement.

Determination of molecular-weight distributions by thermal diffusion has shown considerable promise although the number of systems studied is small. The use of thermal diffusion on a continuous basis to prepare large quantities of fractionated polymer seems feasible but remains undemonstrated. Theoretical predictions of separations have suffered from lack of basic data and, perhaps, from inadequate theory, and have proved ineffective in most cases. See also DIFFUSION.

Turbidimetric Titration

A rapid technique for measuring molecular-weight distributions is that of turbidimetric titration. In outline, this technique works as follows: A dilute solution of the polymer to be characterized is placed in the instrument at a constant temperature with slow stirring, and the turbidity of the solution is recorded continuously. Nonsolvent is then slowly added. The higher molecular-weight polymer precipitates first and increases the turbidity. As more nonsolvent is added, additional lower molecular-weight polymer precipitates, the turbidity continues to increase, and so forth. By calibrating with fractions of known molecular weight, the relationship between amount of nonsolvent added and molecular weight is established. Similarly, the magnitude of the turbidity furnishes a measure of the amount of polymer. Thus, the tedious drying and weighing steps are eliminated by optical weighing.

Variations on this basic procedure exist. For example, the fractionation may be carried out in the inverse order by dissolving a turbid solution. If no solvent–nonsolvent pair is available, temperature may be used to effect precipitation. In this case, the fractionation is started at the higher temperature, the temperature is gradually decreased, and the resulting turbidity measured.

The turbidimetric titration technique for measuring molecular-weight distribution of high polymers was reported by Morey and Tamblyn (108) in 1945. Earlier qualitative uses were described by Adams and Powers (109). Desreux made improvements in both the apparatus and experimental conditions (110,111).

A typical turbidimetric titration apparatus includes a light source, a photocell detector with auxiliary amplification and recording units, a thermostatted test cell with provision for agitation or stirring, and an injection device (112–118). Some generalizations to be considered in constructing or selecting instruments are the following: the light source requires monochromatic light, usually from a high-pressure mercury vapor lamp. In some cases, white light is adequate, but usually a monochromator is preferable. In any case, good control of the light intensity is required, either by use of a double-beam arrangement or by efficient voltage stabilization. Stray light is to be avoided. The rate of stirring in the test cell is important since it must be as low as possible to avoid formation of unstable solutions and yet sufficient to ensure that the sample is homogeneous. A prerequisite of the method is, of course, that as polymer is precipitated, it forms additional particles rather than increasing the size of particles already present. The injection device used to add the nonsolvent has two requirements; the injection must be at a constant reproducible flow rate, and there should be a method of recording flow rate. Some reports advocate the use of a varying rate of nonsolvent addition, with an increasing rate toward the end of the precipitation. This is useful to reduce the time required for the fractionation and imposes only the limitation that the method of adding nonsolvent be reproducible.

The interpretation of the results of turbidimetric titration can be treated either from a theoretical viewpoint or in an empirical manner. In almost every case, the advantage of the turbidimetric titration lies in its simplicity and speed. This encourages the use of an empirical technique. Once a solvent–nonsolvent pair giving a satisfactory range of turbidity versus molecular weight has been found, calibration with reproducible conditions provides an easy technique for determination of molecular-weight distributions for a specific polymer–solvent–nonsolvent system. In a modification of turbidimetric titration higher polymer concentrations are employed and the polymer weight estimated by determining the gel volume which settles out after each nonsolvent addition.

Summative Fractionation

In this technique polymer is first dissolved in a solvent and then a portion of the polymer is precipitated by addition of about one-third the original volume of a partial solvent. The mixture is centrifuged to get rid of the precipitate and the polymer is recovered from the supernatant liquid. This procedure is repeated starting with a fresh polymer sample but using a stronger precipitant to remove more polymer from solution. The polymer from each solution is characterized as to amount and molecular weight to permit construction of molecular-weight-distribution curves. Methods of calculation for summative fractionation have been described in detail (119–122).

Ultracentrifugation

Ultracentrifugation (qv) is a standard method for determining weight-average molecular weight of polymers. Instrumentation is commercially available to make measurements over a wide range of centrifugal forces at temperatures from 0 to 150°C. A variety of optical systems can be used to obtain the experimental data. For a detailed description of various ultracentrifuges, see Refs. 123–125. The study of polydispersity by the centrifuge is not nearly as advanced as that of molecular weight, although polydispersity was one of the original objectives in ultracentrifugal analysis (126,127).

The reason that the ultracentrifuge has not been extensively employed in characterizing molecular-weight distributions lies in the complexity of interpreting the data. For a discussion of basic theory, the reader is referred to the excellent book by Fujita on the *Mathematical Theory of Sedimentation Analysis* (128). The various procedures used will be outlined here, and specific references cited for more detail.

The ultracentrifuge may be operated in two ways. One is the sedimentation–velocity method in which the rate of sedimentation and diffusion are determined during centrifugation at high centrifugal fields as a function of time. The other method is sedimentation equilibrium, in which the sedimentation and diffusion processes are allowed to come to a state of equilibrium. The centrifugal fields employed are usually much lower than in the case of sedimentation–velocity experiments. The majority of molecular-weight-distribution studies have been made by sedimentation–velocity techniques, and only this method will be treated here.

Application of a centrifugal force to a heterogeneous solution causes a separation of the various components of the solute according to their rates of sedimentation. The higher molecular weights have faster rates of sedimentation. Therefore, the centrifugation gives a physical fractionation, represented by a broadening of the sedimentation boundary. Given a physical record representing the various rates of sedimentation and the amount of each species traveling at a given rate of sedimentation, it is in theory possible to construct a molecular-weight distribution. In practice, this is not easy because the broadening of the sedimentation boundary is due not only to heterogeneity in the polymer but to diffusion, concentration, and pressure effects. Each of these requires appropriate mathematical correction before the shape of the boundary can be analyzed for a true molecular-weight distribution. Such a detailed analysis is extremely complex even if all the necessary physical parameters to make such calculations are known. The methods of interpreting the data, therefore, fall into various approximations.

There are two general approaches. One involves measuring certain parameters of the gradient curve. These are then used to compute heterogeneity coefficients for certain assumed forms of the molecular-weight-distribution function. The other general method entails measurement of the distribution of sedimentation coefficients, which can then, provided the sedimentation coefficient–molecular-weight relationship is known, be converted to a molecular-weight-distribution curve.

Consider first the determination of the distribution of sedimentation coefficients, which is the preferred technique if it is feasible. Normalized distributions of sedimentation coefficients can be readily computed from the conventional refractive-index–gradient curves by transformation of coordinates (128–130). The ordinate represents the relative frequency of material having a sedimentation coefficient given by the abscissa. Frequently, this technique is used to compare polymers. It is not a true distribution, again because no corrections have been made for the effects of pressure, diffusion, or concentration. However, this comparative technique can frequently be useful if high accuracy is not required. If the centrifugation is carried out at

Θ conditions, the pressure effect can be reduced to within the order of the experimental accuracy (131); by operating at the Θ temperature, nonideality with respect to concentration is eliminated.

The apparent distribution curves change with time owing to the effect of diffusion, but it is possible to eliminate this effect by extrapolation of measurements made at a series of times to infinity (132–134). This extrapolation is valid because the spreading of the boundary due to differences in sedimentation coefficients is proportional to the square root of the time. Concentration dependence of sedimentation causes a sharpening of the boundary. Again, if measurements are made at several concentrations, graphical extrapolation may be used to eliminate this effect. References 135–137 describe procedures for extrapolation to infinite dilution. Such calculations can be quite laborious. To convert a sedimentation distribution into a molecular-weight distribution, the equation

$$S = KM^a \tag{10}$$

is used, where K and a are constants for each specific polymer–solvent system, S is sedimentation coefficient, and M is molecular weight. Strictly, this relationship is applicable only to linear polymers. Provided the constants in the equation are known, the molecular-weight distribution can be obtained readily from the distribution of sedimentation coefficients. In practice, however, the constants may not be known; if this is true, then a considerable experimental program is required to evaluate them.

Summarizing, uncorrected data may be used for a rough comparison on the degree of polydispersity. If Θ conditions are chosen, pressure effects are greatly reduced. Procedures are available for correcting for diffusion and for concentration, but, if these are employed, the measuring time is greatly increased. Ultracentrifugation will probably be principally used when absolute molecular-weight distributions are not required but a comparison is sufficient.

For comparative purposes, the use of readily calculated parameters and an assumed distribution function can be useful. For example, Gralen (138) describes the width of the curve as defined by the ratio of the area to the maximum height as a useful parameter. Variations on the method have been reported (139,140). Baldwin and Williams (132, 141) used moments as a measure of the position and width of the boundary gradient curve. This technique includes elimination of the diffusion effect but does not permit very detailed measurement of polymer distribution.

Rheological Methods

Studies have been made for many years to develop a means of determining molecular weights and molecular-weight distributions from rheological measurements in order to avoid the more complex precipitation and solution techniques. Plots of apparent viscosity or average shear rate versus shear stress are commonly used. Such non-Newtonian flow curves represent the response of the solution or melt to variable shearing force. The approaches are basically empirical in relating the flow curves to absolute molecular weights and distributions. Most approaches involve relating flow curves to inhomogeneity numbers such as $U = \overline{M}_w/\overline{M}_n - 1$ (142). The constants in the equation $[\eta] = K \cdot \overline{M}_v{}^a$, relating molecular weight to intrinsic viscosity, are sensitive to molecular-weight distribution. With increasing homogeneity, K becomes smaller and a remains unchanged or increases (143). Using equation 11 for viscosity-average molecular weight,

$$\overline{M}_v = \left[\frac{\Sigma N_i M_i^{a+1}}{\Sigma N_i M_i}\right]^{1/a} \tag{11}$$

a comparison of $\overline{M}_v$ obtained in different solvents should provide an index of polydispersity (144–147). In general, however, viscosity measurements do not provide a sensitivity estimation of polydispersity (148).

Usually the decrease in viscosity with increasing shear rate will be less sharp and spread over more decades of viscosity for solutions of a polymer with a broad molecular-weight distribution. Plots of log shear rate versus log shear stress or log viscosity versus log shear rate are widely used. Dynamic measurements also offer a measure of polydispersity. Flow curves are usually obtained for dilute solutions. Non-Newtonian effects are larger in concentrated systems but perhaps more difficult to interpret (149,150). The flow curve for undiluted polymer consists of several contributions: the flow of individual particles, the contribution of mutual interactions, and the effect of distribution of size and shape of particles. The last term, which is a measure of polydispersity, must play a prominent role in the shear dependence of viscosity if molecular-weight distribution information is to be extracted from flow curves.

At the onset of contributions from individual molecules, the non-Newtonian flow will be proportional to some power of molecular weight, and the amount of its contribution proportional to the number of such molecules present in solution. Therefore, low-mw species will generally contribute less than high-mw polymer molecules. Thus, the polydispersity curve will show that the amount of shear dependence at maximum shear rate should be a measure of the height of the distribution curve, whereas the shear rate at the first observation of non-Newtonian flow will be indicative of the largest molecules present. The low-mw tail of the distribution will generally escape detection by flow curve analysis.

For defining polydispersity from shear curves, it has been found that plots of log viscosity versus shear stress at the capillary wall yield straight lines over a range of several orders of magnitude (151,152). An alternative method is to plot log viscosity versus log shear stress, which is consistent with the "power law" expression for non-Newtonian flow. The slope of this line may serve as an index of polydispersity. Alternative interpretations of non-Newtonian shear curves in terms of distribution have also been used (153).

It has been observed that the inflection point in plots of log shear rate, D, versus log shear stress, T, characterized by its value of the mean shear rate, $\hat{D}$, is related to molecular weight of the polymer in solution by the equation $\hat{D} = a \cdot M^{-b}$; a and b are empirical constants which have been determined for a variety of systems (154). A constant value of $\hat{D}$ is achieved only at low concentrations. From this single molecular weight and the shape of the flow curve, information on polydispersity of the polymer in solution can be obtained (155–157).

Flow curves over the various modes of relaxation have been constructed. This summation appears in the equation instead of a closed function for flow behavior (158).

The initial shear stress for non-Newtonian flow, T_i, of polyethylene melts is related to molecular weight by equation 12.

$$1/T_i \sim (1 - M_e/M)^n \tag{12}$$

where M_e is the molecular weight between entanglements for a molecule of such size that entanglements occur. Plots of log $1/T_i$ versus log M/M_e give straight lines which intersect the ordinate and have larger slopes the broader the molecular-weight distribution (159).

The measurement of viscoelastic properties of solutions and melts can provide a spectrum of relaxation times which is also related to molecular-weight distribution. For linear amorphous polymers, the relaxation or creep functions extend over a longer time than theoretically predicted (160,161). This is because the region of rubbery flow depends on both molecular weight and polydispersity (162,163). Monodispersed polymers have clearly narrower distributions of relaxation times. The effect of polydispersity on steady-state elastic compliances is proportional to $\overline{M}_{z+1} \cdot \overline{M}_z / \overline{M}_w$ (164–166) (see also VISCOELASTICITY). Flow birefringence (qv) of polymer solutions also presents a qualitative method of determining molecular-weight distribution (166, 167).

Brownian Diffusion

In a diffusion experiment an artificial boundary is created between a solution of concentration c_0 and a pure solvent. Suitable equipment for this purpose has been described in several papers (168–171). The concentration gradient $\partial c/\partial x$ in the x direction perpendicular to the original boundary tends to equalize the concentration difference between the layers. The progress of this process with time, t, is given by Fick's second law (172):

$$\frac{\partial c}{\partial t} = \frac{\partial}{\partial x}\left(D \frac{\partial c}{\partial x}\right) \tag{13}$$

Here D is the isothermal diffusion constant. The concentration change is observed by the same optical methods as in the case of ultracentrifugation, ie, Lamm's scale method, Philpot-Svensson schlieren optics, or interference optics.

In the solution of equation 13 for a homogeneous solute with a diffusion constant independent of concentration, the concentration gradient is described by a symmetrical Gaussian distribution (173). The corresponding curve for a polydisperse solute is lower and wider. The ratio of the heights, H, of the two normalized curves can be used as a measure of the width of the distribution of the polydisperse polymer (174,175).

More information concerning the shape of the distribution can be obtained by following the time dependence of polymer concentration along the x direction. Methods exist for calculating various averages of the diffusion constant in this manner (176,177). Using these averages, a distribution of diffusion coefficients can be obtained. If the relation between diffusion constant and molecular weight is known for the polymer–solvent system under investigation, the former distribution can be converted into a molecular-weight distribution.

It was assumed thus far that the diffusion coefficient is independent of concentration. However, this has been found to be true only in ideal solutions (178). In all other cases measurements must be carried out at several concentrations and extrapolated to zero concentration.

The method has been applied to a series of natural polymers (174), to poly(vinylpyrrolidone) (175), to poly(vinyl alcohol) (179), and to a styrene–butadiene copolymer (174).

Osmodialysis is the name given to the technique of diffusion of polymer molecules through an imperfect osmotic membrane. Except for the presence of the membrane, the experiment resembles that of free Brownian diffusion, and similar results should be obtained (180,181).

Fractionation of Mixtures and Copolymers

Additional complications are encountered if the polymer under investigation is inhomogeneous not only with respect to molecular weight but also with respect to chemical composition.

In the case of mixtures, it is usually possible to find a solvent–nonsolvent system with which components can be separated by repeated precipitation fractionation. An example of a practical application is the separation of the soluble portion of a polyolefin from the crystalline portion (182).

In the case of a copolymer the fractionation should be carried out by any desirable method in several solvent–nonsolvent systems. Solubility of the sample in each system is affected by molecular weight as well as by chemical composition. Consequently, the chemical nature of fractions with identical molecular weights will vary with the solvent–nonsolvent system employed. The system showing the largest differences should preferably be used. As an illustrative example, a copolymer of vinyl acetate and vinyl chloride with an average chlorine content of 31.3 wt % was fractionated by precipitation in acetone–petroleum ether. The chlorine content of the fractions varied between 29 and 33 wt %. Fractionation in acetone–methanol and in several other systems resulted in a range of 19.6–38.3 wt % chlorine (183).

Chemical inhomogeneity may be characterized by several parameters (184). The weight averages $\bar{E}_w$ of composition of a copolymer obtained from two monomers, 1 and 2, are defined by

$$\bar{E}_{1,w} = \frac{\Sigma c_i E_{1,i}}{\Sigma c_i} \tag{14}$$

$$\bar{E}_{2,w} = \frac{\Sigma c_i E_{2,i}}{\Sigma c_i} \tag{15}$$

where c_i represents the weight of the ith fraction and $E_{1,i}$ and $E_{2,i}$ are the relative amounts of components 1 and 2 in the ith fraction.

The chemical inhomogeneity with respect to component 1 is given by

$$U_1 = U_1^+ + U_1^- \tag{16}$$

The partial inhomogeneities U_1^+ and U_1^- are defined as

$$U_1^+ = \frac{\Sigma c_{1,i}^+ \, \Delta E_{1,i}^+}{\Sigma c_{1,i}^+} \tag{17}$$

$$U_1^- = \frac{\Sigma c_{1,i}^- \, \Delta E_{1,i}^-}{\Sigma c_{1,i}^-} \tag{18}$$

The plus sign refers to fractions that have more than the average value of chemical composition, while the minus sign designates those that have less.

A detailed discussion of this topic with a table of suitable fractionation conditions for a large variety of copolymers is given in Ref. 185.

Treatment of Data

The results of a successful fractionation can be summarized as shown in the first two columns of Table 2. The molecular weight M_i of each fraction has to be determined by a convenient method, in most cases by measuring the intrinsic viscosity $[\eta]$ and using the proper Mark-Houwink equation relating $[\eta]$ and the molecular weight. In the fourth column in Table 2 the weight fraction, w_i, of each fraction has been calculated. The number-average molecular weight, $\bar{M}_n$, and the weight-average molecular weight, $\bar{M}_w$, of the whole polymer can now be obtained using the equations defining these quantities

Table 2. Fractionation of 49.85 g of Polyisobutylene

Fract. no.	Wt, g	Mol wt, M_i	Wt fract., w_i	Cumulative wt fract., $C(M_i)$
1	0.581	1,600	0.0120	0.0060
2	2.162	4,800	0.0448	0.0344
3	1.509	7,600	0.0313	0.0724
4	1.526	9,700	0.0316	0.1039
5	0.914	11,600	0.0189	0.1292
6	0.880	12,400	0.0182	0.1477
7	2.523	14,700	0.0523	0.1829
8	2.638	18,700	0.0546	0.2364
9	2.200	20,900	0.0456	0.2865
10	3.050	23,700	0.0632	0.3409
11	3.215	29,500	0.0666	0.4058
12	4.447	35,300	0.0921	0.4851
13	3.195	43,900	0.0662	0.5643
14	2.428	51,400	0.0503	0.6226
15	2.808	58,900	0.0582	0.6768
16	3.072	66,500	0.0636	0.7377
17	3.157	78,500	0.0654	0.8022
18	2.357	95,400	0.0488	0.8593
19	1.576	110,000	0.0326	0.9000
20	2.446	129,000	0.0507	0.9417
21	1.598	142,000	0.0331	0.9669
total	48.282			

$$\bar{M}_n = \frac{1}{\Sigma w_i/M_i} \tag{19}$$

$$\bar{M}_w = \Sigma w_i M_i \tag{20}$$

The next accessible function is the cumulative molecular-weight distribution $C(M_i)$ defined by equation 21. Its calculation is described below.

$$C(M_i) = \int_0^M W(M_i)dM \tag{21}$$

$W(M_i)$ is the differential molecular-weight distribution, which can be obtained from $C(M_i)$ according to

$$W(M_i) = \frac{dC(M_i)}{dM_i} \tag{22}$$

This function represents the amount in grams present of each molecular-weight species in 1 g of unfractionated polymer. The mole fraction of each species is given by the differential number distribution $N(M_i)$, defined as

$$N(M_i) = \frac{W(M_i)/M_i}{\int_0^\infty \frac{W(M_i)}{M_i}\, dM} \tag{23}$$

The cumulative molecular-weight distribution $C(M_i)$ is most conveniently obtained from the experimental data in Table 2 by a procedure entailing the assumption that each fraction is distributed symmetrically around its average molecular weight, M_i, and that the extreme values are given by the average molecular weights of the neighboring fractions (186). Consequently, the value of $C(M_i)$ for a particular fraction is obtained by summing the weight fractions, w_i, of all previous fractions and adding one-half the weight fraction of that particular fraction:

$$C(M_i) = {}^1/_2 w_i + \sum_1^{i-1} w_i \tag{24}$$

These points are plotted versus the corresponding molecular weights, M_i, and connected by a smooth curve.

The differential molecular-weight distribution is obtained by graphically determining the slope of the cumulative curve at selected molecular weights and plotting these values versus the corresponding molecular weights. This function can again be presented as a smooth continuous curve or in the form of a histogram.

Division of selected ordinate values of the differential weight curve by the corresponding molecular weights and plotting the ratios versus these molecular weights results in a differential number distribution.

Several authors have published theoretically founded or empirical analytical functions to describe molecular-weight distributions (142,187–190). When applicable, the use of these functions can greatly reduce the labor required to evaluate fractionation data and may sometimes make it possible to obtain information regarding the polymerization kinetics of the sample under investigation.

Bibliography

1. H. Tompa, *Polymer Solutions*, Butterworths Publications Ltd., London, 1956.
2. M. J. Voorn, *Advan. Polymer Sci.* **1,** 192 (1959).
3. M. L. Huggins and H. Okamoto, in M. J. R. Cantow, ed., *Polymer Fractionation*, Academic Press, Inc., New York, 1966.
4. M. L. Huggins, *J. Phys. Chem.* **46,** 151 (1942).
5. P. J. Flory, *J. Chem. Phys.* **10,** 51 (1942).
6. M. L. Huggins, *J. Am. Chem. Soc.* **86,** 3535 (1964).
7. P. J. Flory, *Principles of Polymer Chemistry*, Cornell University Press, Ithaca, New York, 1953.
8. H. Staudinger, K. Frey, and W. Starck, *Ber.* **60,** 1782 (1927).
9. E. H. Merz and R. W. Raetz, *J. Polymer Sci.* **5,** 587 (1950).
10. D. R. Morey and J. W. Tamblyn, *J. Phys. Chem.* **51,** 721 (1947).
11. Y. Fujisaki and H. Kobayashi, *Kobunshi Kagaku* (Chemistry of High Polymers) **18,** 305 (1961).
12. Ref. 7, p. 341.
13. L. H. Arond and H. P. Frank, *J. Phys. Chem.* **58,** 953 (1954).
14. S. N. Chinai, *J. Polymer Sci.*, **25,** 413 (1957).

15. W. H. Beattie and C. Booth, *J. Appl. Polymer Sci.* **7,** 507 (1963).
16. R. Koningsveld and A. J. Pennings, *Rec. Trav. Chim.* **5,** 552 (1964).
17. C. A. Baker and R. J. P. Williams, *J. Chem. Soc.* **1956,** 2352.
18. J. L. Jungnickel and F. J. Weiss, *J. Polymer Sci.* **49,** 437 (1961).
19. N. S. Schneider, L. G. Holmes, C. F. Miyal, and J. D. Loconti, *J. Polymer Sci.* **37,** 551 (1959).
20. N. S. Schneider and L. G. Holmes, *J. Polymer Sci.* **38,** 552 (1959).
21. W. F. Haddon, Jr., R. S. Porter, and J. F. Johnson, *J. Appl. Polymer Sci.* **8,** 1371 (1964).
22. C. J. Panton, P. H. Plesch, and P. P. Rutherford, *J. Chem. Soc.* **1964,** 2586.
23. T. J. R. Weakley, R. J. P. Williams, and J. D. Wilson, *J. Chem. Soc.* **1960,** 3963.
24. J. V. Ch'ien and S. N. Chu, *K'o Hsueh T'ung Pao* **1959,** 525; *Chem. Abstr.* **54,** 7216b (1960).
25. J. E. Guillet, R. L. Combs, D. F. Slonaker, and H. W. Coover, Jr., *J. Polymer Sci.* **47,** 307 (1960).
26. J. E. Guillet, R. L. Combs, D. F. Slonaker, J. T. Summers, and H. W. Coover, Jr., *SPE Trans.* **2,** 164 (1964).
27. C. M. Hansen and G. A. Sather, *J. Appl. Polymer Sci.* **8,** 2479 (1964).
28. J. M. Hulme and L. A. McLeod, *Polymer* **3,** 153 (1963).
29. S. R. Caplan, *J. Polymer Sci.* **35,** 409 (1959).
30. N. T. Pope, T. J. Weakley, and R. J. P. Williams, *J. Chem. Soc.* **1959,** 3442.
31. A. Capiro, P. Cordier, J. Jozefowicz, and J. Sebban-Damon, *J. Polymer Sci.* [C] **4,** 491 (1963).
32. W. Cooper, G. Vaughan, and R. W. Madden, *J. Appl. Polymer Sci.* **1,** 329 (1959).
33. W. Cooper, G. Vaughan, and J. Yardley, *J. Polymer Sci.* **59,** S2 (1962).
34. N. S. Schneider, J. D. Loconti, and L. G. Holmes, *J. Appl. Polymer Sci.* **5,** 354 (1961).
35. P. M. Henry, *J. Polymer Sci.* **36,** 3 (1959).
36. D. L. Flowers, W. A. Hewett, and R. D. Mullineau, *J. Polymer Sci.* [A] **2,** 2305 (1964).
37. M. J. R. Cantow, R. S. Porter, and J. F. Johnson, *J. Polymer Sci.* [C] **1,** 187 (1963).
38. D. C. Pepper and P. P. Rutherford, *J. Appl. Polymer Sci.* **2,** 100 (1959).
39. M. F. Vaughan and J. H. S. Green, *Soc. Chem. Ind. (London) Monograph* **17,** 81 (1963).
40. R. S. Alm, R. J. P. Williams, and A. Tiselius, *Acta Chem. Scand.* **6,** 826 (1952).
41. A. Cherkin, F. E. Martinez, and M. S. Dunn, *J. Am. Chem. Soc.* **75,** 1244 (1953).
42. L. M. Marshall, K. O. Donaldson, and F. Friedberg, *Anal. Chem.* **24,** 773 (1952).
43. G. Meyerhoff and J. Romatowski, *Makromol. Chem.* **74,** 222 (1964).
44. N. S. Schneider, *J. Polymer Sci.* [C] **8,** 179 (1965).
45. J. F. Gernert, M. J. R. Cantow, R. S. Porter, and J. F. Johnson, *J. Polymer Sci.* [C] **1,** 195 (1963).
46. N. S. Schneider, *Anal. Chem.* **33,** 1829 (1961).
47. J. Polacek, L. Schulz, and I. Kossler, Paper presented at IUPAC Symposium, Prague, 1965.
48. N. S. Schneider, J. D. Loconti, and L. G. Holmes, *J. Appl. Polymer Sci.* **3,** 251 (1960).
49. G. V. Schulz, P. Deussen, and A. G. R. Scholz, *Makromol. Chem.* **69,** 47 (1963).
50. L. D. Moore, Jr., A. Greear, and J. O. Sharp, *J. Polymer Sci.* **59,** 339 (1962).
51. A. S. Kenyon, I. O. Salyer, J. E. Kurz, and D. R. Brown, *J. Polymer Sci.* [C] **8,** 205 (1965).
52. M. J. R. Cantow, R. S. Porter, and J. F. Johnson, *Nature* **192,** 752 (1961).
53. M. J. R. Cantow, R. S. Porter, and J. F. Johnson, *J. Appl. Polymer Sci.* **8,** 2963 (1964).
54. L. E. Maley, *J. Polymer Sci.* [C] **8,** 253 (1965).
55. E. Ungerer, *Kolloid Z.* **36,** 228 (1925).
56. H. Deuel, J. Solms, and L. Anyas-Weisz, *Helv. Chim. Acta* **33,** 2171 (1950).
57. R. M. Wheaton and W. A. Bauman, *Ann. N. Y. Acad. Sci.* **57,** 159 (1953).
58. J. Porath and P. Flodin, *Nature* **183,** 1657 (1959).
59. K. O. Pedersen, *Arch. Biochem. Biophys. Suppl.* **1,** 157 (1962).
60. J. Porath, *Biochim. Biophys. Acta* **39,** 193 (1960).
61. J. Porath, *Helv. Chim. Acta* **4,** 776 (1959).
62. W. Bjork and J. Porath, *Acta Chem. Scand.* **13,** 1256 (1959).
63. M. F. Vaughan, *Nature* **188,** 55 (1960).
64. M. F. Vaughan, *Nature* **195,** 801 (1962).
65. P. I. Brewer, *Nature* **188,** 934 (1960).
66. P. I. Brewer, *Nature* **190,** 625 (1961).
67. P. I. Brewer, *J. Inst. Petrol.* **48,** 277 (1962).
68. J. C. Moore, *J. Polymer Sci.* [A] **2,** 835 (1964).

69. P. Flodin, "Dextran Gels and Their Application in Gel Filtration," Dissertation, University of Uppsala, Sweden, 1962.
70. J. C. Moore and J. G. Hendrickson, *J. Polymer Sci.* [C] **8,** 233 (1965).
71. K. H. Altgelt, *Makromol. Chem.* **88,** 75 (1965).
72. P. Flodin, *J. Chromatog.* **5,** 106 (1961).
73. J. Porath and E. B. Lindner, *Nature* **191,** 69 (1961).
74. R. E. Jentoft and T. H. Gouw, *Anal. Chem.* **38,** 949 (1966).
75. F. Rodriguez, R. A. Kulakowski, and O. K. Clark, *Ind. Eng. Chem. Prod. Res. Develop.* **5,** 121 (1966).
76. J. Porath and P. Flodin, *Arch. Biochem. Biophys. Suppl.* **1,** 152 (1962).
77. M. J. R. Cantow, R. S. Porter, and J. F. Johnson, *J. Polymer Sci.* [B], in press.
78. G. Meyerhoff, *Ber. Bunsenges. Physik. Chem.* **69,** 866 (1965).
79. J. L. Waters, *Preprints ACS Div. Polymer Chem.* **6,** 1061 (1965).
80. G. Meyerhoff, *Makromol. Chem.* **80,** 282 (1965).
81. D. M. W. Anderson and J. F. Stoddart, *Anal. Chim. Acta* **34,** 401 (1966).
82. P. I. Brewer, *Polymer* **6,** 603 (1965).
83. G. A. Gilbert, *Nature* **210,** 299 (1966).
84. J. G. Hendrickson and J. C. Moore, *J. Polymer Sci.* [A1] **4,** 167 (1966).
85. W. B. Smith and A. Kollmansberger, *J. Phys. Chem.* **69,** 4157 (1965).
86. D. J. Harmon, *J. Polymer Sci.* [C] **8,** 243 (1965).
87. H. E. Adams, K. Farhat, and B. L. Johnson, *Ind. Eng. Chem. Prod. Res. Develop.* **5,** 126 (1966).
88. N. Nakajima, *J. Polymer Sci.* [A2] **5,** 101 (1966).
89. M. J. R. Cantow, R. S. Porter, and J. F. Johnson, *J. Polymer Sci.* [C] (Prague issue), in press.
90. A. R. Schultz, *J. Appl. Polymer Sci.* **10,** 353 (1966).
91. W. B. Smith, J. A. May, and C. W. Kim, *J. Polymer Sci.* [A2] **4,** 365 (1966).
92. G. Heufer and D. Braun, *J. Polymer Sci.* [B] **3,** 495 (1965).
93. H. E. Pickett, M. J. R. Cantow, and J. F. Johnson, *J. Appl. Polymer Sci.* **10,** 917 (1966).
94. L. H. Tung, *J. Appl. Polymer Sci.* **10,** 375 (1966).
94a. H. Benoit, Z. Grubisic, P. Rempp, D. Decker, and J. L. Zilliox, *Gel Permeation Chromatography Measurements on Linear and Branched Polystyrenes of Known Structure,* Waters Associates, Inc.
95. H. L. Berger and A. R. Schultz, *J. Polymer Sci.* [A] **2,** 3643 (1965).
96. F. Rodriguez and O. K. Clark, *Ind. Eng. Chem. Prod. Res. Develop.* **5,** 118 (1966).
97. M. Bohdanecky, P. Kratochvil, and K. Solc, *J. Polymer Sci.* [A] **3,** 4153 (1965).
98. I. Kossler and J. Krejsa, *J. Polymer Sci.* **57,** 509 (1962).
99. P. Debye and A. M. Bueche, in C. H. A. Robinson, ed., *High Polymer Physics,* Chemical Publishing Co., Inc., New York, 1948, p. 497.
100. H. F. Fritzmeier and J. J. Hermans, *Bull. Soc. Chim. Belges* **57,** 136 (1948).
101. G. Langhammer and K. Quitzsch, *Makromol. Chem.* **17,** 74 (1955).
102. G. M. Guzman and J. M. Fatou, *Anales Real Soc. Espan. Fiz. Quim. (Madrid), Ser. B* **54,** 609 (1958).
103. I. Kossler and J. Krejsa, *J. Polymer Sci.* **35,** 308 (1959).
104. G. Langhammer, H. Pfennig, and K. Quitzsch, *Z. Elektrochem.* **62,** 458 (1958).
105. G. Langhammer, *Svensk Kem. Tidskr.* **69,** 328 (1957).
106. J. Krejsa, *Makromol. Chem.* **33,** 244 (1959).
107. I. Kossler and M. Stolka, *J. Polymer Sci.* **44,** 213 (1960).
108. D. R. Morey and J. W. Tamblyn, *J. Appl. Phys.* **16,** 419 (1945).
109. H. E. Adams and P. O. Powers, *Ind. Eng. Chem. Anal. Ed.* **15,** 711 (1943).
110. J. Bischoff and V. Desreux, *Bull. Soc. Chim. Belges* **60,** 137 (1951).
111. A. Oth and V. Desreux, *Bull. Soc. Chim. Belges* **63,** 261 (1954).
112. W. Scholtan, *Makromol. Chem.* **24,** 104 (1957).
113. S. Krozer, N. Vainryb, and L. Silena, *Vysokomolekul. Soedin.* **2,** 1876 (1960).
114. I. E. Climie and E. F. T. White, *J. Polymer Sci.* **47,** 149 (1960).
115. K. P. Shen Kwei, *J. Polymer Sci.* [A] **1,** 2309 (1963).
116. F. D. Hartley, *J. Polymer Sci.* **34,** 397 (1959).
117. G. Gooberman, *J. Polymer Sci.* **40,** 469 (1959).
118. A. R. Mathieson, *J. Colloid Sci.* **15,** 387 (1960).

119. S. Coppick, O. A. Battista, and M. R. Lytton, *Ind. Eng. Chem.* **42,** 2533 (1950).
120. G. Beall, *J. Polymer Sci.* **4,** 483 (1949).
121. H. Mark and R. Simha, *Trans. Faraday Soc.* **36,** 611 (1940).
122. F. W. Billmeyer, Jr., and W. H. Stockmayer, *J. Polymer Sci.* **5,** 121 (1950).
123. T. Svedberg and K. O. Pederson, *The Ultracentrifuge,* Oxford University Press, New York, 1940.
124. J. B. Nichols and E. D. Bailey in A. Weissberger, ed., *Physical Methods of Organic Chemistry,* 3rd ed., Vol. I, Interscience Publishers, Inc., New York, 1960, p. 1007.
125. H. K. Schachman, *Ultracentrifugation in Biochemistry,* Academic Press, Inc., New York, 1959.
126. T. Svedberg and H. Rinde, *J. Am. Chem. Soc.* **45,** 943 (1923).
127. T. Svedberg and J. B. Nichols, *J. Am. Chem. Soc.* **45,** 2910 (1923).
128. H. Fujita, *Mathematical Theory of Sedimentation Analysis,* Academic Press, Inc., New York, 1962.
129. W. B. Bridgman, *J. Am. Chem. Soc.* **64,** 2349 (1942).
130. R. Signer and H. Gross, *Helv. Chim. Acta* **17,** 726 (1934).
131. I. H. Billick, *J. Phys. Chem.* **66,** 1941 (1962).
132. R. L. Baldwin and J. W. Williams, *J. Am. Chem. Soc.* **72,** 4325 (1950).
133. A. F. V. Eriksson, *Acta Chem. Scand.* **10,** 360 (1956).
134. R. L. Baldwin, *J. Phys. Chem.* **63,** 1570 (1960).
135. J. W. Williams and W. M. Saunders, *J. Phys. Chem.* **58,** 854 (1954).
136. R. L. Baldwin, *J. Am. Chem. Soc.* **76,** 402 (1954).
137. N. Gralen and G. Lagermalm, *J. Phys. Chem.* **56,** 514 (1952).
138. N. Gralen, "Sedimentation and Diffusion Measurements on Cellulose Derivatives," Dissertation, Uppsala, 1944.
139. I. Jullander, *J. Polymer Sci.* **2,** 329 (1947).
140. A. F. V. Eriksson, *Acta Chem. Scand.* **7,** 623 (1953).
141. J. W. Williams, R. L. Baldwin, W. M. Saunders, and P. G. Squire, *J. Am. Chem. Soc.* **74,** 1542 (1952).
142. G. V. Schulz, *Z. Physik. Chem.* [B] **43,** 25 (1939).
143. H. P. Frank and J. W. Breitenbach, *J. Polymer Sci.* **6,** 609 (1951).
144. P. F. Onyon, *Nature* **183,** 1670 (1959).
145. H. L. Frisch and J. L. Lundberg, *J. Polymer Sci.* **37,** 123 (1959).
146. J. L. Lundberg, M. Y. Hellman, and H. L. Frisch, *J. Polymer Sci.* **46,** 3 (1960).
147. C. Mussa, *J. Polymer Sci.* **41,** 541 (1959).
148. J. W. Breitenbach, *Makromol. Chem.* **60,** 18 (1963).
149. R. Schwaben, *Einführung in die Viskosimetrie und Rheometrie,* Springer-Verlag, Berlin, 1952.
150. W. Philippoff, *Viskosität der Kolloide,* Th. Steinkopff Verlag, Dresden und Leipzig, 1942.
151. W. Meskat, *Chem. Ingr. Tech.* **24,** 333 (1952).
152. J. Schurz, *Das Papier* **9,** 45 (1955); *Kolloid-Z.* **138,** 149 (1954).
153. K. Edelmann, *Rheol. Acta* **1,** 53 (1958).
154. J. Schurz, *J. Colloid Sci.* **14,** 492 (1959).
155. J. Schurz and H. Streitzig, *Monatsh. Chem.* **87,** 632 (1956); **88,** 325 (1957).
156. J. Schurz, G. Gröblinghoff, and K. Windisch, *Holzforschung* **15,** 8 (1961).
157. J. Schurz and H. Streitzig, *Monatsh. Chem.* **89,** 229 (1958).
158. E. Menefee and W. L. Peticolas, *Nature* **189,** 745 (1961).
159. H. P. Schreiber, E. B. Bagley, and D. C. West, *Polymer* **4,** 365 (1963).
160. H. Leaderman, R. G. Smith, and R. W. Jones, *J. Polymer Sci.* **14,** 47 (1954).
161. J. D. Ferry, *Viscoelastic Properties of Polymers,* John Wiley & Sons, Inc., New York, 1961, pp. 169, 176, 290, 384.
162. A. V. Tobolsky, *J. Appl. Phys.* **27,** 673 (1956).
163. A. V. Tobolsky, A. Mercurio, and K. Murakami, *J. Colloid Sci.* **13,** 196 (1958).
164. F. Bueche, *J. Appl. Phys.* **26,** 738 (1955).
165. J. D. Ferry, L. Jordan, W. Evans, and M. Johnson, *J. Polymer Sci.* **14,** 261 (1954).
166. Ch. Sadron and H. Mosimann, *J. Phys. Radium* **9,** 384 (1938).
167. A. Peterlin, in F. R. Eirich, ed., *Rheology,* Vol. I, Academic Press, Inc., New York, 1956, p. 618.
168. O. Lamm, Dissertation, Uppsala, 1937.
169. H. Neurath, *Science* **93,** 431 (1941).

170. S. Claesson, *Nature* **158,** 834 (1946).
171. G. Meyerhoff, *Makromol. Chem.* **6,** 200 (1957).
172. A. Fick, *Ann. Phys.* **94,** 59 (1855).
173. O. Wiener, *Ann. Phys.* **49,** 105 (1893).
174. N. Gralen, Dissertation, Uppsala, 1944.
175. L. E. Miller and F. A. Hamm, *J. Phys. Chem.* **57,** 110 (1953).
176. M. Daune and L. Freund, *J. Polymer Sci.* **23,** 115 (1957).
177. J. Moacanin, V. F. Felicetta, and J. L. McCarthy, *J. Am. Chem. Soc.* **81,** 2052 (1959).
178. H.-J. Cantow, *Preprints IUPAC Symposium,* Wiesbaden 1952, Vol. II, Chap. 3.
179. L. Freund and M. Daune, *J. Polymer Sci.* **29,** 161 (1958).
180. M. Hoffman and M. Unbehend, *Makromol. Chem.* **88,** 256 (1965).
181. B. E. Hudson, Jr., *Preprints ACS Div. Polymer Chem.* **7,** 467 (1966).
182. O. Fuchs, *Makromol. Chem.* **58,** 247 (1962).
183. O. Fuchs, *Verhandlber. Kolloid-Ges.* **18,** 75 (1958).
184. H.-J. Cantow and O. Fuchs, *Makromol. Chem.* **83,** 244 (1965).
185. O. Fuchs and W. Schmieder, in M. J. R. Cantow, ed., *Polymer Fractionation,* Academic Press, Inc., New York, 1966.
186. G. V. Schulz and A. Dinglinger, *Z. Physik. Chem.* [B] **43,** 47 (1939).
187. W. D. Lansing and E. J. Kraemer, *J. Am. Chem. Soc.* **57,** 1369 (1935).
188. L. H. Tung, *J. Polymer Sci.* **20,** 495 (1956).
189. H. Wesslau, *Makromol. Chem.* **20,** 111 (1956).
190. F. C. Goodrich and M. J. R. Cantow, *J. Polymer Sci.* [C] **8,** 269 (1965).

General References

R. M. Screaton, "Column Fractionation of Polymers," in B. Ke, ed., *Newer Methods of Polymer Characterization,* John Wiley & Sons, Inc., New York, 1964.

P. W. Allen, *Techniques of Polymer Characterization,* Butterworths Publications Ltd., London, 1959.

J. Polymer Sci. [C] **8,** 161–312 (1965), Characterization and Fractionation, Papers presented at Symposiums on Analysis and Fractionation of Polymers, Chicago, Sept. 1964.

G. M. Guzman, in J. C. Robb and F. W. Peaker, eds., *Progress in High Polymers,* Academic Press, Inc., New York, 1961.

E. W. Channen, *Rev. Pure Appl. Chem.* **9,** 225 (1959).

M. J. R. Cantow, ed., *Polymer Fractionation,* Academic Press, Inc., New York, 1966.

Julian F. Johnson and Manfred J. R. Cantow
Chevron Research Company
Roger S. Porter
University of Massachusetts

MOLECULAR WEIGHT DETERMINATION

This article is intended to provide the industrial chemist with background information about the molecular weight determination methods which are most commonly used, as well as a guide to the selection of a method suitable for the sample under study. Selected references are given to reviews or monographs which should be consulted for more complete understanding of an individual technique (1–4). By these means, it is hoped that a chemist with a need for molecular weight information will be equipped either to discuss with a specialist a method which appears suitable for his sample, or to utilize many of the techniques himself.

The determination of molecular weights became possible about one hundred years ago when Avogadro's hypothesis that equal volumes of gases contain the same number of molecules under the same temperature and pressure conditions gained acceptance. This principle is the basis for a method which is still used for gases and volatile liquids and is described below. By 1885 the laws governing dilute solution behavior had been developed by Van't Hoff, in analogy to nonideal gases; this permitted measurement of molecular weights from observations of the colligative properties. Other methods have been developed in this century mostly from observations of physical properties of solutions which approach ideal behavior at infinite dilution.

Molecular weight determinations are truly aids to the identification of monomeric compounds. Preliminary determination of the elemental composition gives the empirical formula which is related to the true molecular formula by an integral number factor which can be estimated once an approximate molecular weight is known. The methods so employed include fairly simple methods, such as the Rast melting point depression method of rather low accuracy, the gas-law methods, and, finally, the tedious x-ray and mass spectrometer techniques which are able to provide highly accurate molecular weights; however, the molecular weights determined by the x-ray and mass spectrometer techniques are usually incidental to the determination of molecular structure.

In polydisperse systems, such as polymers, the molecular weight is not so important in establishing the molecular formula, but is of great value in explaining variations of physical properties when the proportions of components of different chain lengths vary. At very low molecular weights, polymers show no film-forming characteristics, whereas the strength and toughness as measured by tensile strength and elongation at break increase steeply for some polymers in the molecular weight range from 50,000 to 200,000. The presence of broad distributions of molecular weight alters the magnitude of these properties in the directions expected. However, the improvement in many physical properties at higher molecular weights must be balanced against the substantial increase of melt viscosity which makes extrusion and other fabrication more difficult. Knowledge of the molecular weight distribution in these cases may aid in providing the optimum physical properties through control of the conditions of polymerization. Other properties which change with molecular weight are the glass transition temperature and the solubility of a polymer, whose control may be important in applications where exposure to high temperatures or contact with water or other solvents occurs.

NONUNIFORM MOLECULAR WEIGHTS

One trivial source of nonuniform molecular weights is the presence of impurities in a sample, including residual water or other solvents, by-products of synthesis, or adsorbed surfactants. These components are a serious threat to accurate molecular weight measurements because they may contribute largely to the physical or chemical property used in determining molecular weight values. Purification of the sample is called for if such impurities are evident. Often vacuum drying is sufficient to remove solvents, but if they are strongly held, their removal is slow, and a reprecipitation or extraction with less tightly bound solvents is necessary. Recrystallization and extraction are often used to remove nonvolatile impurities; however, the most effective method depends on the particular sample and therefore no generalizations are made here.

The characterization of samples consisting of homologs of differing molecular weights is important both for monomeric organic compounds, such as mixed hydrocarbons of natural origin, and for polymers. Synthetic polymers are the outstanding examples of polydisperse systems, since no practical synthetic process presently produces uniform molecular weights, and in many cases a very wide range of molecular weights results. Heterogeneous systems such as polymers can be characterized most completely by determining the distribution of molecular weights, which indicates the proportions of all species present in the sample. A less complete characterization is provided by the determination of one or more average molecular weights for the sample; this is generally a much easier job, and may provide the information needed for a particular application.

Molecular Weight Distributions

Information on molecular weight distribution is commonly presented as a graph in which the relative number of molecules or the weight of molecules within a narrow range of molecular weights is plotted against the molecular weight, with the curve describing all species present in the sample. Curves representing both types are shown in Figure 1; these have been calculated from an ideal equation representing the distribution expected in some types of polymerization.

Three alternative methods may be employed to obtain curves of these kinds. (a) *Direct.* A method may be chosen which directly presents its results in this form. Only one such technique, *sedimentation equilibrium*, currently provides such results, and this is a complex process with additional difficulties for many polymer solutions. (b) *Fractionation.* The sample may be separated into components each having a narrow range of molecular weights which can be characterized by some average molecular weight determination. The results for each component are collected to produce

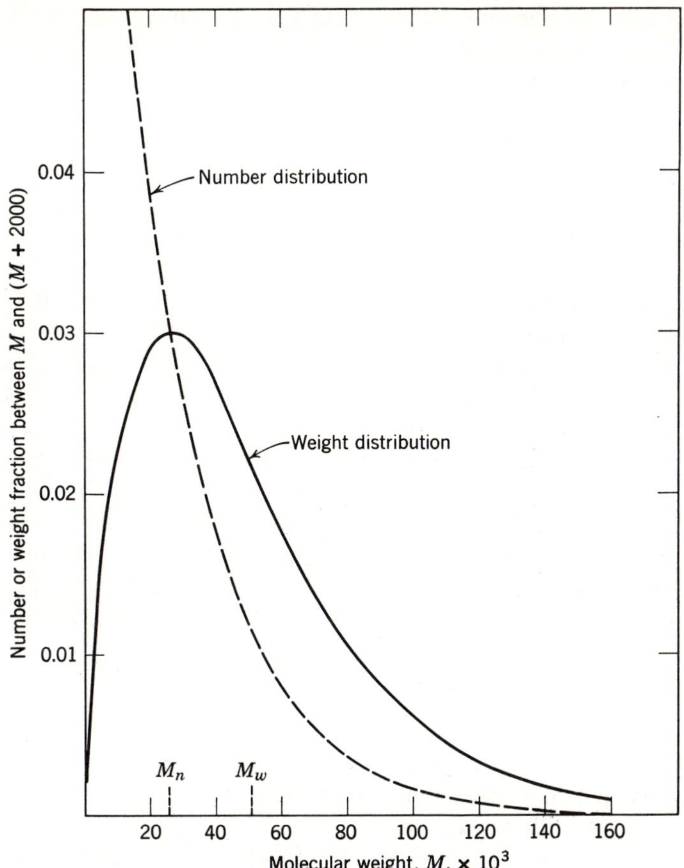

Fig. 1. Most probable molecular weight distribution, calculated for a polymer having $M_w = 50{,}000$ and $M_w/M_n = 2$.

the distribution curve. This is the process most commonly used since moderately efficient fractionation of polymers is not difficult, and quite small weights of fractions can be characterized easily by measurements of limiting viscosity numbers. These results can be converted to molecular weights after calibration as will be described later, and the distribution curve may then be constructed, first as an integral curve, which is then differentiated to produce a result somewhat resembling Figure 1. Fractionation methods have been reviewed by Hall (5) and by Cragg and Hammerschlag (6). Some doubt exists about the exactness of the resulting distribution curves, because the fractions obtained by any procedure are known to be unsharp in most cases, and corrections for this spread in molecular weights have had to be somewhat arbitrary. Nevertheless, it is the best method generally available. (c) *Calculation from averages.* If two average molecular weights of unequal weighting, such as M_n and M_w, are determined for the sample, and a reasonable distribution function can be assumed from prior knowledge of the method of sample preparation, then the approximate distribution curve may be calculated by insertion of the average values. Distribution functions suitable for many polydisperse polymer systems include that of Lansing and Kraemer (7) and the "most probable" distribution function of Schulz

(8,9), which was used for Figure 1. This method is sometimes used, but it requires good accuracy for the two average molecular weights, as well as the often dubious assumption regarding the distribution function.

Average Molecular Weights

The determination of a single value for the molecular weight of a heterogeneous sample is the easiest and the most frequently employed procedure for characterization. Despite the very incomplete knowledge of the sample which it provides, such a method can be quite adequate for many purposes. It is essential that the method for determining the molecular weight be properly chosen so that the results will correlate as closely as possible with whatever properties are under study, since these properties are average values also. The kinds of average to which several important physical properties are related are indicated below.

Methods which produce *number-average molecular weights*, M_n, essentially count the number of molecules present in a given weight of a sample. A mathematical expression of M_n is given as a summation of the product of each molecular weight value, M_i, times the number, n_i, of such molecules present, divided by the total number of all molecules present. This may also be expressed in terms of the weight, w_i, of each component as follows:

$$M_n = \frac{\sum n_i M_i}{\sum n_i} = \frac{\sum w_i}{\sum w_i / M_i} \tag{1}$$

This type of molecular weight is especially sensitive to small molecules, since there can be more present per unit weight, and so care must be taken with these methods to avoid low molecular weight impurities.

The *weight-average molecular weight*, M_w, in contrast, sums the contributions on the basis of the weight fraction of each species, leading to the following equation:

$$M_w = \frac{\sum w_i M_i}{\sum w_i} \tag{2}$$

Thus, this kind of average will be much more sensitive to components of high molecular weight.

Molecular weights may be averaged with various other types of weighting which have value in some applications. Only one additional average will be described here, because of its importance in molecular weight determinations.

The *viscosity-average molecular weight*, M_v, is that derived from measurements of the limiting viscosity number, $[\eta]$, which are calibrated in terms of known molecular weights by the following equation:

$$[\eta] = KM^a \tag{3}$$

where K is a constant and the exponent a is found to range from 0.5 to 1.0 for polymer molecules. When this expression is used to calculate a molecular weight from a measured viscosity number, the result is M_v, which is weighted as follows:

$$M_v = \left[\frac{\sum w_i M_i^a}{\sum w_i}\right]^{1/a} \tag{4}$$

This average is equal to the weight average when $a = 1$, but falls between the weight average and number average when a is smaller. It is always closer to a weight-average quantity, and so should always be associated with other weight-average values.

In order to illustrate the difference in these two averages, their values are indicated in Figure 1 for the "most probable" distribution. In this case the ratio M_w/M_n equals 2, whereas in some types of polymerization a ratio as high as 5 is found. Also, polymers having distribution curves consisting of two peaks may be synthesized under some conditions, or may result from mixing of polymers of widely different molecular weight. In contrast, polymers from which some components have been removed by extraction or fractional precipitation may have ratios of M_w/M_n approaching unity. It is the occurrence of such variability that influences the physical properties of polymers and thus requires a study of molecular weights. So long as the distribution curve does not depart radically from typical curve shapes, the measurement of two averages, such as M_w or M_v and M_n, for a series of samples often gives sufficient information to show how the differences in a physical property arise. When the results are inconsistent, then a more complete measurement of the molecular weight distribution, such as is provided by fractionation and characterization of the fractions, is required.

Molecular Weights and Physical Properties

The relations between molecular weight and physical properties have been established for some polymers (10), but disagreements or uncertainty exists in some cases. The more generally recognized relations include the following: (a) The melt viscosity is related to the weight-average molecular weight, being directly proportional to $M_w^{3.4}$. (b) The tensile strength appears related to an average between the number average and the weight average. However, tensile impact strength appears most closely related to the weight average. (c) The elongation at break increases with molecular weight, but is not clearly related to the number average. (d) The glass transition temperature, T_g, is related to molecular weight (11) through the following equation:

$$T_g = T_{g\infty} - (K/M) \tag{5}$$

(e) Stress relaxation has been shown to be related to the molecular weight distribution of polymers (12), but with different types of dependence for various regions of behavior.

Many of these relationships have not been fully clarified because of the broad molecular weight distributions present in the samples tested. The recent availability of polymers representing very narrow ranges of molecular weight, particularly vinyl polymers prepared by anionic polymerization, offers greatly improved possibilities for understanding these relationships, first for uniform molecular weights and then for broad distributions. Some early work in this area has appeared (10), and further work is in progress.

METHOD SELECTION

Table 1 has been designed to aid the chemist in selecting a molecular weight method suitable for his problem. If a polymer is involved, a choice of the type of average molecular weight needed must first be made, since various physical properties

Table 1. Selection of Molecular Weight Methods

Method	Type of result[a]	Best molecular weight range	Time required: measurement	Time required: calculation	Apparatus: source[b]	Apparatus: cost	Supporting measurement[c]	Usual limitations
gas density	M_n	gases	1 hr	short	lab, com	low		gas samples
vapor density	M_n	up to 200	short	short	lab, com	low		sample bp < 200°C
cryoscopic	M_n	up to 10^4	short	short	lab, com	low, medium	calib	insensitive at high mol wt
ebulliometric	M_n	up to 10^4	short	short	com	medium	calib	insensitive at high mol wt
vapor pressure thermoelectric	M_n	up to 10^4	short	short	com	high	calib	insensitive at high mol wt
isopiestic	M_n	up to 10^3	days	short	lab	low	calib	insensitive at high mol wt
osmotic pressure	M_n	2×10^4–5×10^5	short–18 hr	short	com, lab	medium, high, low		membrane porosity
light-scattering dissymmetry	M_w	10^4–10^5	1 hr[d]	short[d]	com	high	dn	single solvent, clarify
angular extrapolation	M_w	5×10^4–10^7	2 hr[d]	4 hr[d]	com	high	dn	single solvent, clarify
sedimentation velocity	M_z	10^4–10^7	2 hr[d]	8 hr[d]	com	very high	dn, V, ρ	nearly ideal solvent
sedimentation equilibrium	M_w, M_z	500 up	days–weeks	8 hr	com	very high	dn, V, ρ	nearly ideal solvent
sedimentation Archibald	M_w	500 up	1 hr	8 hr	com	very high	dn, V, ρ	
solution viscosity	M_v	10^3–10^7	short	short	com	low	calib	
end-group analysis	M_n	up to 2×10^4	1–4 hr	short	lab	low		condensation polymers
x-ray diffraction	M	any crystal	1–8 hr	varies	com	very high		high crystallinity
surface pressure	M_n	2×10^3–10^5	1 hr	short	lab	medium		surface-active sample and clean surfaces
mass spectrometer	M	up to 700	1–4 hr	short	com	very high	calib	volatile sample at 200°C or less

[a] M = true molecular weight; M_n = number-average molecular weight; M_w = weight-average molecular weight; M_v = viscosity-average molecular weight; M_z = molecular weight distribution.
[b] lab = apparatus may be constructed or modified in the laboratory; com = commercially available.
[c] calib = calibration with a suitable standard; dn = determination of the refractive increment; V = determination of partial specific volume; ρ = determination of density.
[d] Includes measurements of concentration series.

of polymers are strongly affected by low or high molecular weight components. For samples of uniform molecular weight, all methods should yield the same value, so no preference of method is required. The best molecular weight range listed in the table is only approximate and is based on the majority of literature reports. Some applications considerably beyond these ranges have been found, but they usually require refined apparatus and techniques.

The time required for a determination is based on the author's experience with some methods, and on literature reports for others. It has been assumed that the apparatus is in regular operation and that a proper sample is ready. Preparation of the sample often requires more time than the measurement, but the time is too variable to estimate. Most times given are for a single measurement which often must be repeated at a series of concentrations before a molecular weight calculation is possible. However, in some cases, considerable time economies may be made in successive tests, or several tests may be run simultaneously; for example, in the velocity ultracentrifuge, four different samples or concentrations may be recorded at once by use of double-sector cells. The time is indicated as "short" when less than 1 hr is required. The time estimated for calculations assumes that calibrations have been completed and the mathematical procedures are set up.

In the choice of apparatus, the relative importance of cost and time must be evaluated for each proposed application. Where rapid routine measurements are desired, the high cost of commercial apparatus may be justified by the greater convenience and probable freedom from frequent servicing by expert personnel. In some cases, instruments may be constructed in the laboratory from detailed literature reports, or may be modified for a specialized use, though either usually requires substantial expenditure of time before reliable operation can be expected. The commercial apparatus may be available either as components or as fully operational units. In some cases thermostatting or other facilities normally expected in laboratory experimentation will be required.

GAS DENSITY METHODS

Molecular weights of real gases may be determined by use of the ideal gas law, shown in Equation 6, if an extrapolation is made to low pressures where this law is applicable.

$$PV = (w/M)\,(RT) \tag{6}$$

where P is the pressure, w is the weight, M is the molecular weight, R is the gas constant, and T is absolute temperature. The gas density, $d = w/V$, divided by the pressure should provide the molecular weight by the following equation:

$$\frac{d}{P} = \frac{M}{RT} \tag{7}$$

The *limiting density method* is based upon extrapolation of the ratio d/P to ideal conditions. The weight of gas held in a glass bulb fitted with a stopcock is determined from the difference in weight of the evacuated and filled bulb, measuring the temperature and pressure carefully. Such measurements at constant temperature and covering the range 0.2 to 1 atm are adequate to permit extrapolation to ideal conditions by a linear plot of d/P vs P. The resulting molecular weights can be of high accuracy, sometimes within 0.01%. Although simple equipment can be used, highly specialized apparatus has been developed to avoid various errors (13).

In the *limiting pressure method*, measurements are made of the pressure values at which the densities of two gas samples are equal. This pressure ratio approaches the inverse ratio of molecular weights if deviations from ideal behavior are small. A commercially available instrument, the Edwards gas density balance, uses a torsion balance on which a sealed bulb is successively balanced in atmospheres containing known and unknown gas samples at adjustable pressures. The deviations from ideal behavior are often substantial, but can be minimized (13).

These two methods are used for precise gas and isotope analysis, but many samples can be studied more conveniently if they can be liquefied and treated by the methods below. See also GAS ANALYSIS.

VAPOR DENSITY METHODS

Both methods described here determine the volume occupied by a known weight of liquid when it is vaporized. Several modifications of the *Victor Meyer method* are described in laboratory manuals for physical chemistry courses, and detailed procedures are provided for an apparatus suitable for accurate industrial measurements by Bonnar et al. (2). In all variations, a large container filled initially with nitrogen or air and thermostatted at a temperature at least 20°C above the boiling point of the unknown is attached to a gas buret to measure volume without pressure rise. The increase in volume which occurs when the weighed unknown liquid sample in a sealed vial is released and vaporized can be used to calculate the molecular weight from the ideal gas law (Equation 6) by inserting the measured values of P in mm, V in ml, w in g, and T, the absolute temperature, and using the gas constant $R = 6.237 \times 10^4$. Corrections may be applied for the vapor pressure of the fluid in the gas buret, if substantial, and for departures from ideal gas law behavior, which may approach 2%. This method is limited normally to samples of molecular weight below about 200 with thermostatted baths at 200°C or less. However, with specialized apparatus, the method has been used at temperatures as high as 2000°C. Equipment for measurements on a micro scale has been used on a routine basis (2).

The procedure to be followed in the Victor Meyer vapor density method is described below.

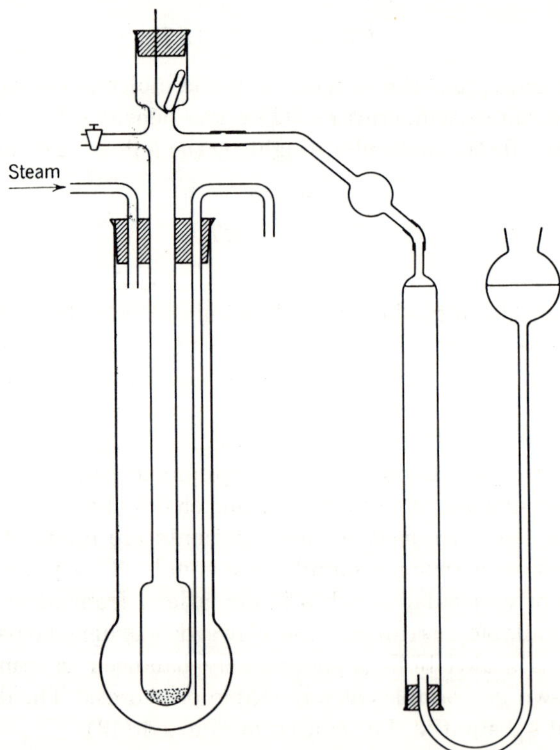

Fig. 2. Victor Meyer vapor density apparatus.

EXPERIMENTAL DETAILS

Apparatus

The Victor Meyer apparatus available from several laboratory supply houses consists of an inner tube immersed in a bath held at a constant temperature above the boiling point of the unknown. Steam from a separate generator is often used, as shown in Figure 2. The inner tube is connected to a buret filled with water to measure the volume increase upon vaporization of the sample.

Procedure

Transfer and weigh about 0.1 g of the liquid sample in the small stoppered vial, and support this in the cool upper section of the inner tube, preventing release of the vial with a rod which can be deflected when desired. With the heated jacket at constant temperature, release pressure inside the apparatus and adjust the leveling bulb to give a zero buret reading. Check for any change in volume, and when stable, release the sample vial, permitting it to fall onto a small pile of sand or glass wool in the lower bulb. As the vapor is released, lower the leveling bulb and read the buret when the volume has become constant. Calculate the molecular weight of the sample as described above, using the absolute temperature, T, of the gas collected in the buret at the measured atmospheric pressure, P. Before further experiments are carried out, flush the inner tube with air to avoid condensation of vapor from previous samples in cool regions of the apparatus.

The *Dumas method* utilizes a large bulb with a small opening which can be sealed after complete vaporization of a liquid sample under controlled temperature and pressure conditions. The weight of vapor and the volume of the container are determined, and the molecular weight is calculated from the ideal gas law. This method is usually less accurate than the Victor Meyer method because only small differences in weight are observed. The procedure follows.

EXPERIMENTAL DETAILS

Apparatus

Use the apparatus shown in Figure 3.

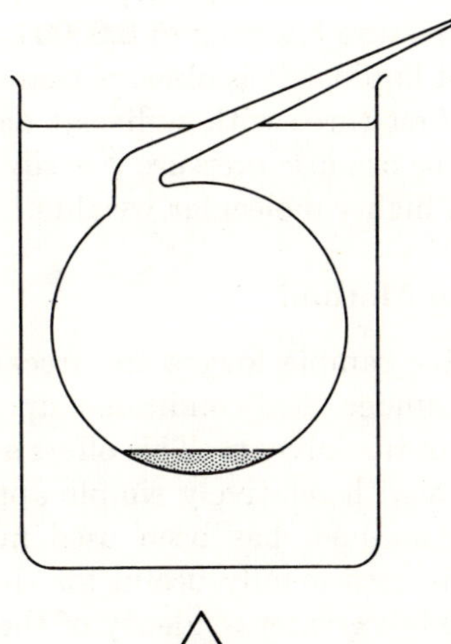

Fig. 3. Dumas vapor density apparatus.

Procedure

Weigh the dry thin-walled bulb of about 200-ml capacity, then add about 5 g of the liquid sample and immerse all but the thin neck of the bulb in a constant temperature bath above the boiling point of the sample. When all the sample has vaporized, seal the thin neck and reweigh the bulb to determine the weight of sample remaining. Then fill the bulb with water and reweigh; the difference between this weight and the weight of the air-filled bulb when divided by the density of water at the temperature of the water gives the volume of the bulb.

Calculation

Calculate the molecular weight by using the gas-law equation (Equation 6), inserting the weight, w, of sample occupying the bulb volume, V, at the absolute temperature, T.

COLLIGATIVE PROPERTY METHODS

Cryoscopic, ebullioscopic, vapor pressure lowering, and osmotic pressure methods are often classed together as colligative methods because they all arise from the decrease in activity of a solvent upon addition of a soluble sample. Activity is the effective concentration, that is, the product of the molar concentration and the activity coefficient. This coefficient usually has values somewhat less than unity and its magnitude can be calculated when the characteristics of the solvent and solute molecules are known. This decrease in activity of the solvent lowers its vapor pressure and thus also increases the boiling point of the solution, provided the solute is nonvolatile. The depression of the freezing point is also a direct consequence of the lowered vapor pressure. The osmotic pressure is developed when the solvent diffuses through a semipermeable membrane to dilute the solution to produce more nearly equal activities.

Measurements of the magnitude of these changes permit the determination of the molecular weight of the dissolved sample under proper conditions. A comparison of the magnitude of the effects is given in the following example. A benzene solution containing 10 g/liter of a solute whose molecular weight is 20,000 will show an elevation of the boiling point of about 0.0012°C, a depression of its freezing point of about 0.0025°C, a relative vapor pressure lowering of 0.00004, and an osmotic pressure corresponding to about 12 cm of liquid. It is obvious that very sensitive means must be employed to determine the first three with sufficient accuracy to provide significant molecular weights, whereas the osmotic pressure is easily measurable, and this method should be applicable to even higher molecular weights.

Freezing-Point Depression Method

The addition of a soluble sample lowers the freezing point of a solvent by an amount which is dependent, under ideal conditions, upon the concentration of solute and the latent heat of fusion of the solvent. This offers a quick method of determining molecular weights of samples with relatively simple apparatus. The method, sometimes called the Beckmann method, has been used most commonly for molecular weights below 1000, but it is occasionally useful for the determination of molecular weights up to 50,000. There is a greater tendency of the solute molecules to associate than is the case with determinations based upon other colligative properties. The theory and applications are discussed in detail by Bonnar et al. (14), Skau et al. (15) and by Rushman (4).

Apparatus suitable for measurements of moderate accuracy is shown in Figure 4, and is available from many laboratory supply houses. The outer jar contains a mixture which maintains a temperature a few degrees below the freezing point of the sol-

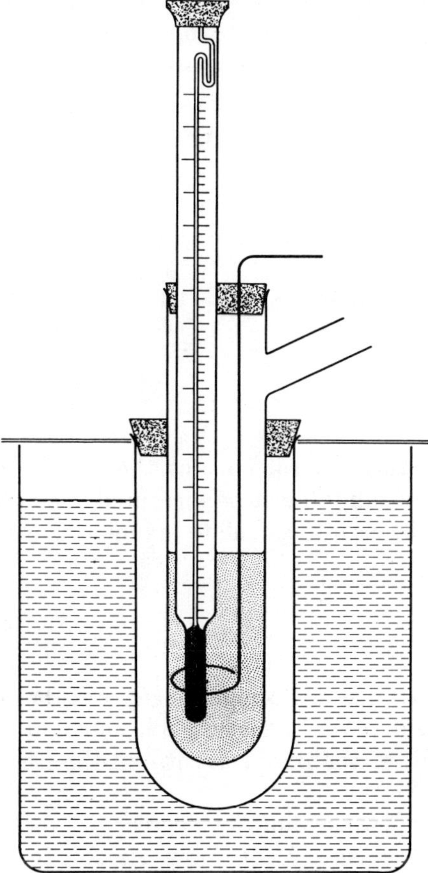

Fig. 4. Apparatus for freezing point depression method.

vent studied, for example, ice in salt water for measurement of aqueous solutions. The inner freezing point tube is separated by an air jacket from the outer bath to prevent too rapid cooling. A Beckmann thermometer and a wire stirrer are placed in the freezing point tube containing the solvent, and successive solute samples may be added through the side arm shown.

More refined apparatus may utilize a thermistor to provide sensitivity as high as 10^{-4}°C, and an evacuated jacket may be used for slower cooling through the freezing range. Apparatus has also been developed for measurements on a micro scale (15), often permitting measurements with a sample of 1 mg, or with the most delicate equipment a total solution weight of 1 μg or less may be used while still retaining a sensitivity of 0.002°C. American Instrument Company, Silver Springs, Md., and Precision Scientific Company, Chicago, Ill., market apparatus designed for routine measurements. One of these utilizes a Beckmann thermometer; the other is designed for small (1 ml) samples and a thermistor sensing element is used.

Solvents which have good solvent action for a variety of samples are preferred for this method, in order to avoid association of solute molecules which is often found at the freezing point of solvents. Polar samples containing hydroxyl, carboxyl, and amino groups should be tested in the more polar solvents in order to avoid association. Table 2 includes freezing points and approximate cryoscopic constants, K_f, for commonly used solvents, listed in order of decreasing polarity.

Table 2. Solvents for Cryoscopic Measurements

Solvent	Freezing point, °C	Cryoscopic constant, K_f
water	0	1.86
acetic acid	16.5	3.7
dioxane	11.8	4.6
nitrobenzene	5.6	6.8
diphenylamine	53.2	7.7
benzene	5.5	5.08

The freezing point of the pure solvent and the freezing point of the solvent containing a known amount of solute must be measured very precisely. The depression of the freezing point may be used to calculate the molecular weight using the following equation:

$$M_n = \frac{c\,K_f}{\Delta T} \tag{8}$$

where M_n is the number-average molecular weight, c is the concentration in grams of solute per 1000 g of solvent, K_f is the cryoscopic constant, and ΔT is the difference between the freezing point of the pure solvent and the freezing point of the solvent containing the solute.

Although the cryoscopic constants in degrees per mole have been measured or calculated for a variety of solvents (2,15), it is preferable to determine the value of the constant with a purified substance of known molecular weight under the conditions used for the unknown. The procedures for the determinations of the cryoscopic constant and the molecular weight of the unknown sample are described below.

EXPERIMENTAL DETAILS

Apparatus

Use the apparatus shown in Figure 4 or similar equipment.

Materials

SOLVENT. Select a purified grade of a suitable solvent. For purposes of illustration, reagent-grade benzene with a specific gravity of 0.879 is used.

STANDARD. Use purified naphthalene or other suitable standard material. The molecular weight of naphthalene may be taken as 128.18.

Procedure

Pipet 25.0 ml of reagent-grade benzene into the freezing point tube. This corresponds to 22.0 g of benzene. Determine the freezing point of the pure benzene by plotting the cooling curve at slow rates of cooling and selecting the region of nearly constant temperature produced by heat liberated during crystallization. See also FREEZING POINT DETERMINATION; THERMOMETRIC ANALYSIS. Remelt the solvent by removing the freezing point tube from the apparatus.

Pelletize and weigh accurately a sample of naphthalene of about 0.140 g. Transfer to the freezing point tube and place the tube in the apparatus. Stir to effect solution and determine the freezing point in the same manner as described for the pure solvent. The difference in freezing point is ΔT.

Repeat the procedure for the unknown using a sample size which will give a ΔT value between 0.200 and 0.300. Avoid residual solvents or adsorbed water if accurate molecular weights are desired. Preliminary desiccation is usually sufficient, but for removal of tightly bound solvents, it is sometimes necessary to treat the sample with the solvent to be used for the freezing point measurement and then remove the solvent by desiccation.

Calculations

To determine the cryoscopic constant, it is convenient to rearrange Equation 8 to the following form:

$$K_f = M_n \, \Delta T/c$$

Using a sample of 0.140 g in 25 ml of benzene, c becomes 6.36. If a ΔT value of 0.250°C is obtained, then it follows that

$$K_f = \frac{(128.18)(0.250)}{6.36} = 5.06$$

Use Equation 8 to calculate the molecular weight of the sample directly.

If the apparent molecular weight determined by this method varies with the concentration present, it is advisable to plot several apparent values versus concentration and extrapolate to zero concentration to obtain the true molecular weight. This extrapolation minimizes nonideal solution behavior as well as reversible dissociation which diminishes at low concentrations.

Melting Point Depression Method

The *Rast* method, often used in the identification of organic compounds, involves rapid measurement of the melting point depression with small samples and simple equipment. Camphor is usually employed as solvent since it has a very large cryoscopic constant, 39.7, and dissolves many types of compounds. By the use of this solvent and relatively high solute concentrations, the melting point depressions can be measured easily with simple mercury thermometers in a capillary tube melting point apparatus.

EXPERIMENTAL DETAILS

Procedure

Weigh accurately about 0.05 g of sample and place it in a small test tube with a weighed piece of camphor (about 0.5 g). Melt the camphor over a low heat, and quickly mix the two components to form a clear liquid. Cool the melt and crush the solid to a powder. Introduce a small sample of the powder into a glass capillary tube of the type used in melting point determinations, and pack the powder into a layer about 1 mm thick. Suspend the tube in a stirred oil bath and heat the bath so as to raise the temperature about 2°C/min as the melting point is approached. Record the temperature at which the last solid disappears as the melting point. Determine the melting point of the pure camphor in a similar way.

Calculation

Calculate the molecular weight from the following equation:

$$M = \frac{39{,}700\ U}{S(T_s - T_u)}$$

where U and S are the weights of unknown and solvent, and T_s and T_u are the melting points of solvent and solution. The sample concentration must be between 0.2 and 0.5 M, which is equivalent to a melting point depression between about 8 and 20°C, in order to conform to this equation. Carry out a second experiment with a sample of suitable weight if the first was not within the proper range.

Ebulliometric Method

The addition of a nonvolatile, soluble sample increases the boiling point of a solvent in direct proportion to the number of molecules added per unit volume. Precise measurements of this elevation are most often used for determining molecular weights below 1000. Some applications have been made up to a molecular weight of 30,000, but since the number of molecules added for a fixed weight concentration decreases greatly at high molecular weight, high concentrations or very high temperature sensitivity are needed. The method requires more complex apparatus than cryoscopic determinations, but is less subject to errors due to association of solute molecules because of the greater solvent power under the conditions of measurement. A detailed treatment of the method is given by Bonnar et al. (2) and reviews are presented by Rushman (4) and by Swietoslawski and Anderson (16). See also BOILING POINT DETERMINATION; TEMPERATURE MEASUREMENT AND CONTROL.

The elevations, ΔT, of the solvent boiling point upon successive additions of small solute concentrations, c, in g/1000 g of solvent, are used to calculate the apparent molecular weight of the solute from the following equation:

$$M_{app} = \frac{cK_b}{\Delta T} \tag{9}$$

where K_b, the ebullioscopic constant is determined, usually in the same apparatus by calibration with a standard of known molecular weight. The resulting apparent values of the molecular weight are then plotted as a function of the concentration used in their measurement, and the curve, usually a straight line, is extrapolated to zero concentration to give the true molecular weight.

In most applications, simultaneous measurement of the boiling points of the solvent and solution are made in order to avoid errors caused by fluctuations in atmospheric pressure which may be sufficiently rapid to affect successive measurements of boiling points. Two alternative methods for simultaneous measurements are used; two similar ebulliometers containing solvent and solution may be used, or one instrument may be arranged so that both the boiling point of the solution and the condensation temperature of the vapor are observed. The latter method is employed in the Menzies–Wright ebulliometer which uses a differential vapor pressure thermometer to measure only the difference in the boiling and condensation temperatures; see BOILING POINT DETERMINATION. This type of instrument has been widely used, but must be carefully designed and operated to provide the true differential. The use of twin instruments is now more common, particularly since thermocouples or electrical resistance thermometers and associated measuring devices can provide the high temperature sensitivity needed, and permit direct measurement of the difference in boiling points of solvent and solution.

Apparatus for this method must be designed to avoid the effect of hydrostatic pressure on the boiling point, and more refined apparatus operates within an outer jacket which is well thermostatted, often using a vapor bath of the same solvent. A variety of designs have been successfully used (2,4,16). Routine measurements with a series of similar instruments have been described (2); both Beckmann thermometers and thermistors were found suitable for measurements on macro-size samples with molecular weights up to about 3000. Micro instruments requiring samples of 10 mg have also been developed using thermistors sensitive to less than 10^{-4}°C differentials. Alternatively, the high thermal sensitivity of thermistors or thermopiles has been used by several workers to determine molecular weights up to 30,000 (4).

The choice of solvents for ebulliometric measurements is based on their stability and freedom from impurities or tendency to absorb atmospheric moisture, as well as their ability to dissolve the variety of samples to be studied. Solvents are usually selected which have a polarity somewhat similar to that of the samples in order to avoid association of the solute molecules and to approach ideal solution behavior. Suitable solvents, together with their boiling points and ebullioscopic constants, are listed in Table 3, in approximate order of decreasing polarity.

Table 3. Solvents for Ebulliometric Measurements

Solvent	Boiling point, °C	Ebullioscopic constant, K_b
water	100	0.51
acetic acid	118	3.07
ethanol	78	1.19
2-butanone	80	2.56
tetrahydrofuran	63	2.50
dichlorethane	83	1.90
benzene	80	2.53
toluene	111	3.33
cyclohexane	81	2.79

Both the sample and the standard substances should be thoroughly dried by evacuation, since even small amounts of water or residual solvent may affect the apparent molecular weight greatly. Before molecular weight measurements of an unknown are attempted, its complete solubility in the chosen solvent should be checked, and a more suitable solvent should be chosen if the solubility is very limited or slow. The sample should have negligible vapor pressure at the boiling point of the solvent; a difference of 100°C in the boiling points of the sample and solvent is sufficient to assure this condition. Any instability of the sample can be detected by the drift of the boiling point of the solution for longer times than are normal for the apparatus.

EXPERIMENTAL DETAILS

Apparatus

This method utilizes two identical ebulliometers of the Cottrell or Washburn and Read designs as sketched in Figure 5. Instruments of this type are available

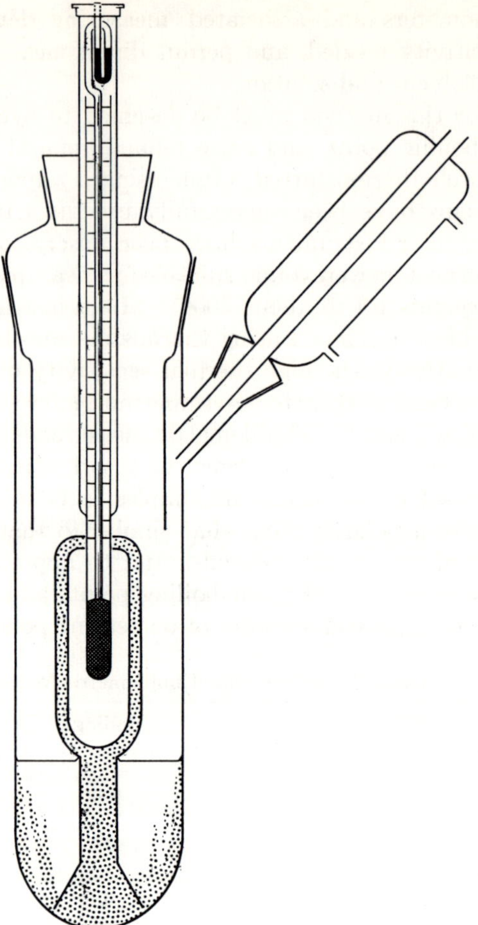

Fig. 5. Cottrell apparatus for boiling point elevation measurements.

from many laboratory supply houses. The lower section of each instrument should be heated with a smoothly variable source, such as an electrically heated aluminum block or a shallow oil bath on a thermostatted hot plate. The upper section of the apparatus should be insulated from drafts with asbestos or glass wool. Two Beckmann thermometers are recommended to measure the boiling points of solvent and solution simultaneously.

Procedure

Fill both clean instruments with equal volumes of solvent, from 20 to 50 ml depending on the design, and insert the thermometers which have been set to the temperature range to be studied. Adjust the heating rate to provide steady boiling and pumping of liquid over the thermometer bulb. Adjust the heat input so that the rate at which condensate drops from the condenser is at about midpoint of the range over which the thermometer reading shows no variation. At this optimum heating rate, read the thermometer over a period of 15 min or until no further drift (within about 0.002°C) is observed. Continue heating and reading one instrument, while cooling the other preparatory to addition of a sample. Remove the condenser and insert a weighed sample sufficient to raise the boiling point by about

0.2°C. The necessary weight may be estimated by use of Equation 9 together with the value of K_b for the solvent and the approximate molecular weight. For a 30-ml volume of benzene, for example, a 0.5-g sample having a molecular weight of 200 would raise the temperature about 0.2°C. Now determine the stable boiling temperature of the solution at the same rate of condensation. The temperature rise observed must be corrected for any change which has been noted in the boiling point of solvent in the other apparatus. This can usually be attributed to change in barometric pressure; for example, only 1 mm change in pressure alters the boiling point of water by 0.04°C. Measure the further elevation of the boiling point upon addition of three more samples, each having about the same weight as the first. Determine the value of K_b for the solvent by successive additions of a standard compound of known molecular weight to a fresh batch of solvent in the same way and in the instrument used for the unknown. Suitable standard substances for solvents of low or moderate polarity are diphenylbenzil, triphenylmethane and methyl stearate; sucrose has been used as a standard for highly polar solvents.

Calculations

Calculate the four apparent molecular weights by inserting the elevations caused by each addition into Equation 9 together with the value of K_b determined with a standard substance and extrapolated to the ordinate to obtain the value at infinite dilution. Construct a graph of apparent molecular weight values versus the total sample weight present for each, and extrapolate this to the ordinate to obtain the molecular weight at infinite dilution.

Vapor Pressure Methods

These methods all utilize the decrease of vapor pressure of a solvent upon addition of a soluble sample in order to determine the molecular weight of the solute. Direct manometric observation of this depression has been accomplished, but is generally too complicated except for quite low molecular weights. Instead, several effects resulting from vapor pressure differences are most commonly used. Detailed discussions of several of these are given by Bonnar et al. (2), Wagner and Moore (17), and Rushman (4). See also PRESSURE MEASUREMENT AND CONTROL.

THERMOELECTRIC METHOD

In this technique, measurements are made of the temperature rise resulting from condensation of solvent vapor from a saturated atmosphere into a solution. In order to measure the tiny thermal effect quickly, small liquid samples, often single drops, are placed in contact with a highly temperature-sensitive element of low mass. Two of these units, one containing only solvent and the other the solution for test, are isolated in an atmosphere saturated with solvent vapor. The equilibrium temperature difference between the two samples is related to the molal solute concentration of the solution, and so gives a measure of molecular weight. The method is not absolute, since heat losses, incomplete saturation of the vapor, and inefficient condensation on the liquid surfaces depend on the sample and equipment used; thus calibration is required with known standards in the same solvent. The advantages of the method are its high speed in certain instrument designs, the very small samples required, since only drops of solution are needed, and the ability to work at moderate temperatures where association or decomposition of samples are avoided. The disadvantages are

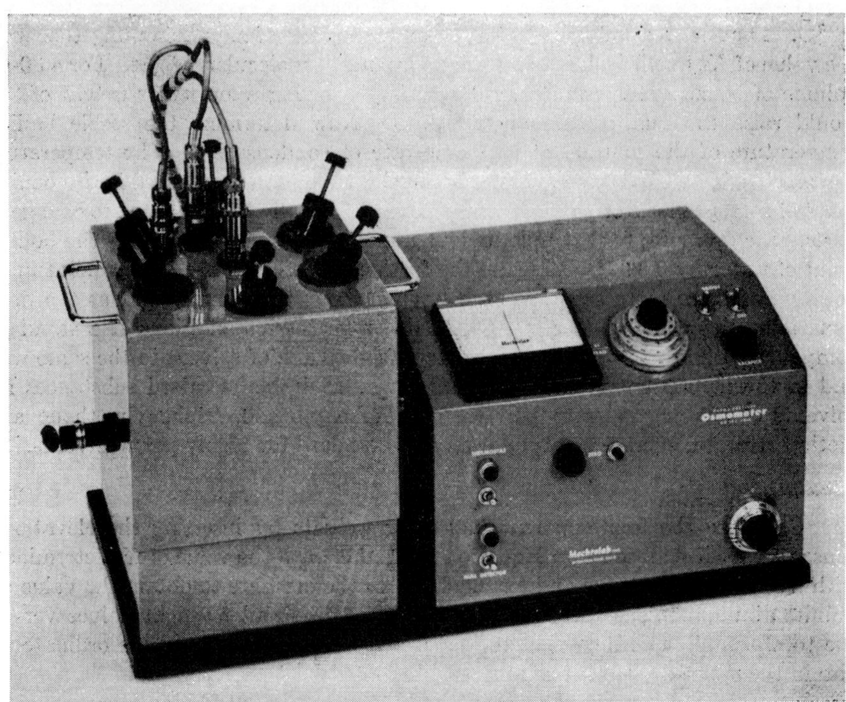

Fig. 6. Vapor pressure osmometer. Courtesy Mechrolab Inc.

the need for carefully designed equipment and very good thermostatting of the atmosphere.

Apparatus of quite varied types utilizes thermocouples or thermopiles, and more recently thermistors to provide thermal sensitivities between 0.001 and 0.0001°C. Quite accurate values are reported up to molecular weights of 1000, but some applications have been made at molecular weights as high as 5000 or more with decreasing precision. One instrument utilizing thermistors is now available commercially from Mechrolab Inc., Mountain View, Calif. It has been shown to produce molecular weights accurate to 1% up to values of 600, and is said to allow as many as eighty such measurements within eight hours. The instrument is shown in Figure 6. Equipment of somewhat similar type can be built from published designs, but the cost of such a commercial instrument may be justified by its compactness and reliable operation at high speed.

The procedure used with any of the modern designs is to establish a zero reading with droplets of solvent applied to both the sensing elements after temperature equilibrium and vapor saturation have been achieved. Then a solution containing from 10^{-3} to 3×10^{-2} g/ml of sample, depending on its molecular weight range, is added to one element, and the stable temperature difference developed is measured. Similar measurements are usually made for at least two concentrations as a check and to observe any concentration dependence which may indicate association or nonideal behavior. A calibration curve for the solvent system is constructed from measurements with a standard substance covering the concentration range to be studied. A simple plot of the observed current or resistance change versus the molal concentration of the solute is adequate for many purposes, though other plotting methods may be used (2). The apparent molal concentration of the unknown is then read off such a calibration curve for each experimental point, and the apparent molecular weights calculated from the known weight concentrations. If the apparent molecular weight values at all concentrations do not agree, then a plot of these versus concentration may be extrapolated to give the true molecular weight, or may indicate some association or dissociation process. In this event, measurements in a better solvent or one in which dissociation does not occur may be used to confirm the value. The samples and standards must be free of impurities as already indicated for the boiling point method.

EXPERIMENTAL DETAILS

Apparatus

Use the Mechrolab vapor pressure osmometer illustrated in Figure 6.

Procedure

Fill the solvent cup with the selected solvent and locate the vapor wicks in place to produce a vapor-saturated atmosphere. Replace the upper housing and allow the apparatus to reach thermal equilibrium, which takes about 1 hr if the block has been thermostatted previously. Prepare a stock solution of the unknown which is estimated to be about 0.1 M or less for low molecular weight compounds, or about 0.025 M for polymers. Also prepare dilutions from this stock which are about one-half, one-fourth, and one-eighth of the stock concentration. Use the same batch of solvent in all operations. Fill four syringes with the solutions, as little as 0.1 ml is sufficient, and insert the syringes in the thermal block. Fill two syringes with solvent, and locate them in their holes in the block. Deposit one drop of solvent on each thermistor bead, after rinsing, and adjust the bridge to zero balance after 2 min. Rinse one thermistor with the most dilute solution and leave one drop in place. Measure the change of resistance, ΔR, to balance the bridge after 2 min. Repeat the last two steps with each sample solution.

Calculations

Prepare a calibration curve for the solvent by repeating the measurements above with a substance of known molecular weight and purity, and plotting the observed values of ΔR versus the molar concentration of the standard. From the ΔR values, determine the apparent molar concentration of each solution from the calibration curve. Calculate the apparent molecular weight at each concentration by dividing the molar concentration by the weight concentration of the sample in g/liter. If the results are not constant, prepare a graph of the apparent molecular weight values against concentration and extrapolate to zero concentration to obtain the true molecular weight.

ISOPIESTIC METHOD

This is generally a much slower process in which solvent is allowed to distill between two neighboring volumes of solution, one containing an unknown solute and the other a standard substance, until equilibrium is reached. Then the weight concentration in each solution is determined from the weight or volume remaining in each, and from this the molecular weight of the unknown may be calculated, since the molal concentrations in each are presumably equal.

The experimental equipment designed for operation at reduced pressure speeds the distillation and therefore the attainment of equilibrium, but is generally complex.

Very simple devices, such as slugs of sample and reference solutions in capillaries, in which the lengths of the slugs are measured, are generally very slow. A compromise in which the solutions in platinum crucibles are held in a desiccator at somewhat reduced pressure is described in Reference 2. Both micro-size and macro samples have been studied, but the equilibration still requires between 1 and 3 days. Despite the long time, such methods may be of value where a number of samples can be set up and given no attention for the required time. This method has produced some of the most precise vapor pressure data when great care was taken to maintain uniform temperatures.

Osmotic Pressure Method

Osmotic pressure is the most sensitive of the colligative properties. Because of this sensitivity and membrane limitations, this method is most commonly used for measuring molecular weights between 20,000 and 500,000. Detailed descriptions of the method are given by Moore and Wagner (17), by Bonnar et al. (2), and by Doty and Spurlin (3).

Measurements are made of the pressure exerted by solvent molecules diffusing through a semipermeable membrane into a solution in which the activity of the solvent molecules is lower. If the solution compartment is closed, the pressure builds up until an equilibrium pressure is reached, after which no net solvent flow occurs. This osmotic pressure, π, is related to the number-average molecular weight, M_n, and concentration, c, of the solute by the following relationship:

$$\frac{\pi}{c} = RT\left(\frac{1}{M_n} + A_2c + A_3c^2 \ldots\right) \tag{10}$$

where R is the ideal gas-law constant in appropriate units, T is the absolute temperature, and A_2 and A_3 are constants for the solvent and solute system. Normally, the osmotic pressure is measured as the height of solution in a capillary tube, corrected for capillary rise of solvent in the same tube; this is shown in Equation 11. The practical units then employed in calculating molecular weights are given in terms of the height, h, in cm of liquid of density ρ, and with solute concentrations, c, given in g/ml.

$$\frac{h}{c} = \frac{8.48 \times 10^4 T}{\rho}\left(\frac{1}{M_n} + A_2c \ldots\right) \tag{11}$$

At 25°C, and for a liquid of unit density, the following equation applies:

$$\frac{h}{c} = 2.53 \times 10^7\left(\frac{1}{M_n} + A_2c \ldots\right) \tag{12}$$

In order to determine the molecular weight of a sample, values of h/c are found at several concentrations, and are plotted against concentration. Such a plot is often linear, and the intercept at the h/c axis equals $8.48 \times 10^4T/dM_n$, from which M_n may be calculated. The slope of the line provides the constant A_2, which is related to the strength of interaction of the solute with the solvent, being smallest for poor solvents.

The apparatus for measuring osmotic pressures usually utilizes a flat sheet of membrane clamped between two compartments, one containing the solvent and the other the solution. Each compartment is connected to a vertical glass tube or capillary in which the progressive development of osmotic pressure can be measured. For most rapid attainment of equilibrium, the exposed area of the membrane should be large and the volume change necessary for pressure development should be small, suggesting that capillaries be of smallest practical diameter. In addition, it is desirable to clamp the membrane to minimize bulging, and it is best to minimize the enclosed volumes to decrease temperature sensitivity. A variety of instruments have been utilized in recent literature, and the choice for a given study will depend on how extensive the work and how rapidly one wishes to collect results. Table 4 shows a comparison of the characteristics governing the rate of approach to equilibrium of several instruments representative of the large number of types which might be selected. The ratio of the exposed membrane area to the sensitivity of the measuring method is used as a guide to the speed of approach to equilibrium for membranes of the same porosity. The last column in Table 4 shows reported times to complete a single pressure measurement using membranes suitable for polymers of moderate molecular weight.

Table 4. Comparison of Speed of Various Osmometers

Instrument	Effective area, cm^2	Pressure sensitivity, ml/cm head	Speed factor, area/sensitivity	Measurement time for a single determination
Schulz-Wagner[a]	3	0.0038	800	18 hr
Zimm-Myerson[b]	6	0.0019	3100	60 min
Fuoss-Mead[c]	75	0.0030	25,000	15 min
Mechrolab Series 500	approx 1.4[d]	$<10^{-8}$	$>10^8$	3-10 min
Hallikainen Model 1361B	approx 8[e]	10^{-7}	approx 10^8	5-10 min

[a] G. V. Schulz, *Z. Physik. Chem. (Leipzig) A* **176,** 317 (1936); R. H. Wagner, *Ind. Eng. Chem., Anal. Ed.* **16,** 520 (1944).

[b] B. H. Zimm and I. Myerson, *J. Am. Chem. Soc.* **68,** 911 (1946).

[c] R. M. Fuoss and D. J. Mead, *J. Phys. Chem.* **47,** 59 (1943).

[d] Calculated from estimates of the area of grooves.

[e] Information from manufacturer.

The first two instruments listed are representative of many designs having relatively small membrane area and conventional pressure measurement in capillaries. They are many times slower than the Fuoss-Mead type, though they are less complicated and much less expensive so that a series of measurements with multiple units can be under way at one time. Several modifications of the Fuoss-Mead instrument have been described; they have in common a relatively large membrane clamped between two matching grooved plates to minimize membrane ballooning, and grooves acting as the solution and solvent compartments. Typical measurements with this design are described below.

Three automatic osmometers have been marketed, and these combine the advantages of clamped membranes and small compartments with very sensitive devices for pressure sensing and balancing by servo mechanisms. The osmometers manufactured by Hallikainen Instruments, Berkeley, Calif., and by Dohrmann Instrument Co., San Carlos, Calif., are based on a design originated at Shell Development Co. (18). The photograph of the Hallikainen design shown in Figure 7 indicates the compactness of these new tools. The former two instruments utilize the deflection of

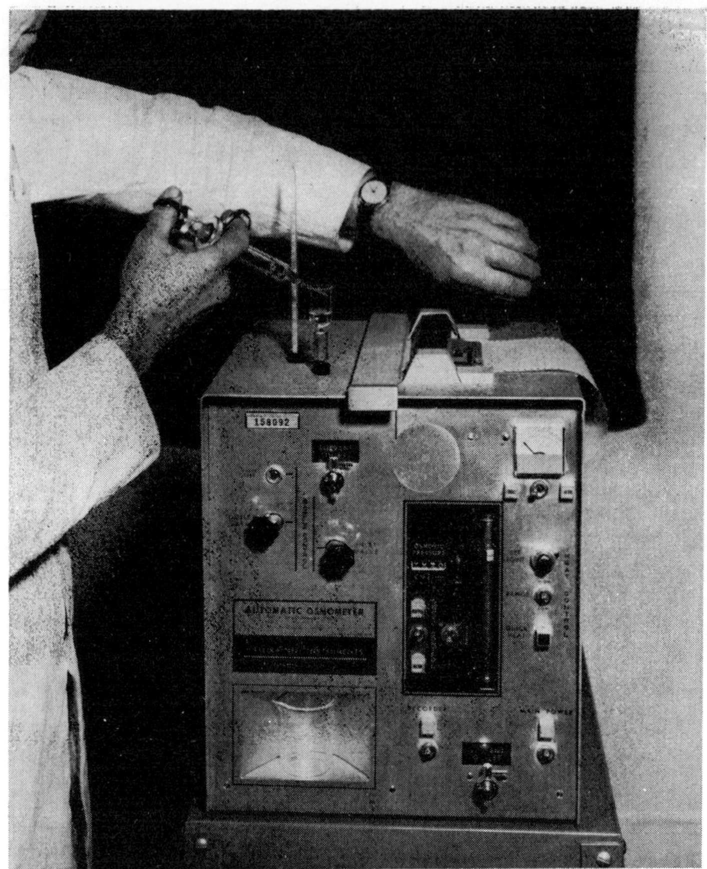

Fig. 7. Hallikainen Model 1361B automatic osmometer. Courtesy Hallikainen Instruments.

a thin metal diaphragm to detect pressure changes which are balanced by a servo-driven arrangement for increasing the head in the solution compartment. The instrument of Mechrolab, Inc., Mountain View, Calif., has been described and some tests reported (19). This instrument detects osmotic flow by movement of an air bubble, with adjustment by applying a negative head to the solvent compartment. This design permits measurements with unusually small volumes of solution (0.3 ml), although the 2-ml volume required by the other instrument is a great improvement over conventional designs. The sketches shown in Figure 8 compare the principles of these two instrument types. Both types detect very small volume changes (see Table 4) and balance these automatically so that apparent equilibrium can be reached in a few minutes, and the entire process can be displayed as a recorded trace as illustrated in Figure 9. An additional advantage of the rapid measurement is the minimum loss of low-molecular weight components by diffusion through the membrane pores, and the ability to detect and utilize the changes in osmotic pressure, which result from such diffusion on longer standing, to estimate the proportion of diffusible components present. The performance of both types of automatic osmometer has been reported (18,19). For polystyrene and poly-α-methylstyrene, molecular weights in the normal range agreed with independent measurements within 5%; fairly accurate results were obtained for fractions of molecular weight as low as 3000 even though

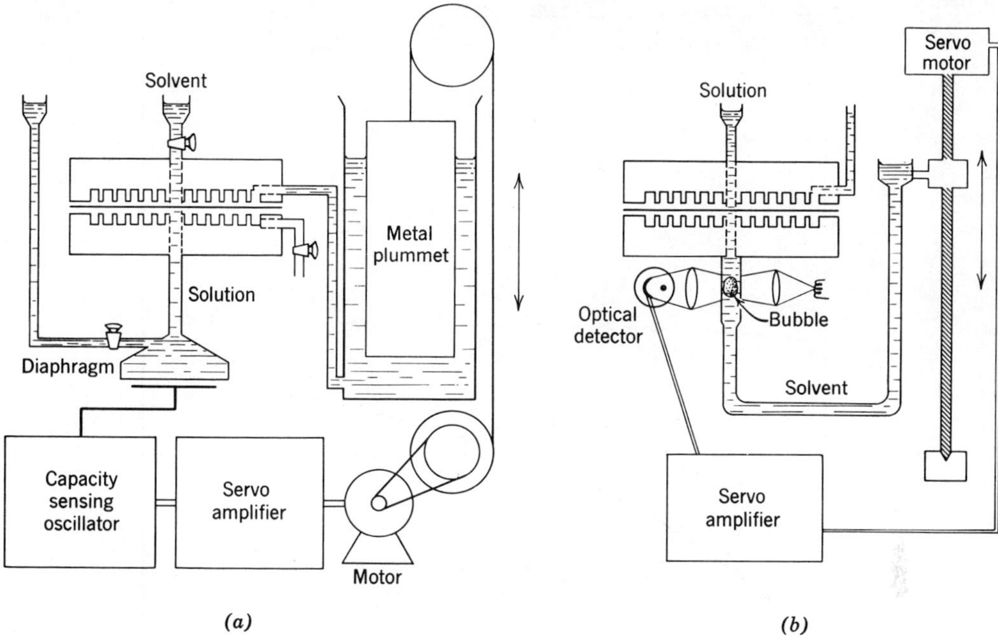

Fig. 8. (*a*) Rapid-recording osmometer with pressure sensing. (*b*) Rapid-recording osmometer with flow sensing.

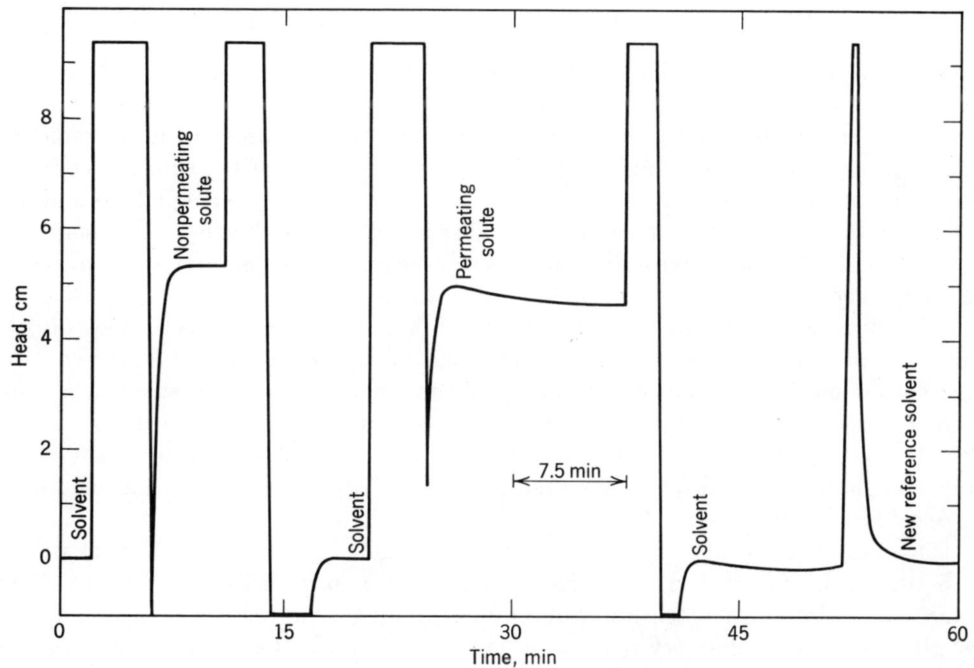

Fig. 9. Osmotic pressure measurement obtained with Hallikainen Model 1361B automatic osmometer. Stable pressures with a nonpermeating sample and drift with a permeating sample are shown. Courtesy Hallikainen Instruments.

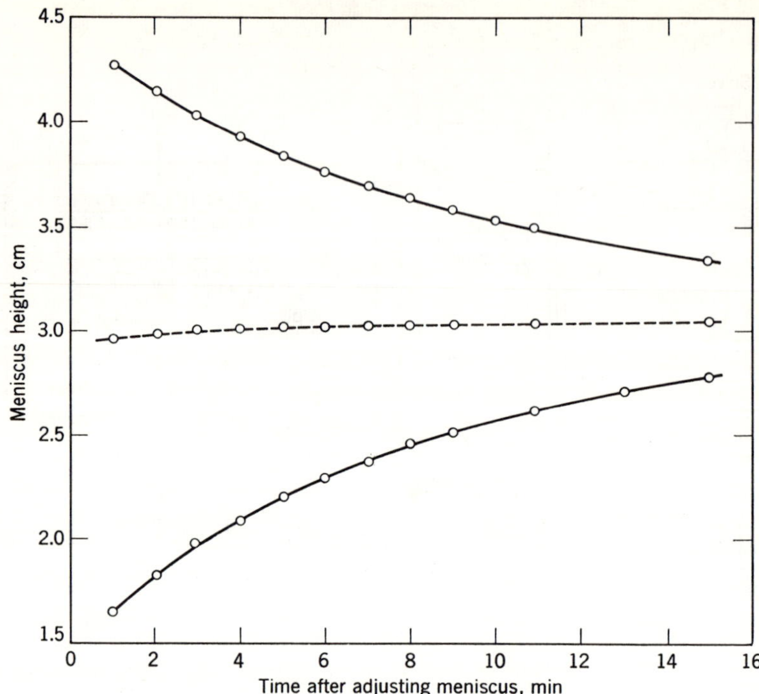

Fig. 10. Osmotic pressure by the half-sum method, using the Fuoss-Mead type of osmometer, with 0.01 g/ml of gelatin in aqueous salt solution.

diffusion was substantial. The current widespread use of these instruments should soon provide literature on other polymers.

The degree of temperature control required for osmotic pressure measurements depends on the apparatus, being minimized for compartments of the smallest volume, and for symmetrical designs in which both compartments are of equal volume so that thermal expansion affects capillary heights equally on both sides. Temperature control to 0.01°C is then often satisfactory, since the mass of the instrument damps out fluctuations effectively.

Membranes suitable for osmometry can be prepared by coating solutions of cellulose derivatives, followed by reconversion to insoluble cellulose; however, the control of pore sizes and the avoidance of leaks make the process tedious. Instead, it is now common practice to purchase cellulose membranes of various controlled pore sizes, such as gel-cellophane, available in wet form from cellophane manufacturers, or filters especially prepared for osmometry (17), which are then conditioned for use in the selected solvent.

The measurement of osmotic pressure can be made by a static method, in which the liquid level is allowed to rise to its equilibrium height without adjustment. This requires a minimum of attention, but requires a long time for slower instruments, though it is quite satisfactory for fast-acting instruments. Results for many instruments can be obtained faster and more reliably by the *half-sum* method, as shown in Figure 10, which requires that the liquid height in the capillary be adjustable. The approach to equilibrium is followed with time both from a height set somewhat above the expected final value, and then from a height below the final value by about the same amount, as estimated from the first test. Both approaches are then plotted in terms of time, with zero time being taken when the position was adjusted. Average heights at various times are then computed from the two curves, and these average values are seen to approach a constant height, equal to the equilibrium value. In some osmometers an external pressure is applied to counteract the osmotic pressure, and a *dynamic* method of measurement can then be used. The external pressure is successively adjusted to various values and the rate of flow through the membrane is determined for each pressure. These data are then extrapolated or interpolated to the position of zero rate of movement which yields the osmotic pressure.

The diffusion of some sample components through the pores of the membrane is often observed with many polymer samples and is the cause of greatest concern in osmotic pressure measurements. The pressure decrease resulting from this diffusion may not be detected in slow-acting osmometers, and the lower osmotic pressure found may represent only the high molecular weight components, and may thus lead to serious error. On the other hand, Staverman et al. (20,21) have provided theoretical and experimental evidence that the osmotic pressure measured rapidly with a leaky membrane will also be in error, because diffusible components are less than ideally effective. Nevertheless, rapid measurements offer the result closest to the true osmotic pressure, and become more accurate the less chance there is for diffusion, either by using less porous membranes or by using samples having fewer low molecular weight components. Bruss and Stross (22) have compared number-average molecular weights determined with porous membranes by rapid dynamic methods and a slow static approach, and found the results in good agreement for clean high molecular weight samples, but in serious disagreement when components with molecular weights below about 10,000 were present. It was found that molecular weights measured quickly under these conditions were in fair agreement with those measured by the boiling point elevation method, if the true molecular weight was over 6000. Donnet and Roth (23) compared the apparent molecular weights found for one commercial unfractionated polymer when osmotic pressures were measured with membranes whose limit of permeability ranged from 1000 to 30,000. The resulting molecular weights ranged from 7000 to 220,000, whereas when the same polymer had been purified by a preliminary single-stage precipitation, the molecular weights were more uniform whatever the permeability of the membrane.

These results and many other publications indicate that past osmotic pressure measurements have not generally been carried out with sufficient care, first to remove unwanted components, and then to insure that the pressures measured represent all the components of the sample. This earlier lack of attention was primarily due to the difficulty of measuring pressures with sufficient speed to provide results in a reasonable time when membranes of low permeability were needed, or to avoid loss by diffusion if porous membranes were used. The increasing use of rapid and highly sensitive osmometers, together with membranes of more consistent quality, promises to overcome these deficiencies.

Ionized macromolecules, such as proteins and various natural and synthetic polyelectrolytes, produce osmotic pressures which are larger than for comparable nonionized polymers because of binding of counterions whose osmotic effect may greatly exceed that of the polymeric component. The addition of substantial concentrations (0.1–0.5 ionic strength) of diffusible salt suppresses this effect so that the value of the molecular weight obtained by extrapolation of osmotic pressure data is that of the macro ion only. The slope of the plot of π/c versus c is greatly affected by the charge on the macro ion, and it is best practice to minimize this charge if possible, for example, by adjusting the pH of protein solutions to near the appropriate isoelectric point.

The following abbreviated procedure for operation of the Hallikainen automatic osmometer is similar in many respects to that used for other servo-balanced instruments. It is assumed that the semipermeable membrane is already installed, since one membrane may be used for as long as a year.

EXPERIMENTAL DETAILS

Procedure

Turn on the heater switch and allow the cell to reach the selected operating temperature. Fill the upper compartment and the manometer with solvent, flushing out any bubbles by use of a syringe. For a preliminary check of the zero position, fill the lower (sample) compartment with solvent, then close both the inlet and outlet valves. After about 5 min, drain the solvent from the upper compartment until the digital readout indicates zero. Prepare a stock solution of the unknown polymer containing 0.005–0.01 g/ml, using the same batch of solvent as previously used. Also prepare dilutions having, for example, three-fourths, one-half, and one-fourth of the stock concentration. As little as 25 ml of stock may be required to provide 10 ml of each concentration for the tests. Now replace the solvent in the lower compartment with a solution of the unknown, using a total of 10 ml, of which most flushes out previous contents of the compartment. Close both the inlet and outlet valves. After about 5 min, observe the recorder trace for indications of pressure which is stable for several minutes. Record the pressure, in cm of solvent, indicated by the digital readout. Replace this sample solution with another dilution and repeat the procedure for each dilution.

Calculations

Divide each equilibrium height, h, by the sample concentration, c, in g/ml, and plot the values against the sample concentration. Draw a straight line or slight curve through the points and extrapolate it to zero concentration. Calculate the molecular weight from the value of h/c at the intercept by use of Equation 11.

LIGHT-SCATTERING METHODS

Dissolved polymer molecules act as discontinuities of sufficient size that dilute solutions, which are usually less than 0.01 g/ml, scatter sufficient light from a strong beam to be measured easily with a photometer of special design. The method is absolute and is presently the most convenient method for determining weight-average molecular weights of polydisperse polymers, which values are related to certain physical properties. The normal range of the method is from about 10,000 molecular weight upwards, with the sensitivity increasing the higher the molecular weight. A detailed study of the method and its applications is given by Stacey (24); reviews are presented by Peaker (25), McIntyre (26), and Oster (27).

The scattered intensity is related to the concentration and molecular weight of the sample, and also to the difference in refractive index of the polymer and solvent. This requires that an independent measurement be made of the change of refractive index, n, with solute concentration, the gradient dn/dc. One important criterion in the choice of a solvent is that its refractive index differs substantially, and preferably more than 0.1, from that of the sample.

The method requires the measurement of the light intensity scattered by at least four concentrations of the sample, often between 2×10^{-3} and 10^{-2} g/ml. The illuminating light should be an intense, nearly paralleled beam of monochromatic light, such as can be obtained by isolating one of the lines from a medium-pressure mercury arc. The intensity of the weak scattered light is measured with a phototube shielded so as to receive a fairly limited angular range. The receiver should be arranged so as to measure the light scattered in the angular range from about 40 to 135° with respect to the incident beam. Many designs suitable for this purpose have been described. See LIGHT SCATTERING.

The basic equation from which the weight-average molecular weight, M_w, is calculated is

$$\frac{Kc}{R_\theta} = \frac{1}{M_w P(\theta)} + \frac{2A_2 c}{RT} \tag{13}$$

where K is a constant for a given series of measurements and is calculated from the refractive index of the solvent, its concentration gradient, and the wavelength of the light; c is the solute concentration, in g/ml; R_θ is the experimentally observed intensity of scattered light at the angle θ; $P(\theta)$ is a function of the angle of measurement and the size of the dissolved molecules, having a value of unity at zero angle and progressively smaller values which can be calculated for larger angles and increasing molecular size. The virial coefficient, A_2, is the same as that appearing in the osmotic pressure equation, as are the gas-law constant, R, and absolute temperature, T.

The determination of molecular weight under the simplest conditions then involves the measurement of the scattered light intensity at one angle (usually 90°) for a series of sample concentrations. After K is evaluated, a plot of the first term of the equation against concentration will generally produce a straight line which can be extrapolated to zero concentration. This intercept equals the value of $1/M_w P(\theta)$, since the last term of the equation is eliminated. If it is certain that all components of the sample are small, below about a molecular weight of 20,000, then the value of $P(\theta)$ has a fixed value, so that the molecular weight can be calculated at once. However, in most cases some components may be larger, and the value of $P(\theta)$ must be determined by further measurements of the scattered light intensity at other angles. Two methods for this determination have been in common use.

DISSYMMETRY METHOD. This is the easier of the two methods for evaluating $P(\theta)$, and is applicable to polymers with molecular weights up to about 100,000. Measurements of the scattered intensity are made at two angles symmetrical with respect to 90°, often 45° and 135°, in addition to the usual measurements at 90°. The ratio of the intensities at 45 to 135° is called the dissymmetry coefficient, z. Measurements of z are made at each solute concentration and are extrapolated to zero concentration; the intercept is then used to determine the factor $P(\theta)$ and also to derive an average dimension of the molecules, both values being read from published graphs relating these quantities. Before either of these results can be derived, the shape of the molecules must be known, since this affects the values. Most polymers in solution approach the random-coil configuration, and some proteins have spherical shapes; these are commonly assumed until more complete study indicates otherwise. Once the value of $P(\theta)$ is known, it may be substituted into the general equation to

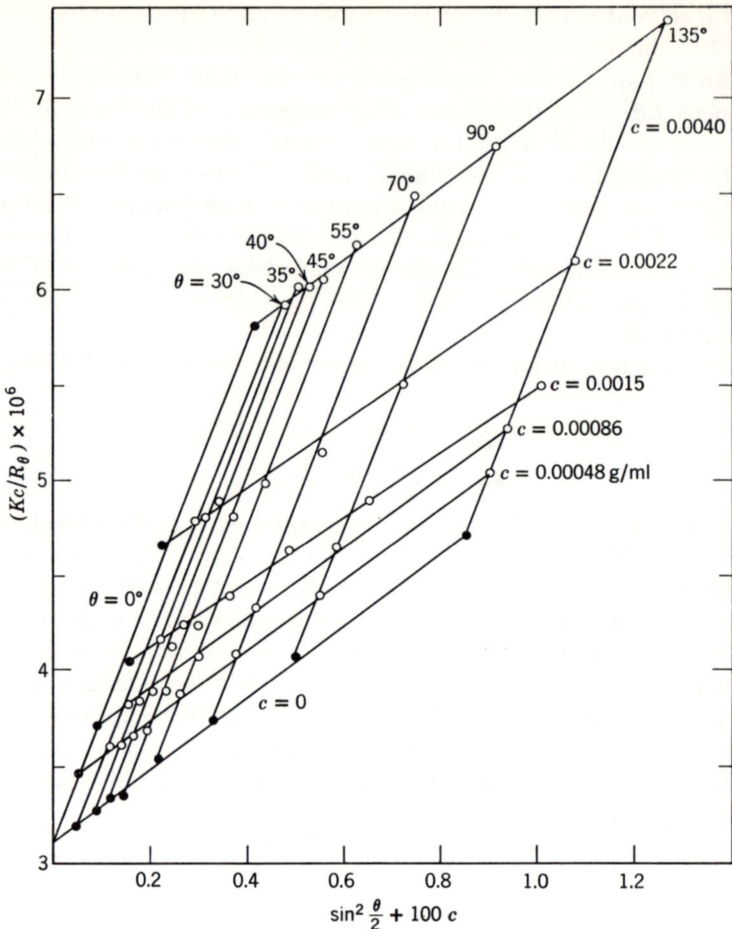

Fig. 11. Grid plot of light-scattering data, with gelatin, M_w = 320,000, in aqueous salt solution. Solid dots are values extrapolated to zero angle or zero concentration.

produce the molecular weight. The additional measurements required for the dissymmetry method occupy little extra time, provided the instrument and sample cells are designed for observations at three angles. Therefore, the method is used as a preliminary check of all samples, even when no correction is required. When the values of z exceed about 1.2, the more complex methods described below are required, and are often used even for smaller values.

ANGULAR EXTRAPOLATION METHODS. The most complete of these methods, the grid method, was originated by Zimm. It utilizes measurements of scattered intensity for a series of angles, often between 30 and 135°, but concentrating on the smaller angles. The intensities, expressed as $Kc/R(\theta)$ values, are plotted versus a function of the angle, $\sin^2\theta/2$. These points form a curve which can then be extrapolated to zero angle, that is $\sin^2\theta/2 = 0$. Similar curves for each of several solute concentrations are plotted in the same way, but are displaced from each other by using a compound abscissa, $(\sin^2\theta/2) + kc$, in which the value of k is arbitrarily selected to provide good spacing of the curve, as shown in Figure 11. The curves formed at each concentration can be extrapolated to give a point at zero angle. A second extrapolation of the curves formed of all zero angle points can then be made which intersect the ordinate axis. Similarly, the results at each angle can be extrapolated to give a point at zero concentration, and a second extrapolation of all zero concentration points should intersect the ordinate axis at the same point as the previous double extrapolation. This intercept, Kc/R, at zero angle and zero concentration needs no correction since $P(\theta)$ for zero angle is unity; thus, the molecular weight, M_w, is the reciprocal of the intercept. An average dimension of the molecules may also be derived from the graph, and no assumption need be made regarding the shape of the molecules.

Results which are equivalent to the grid method may be obtained by separate plots as a function of the angle and of the concentration, followed by the same kinds of extrapolation. Some simplification to permit measurements at only a few angles may be made for systems in which earlier measurements of similar samples have shown that the angular extrapolation produces straight lines. This occurs most often for polymer fractions or other samples in which a narrow distribution of sizes is present.

Photometers for light-scattering measurements are specialized designs for this application, as described in the article LIGHT SCATTERING. These require careful adjustment for symmetry and frequent calibration during use. When the instrument is stable and proper samples are ready, the time required for measurements is relatively short (1–2 hr) since a series of concentrations may be prepared by dilution in a single cell.

The clarification of solvent and solution to remove dust or other contaminants is usually accomplished by pressure filtration through very fine sintered glass or some types of membrane filters; the smaller the scattered intensity, the more care is required in filtration. This treatment works well for many polymers and some proteins, but some samples of natural origin and certain high molecular weight polymers include suspended colloidal components which may be quite difficult to remove. Centrifugation at high speed, partial precipitation, or filtration in the presence of filter aids may then be required before a sample suitable for measurement is obtained. It is essential to determine the effect which any of these treatments, including filtration, have on the sample; the most cogent observation is the solution viscosity which will indicate any significant change in the molecular weight, but concentration measurements may also be useful.

The *refractive index gradient*, dn/dc, must be measured in a sensitive (about 10^{-5} unit) refractometer which utilizes the same wavelength of light as that used in the scattering measurements. Either absolute or differential instruments may be used, but the latter are faster, since the gradient is provided with a single measurement. Both the instruments and techniques are described in the article REFRACTOMETRY.

Single solvents are usually required for light-scattering measurements because the selective attraction of one solvent component of a mixture alters the refractive index near the solute molecules and thus affects the intensity of scattered light.

ULTRACENTRIFUGAL SEDIMENTATION

At the very high centrifugal fields, often more than 100,000 times gravity, developed in the ultracentrifuge, dissolved macro molecules sediment through the solvent medium in a relatively short time if their density is appreciably different from that of the solvent. The molecular weight of the solute can be determined either from the rate of sedimentation or the equilibrium position of the molecules after a balance between sedimentation and diffusion is reached. Both these methods depend on observations of the extent of movement of solute molecules at various times of sedimentation, and require optical measurement of the solute concentration at each level in the cell while it is rotating at high speed in a vacuum. The need for refined optical systems, freedom from distortion of the cell at high internal pressures, and the precise control of rotational speed, together with expert technical skill at one time limited the use of this technique to a few laboratories. Recently, however, these problems have been solved or simplified by newer approaches, and have resulted in commercially available apparatus which reliably provides sedimentation data of high quality. Therefore, the use of sedimentation methods has become more widespread in polymer and protein laboratories.

Sedimentation equilibrium provides the only direct means of determining molecular weight distributions of heterogeneous polymers on an absolute basis, although it is a time-consuming process. Sedimentation velocity provides both average molecular weights and distributions, when diffusion constants are also available, but it is possible to obtain useful curves related to the distribution in a shorter time than by any other method. The Archibald method provides weight-average molecular weights in a short time without knowledge of diffusion constants. The various methods and apparatus are thoroughly described by Nichols and Bailey (28), and shorter reviews are given by Frith and Tuckett (29), by Tanford (30), and by Weissberg et al. (31).

Sedimentation Velocity

At high centrifugal fields, a dilute solution of a sample of uniform molecular weight separates into an upper layer of pure solvent and a lower layer of solution, with the boundary moving down the cell with time; diffusion causes some spreading of the initially sharp boundary. The sedimentation constant, s, of the sample is calculated from measurements of the boundary position at two times, t_1 and t_2, by the following equation:

$$s = \frac{2.303}{\omega^2(t_2 - t_1)} \log \frac{x_2}{x_1} \tag{14}$$

in which x_1 and x_2 are distances from the center of rotation, and ω is the angular velocity of the rotor. This sedimentation constant is dependent on solute concentration, particularly in the case of polymers, and therefore values are determined at several concentrations and are extrapolated to give an infinite dilution value which approaches ideal behavior. The molecular weight of the sample may be determined from the following equation:

$$M = \frac{RTs}{D(1 - V\rho)} \tag{15}$$

in which R is the gas constant; T is the absolute temperature; D is the diffusion constant for the sample, used as a measure of the frictional resistance to sedimentation; V is the partial specific volume of the sample; and ρ is the density of the solvent.

Polydisperse samples have broader boundaries because of differences in the sedimentation rates of various components, as well as the diffusion spreading. If only an average molecular weight is to be found, the calculations are as for monodisperse samples, using the steepest concentration gradient in the boundary. Typically, photographs may be taken between 8 and 60 min after the start of sedimentation. If the distribution of molecular weights is desired, then the sedimentation constants of many components in the boundary gradient are calculated at several times of sedimentation. In order to remove the effect of broadening by diffusion, these data at each time may be extrapolated to infinite time since broadening by sedimentation is more rapid than diffusion-broadening. Finally, the distribution of sedimentation constants at infinite time is determined at several concentrations, and an extrapolation of each point on each curve is made to infinite dilution. Further complexity is added by changes in centrifugal field and in concentration with distance in the cell, which must be considered. Procedures for this type of analysis have been systematized, but remain laborious.

In order to convert the resulting distribution of sedimentation constants into a molecular weight distribution, it is necessary to obtain a corresponding distribution of diffusion constants at infinite dilution by independent measurements. For compact molecules of uniform size, such as many proteins, these diffusion constants can be obtained at several concentrations and readily extrapolated to infinite dilution. Problems are encountered with molecules, including most polymers, which are expanded in solution and thus interact strongly at the concentrations usable for diffusion measurements; in these cases the extrapolation of diffusion constants to infinite dilution is inexact because of the large concentration dependence. One promising new approach is the estimation of the frictional coefficients from viscosity measurements, and this has been shown to give consistent results in certain systems (31).

The alternative which is now practiced is to establish a relationship between sedimentation constants and molecular weights by some means and use this to convert to a molecular weight distribution. This relationship may be found by determining the peak sedimentation constants for a series of fractions of known molecular weight; then construct a log s versus log M graph which provides a continuous relationship. A second approach is to determine the complete molecular weight distribution of one sample in the equilibrium ultracentrifuge, and compare the integral distribution curves of molecular weight and sedimentation constant so as to interrelate individual values of each. These are the best methods presently available for interpreting sedimentation velocity data, and are complicated by their dependence on separate diffusion measurements or on evaluation of the molecular weight of fractions or complete knowledge of the distribution curve of a similar sample.

Sedimentation curves can be obtained rapidly at concentrations low enough (about 0.002 g/ml) to distinguish features of the molecular weight distribution of a polymer sample, even without detailed analysis. Therefore, a qualitative comparison can be made of samples in which significant differences in distribution are suspected, by simply superimposing sedimentation curves obtained under similar conditions. If separate calibration curves have been made with fractions of known molecular weight, the approximate assignment of molecular weight to the regions in which samples differ may then be made.

In the choice of a solvent for sedimentation measurements, the following characteristics should be considered: (*1*) the density difference between the solvent and the solute should be at least 0.1 in order to provide rapid sedimentation; (*2*) if the sedimentation patterns are observed by a refractometric method, then the refractive indexes of the solvent and the solute should differ by at least 0.1, otherwise the sample should have an absorption band with which the solvent does not interfere; (*3*) the solvent should approach the "ideal" solvent condition, that is, it should be a poor solvent, since interactions between solute molecules are less, with resulting better separation of components and smaller concentration dependence; (*4*) solvent mixtures are generally not safe because of selective attraction of some components to the solute.

Sedimentation Equilibrium

At much slower centrifuge speeds and with longer times than are used in sedimentation velocity measurements, molecules will settle until they reach a position where their rate of sedimentation just balances the rate of diffusion into upper levels. If the sample is monodisperse, its apparent molecular weight can be calculated from the ratio of concentrations at two positions in the cell, by the following equation:

$$M = \frac{4.606\,RT}{\omega^2(1 - V\rho)(x_2^2 - x_1^2)} \log \frac{c_2}{c_1} \tag{16}$$

If these apparent values show dependence on the solute concentration, then extrapolation of the data to infinite dilution provides the true molecular weight.

For polydisperse samples, the equilibrium sedimentation position is different for each molecular species, so measurement of the concentratiod present at each level provides information from which the molecular weight distribution of the sample may be calculated under favorable conditions. If the solution is ideal, that is, with little or no dependence of the sedimentation constant on concentration, then a detailed analysis of the sedimentation pattern can yield various average molecular weights with different weighting, such as M_n, M_w, and M_z, which are useful measures of the breadth of the distribution. These average molecular weights can then be used to calculate the molecular weight distribution by assuming a distribution function, or by a sufficiently complete analysis of the averages. No good solution to the problem has been found where the concentration dependence of the sedimentation constants is substantial, and so the solvent should be selected to approach ideal conditions.

In addition to the mathematical complications of this method, the approach to equilibrium also requires a long centrifuging time (from a few days to several weeks) with good uniformity of speed and temperature to avoid convection disturbances. The results of one equilibrium experiment may, however, be utilized with the rapid sedimentation velocity experiments to determine molecular weight distributions of samples of similar nature.

Archibald Method

At moderate speeds of centrifuging, the initial separation of sample components from the air–solution meniscus, and also the concentration change at the bottom of the cell have been shown theoretically to approach the conditions at sedimentation equilibrium. This has been used to determine molecular weights rapidly and with no dependence on separate measurements of diffusion constants, according to the differential form of the equilibrium equation:

$$M = \left(\frac{RT}{(1 - V\rho)\omega^2 xc}\right)\left(\frac{dc}{dx}\right) \tag{17}$$

Measurements are required of the linear section of the concentration gradient close to each end of the cell at times soon after the start of sedimentation. These are available with careful use of the schlieren optics system and microphotometer reading of the images. Early tests with samples of known uniform molecular weight demonstrated the value of the method in the range from 131 to high molecular weights. More recent applications to polydisperse polymer systems have shown the usefulness of the method for determining weight-average molecular weights in several systems. Because these applications have occurred since compilation of the general references cited above, two specific examples will be listed. Fujita et al. (32) have found molecular weights by this method for two polystyrene samples in fairly good agreement with results of light-scattering measurements. Weston and Billmeyer (33) have reported fair agreement for one sample of polyethylene of the sedimentation and light-scattering molecular weights. It is apparent that this method may have some advantages where sample preparation is difficult for light-scattering measurements, but the analysis of sedimentation patterns requires great care for accurate results.

SOLUTION VISCOSITY METHOD

This is the method most widely used for the characterization of polymers because of its speed and relative freedom from experimental complications. It is, however, a relative method which at present requires calibration for each type of polymer and each solvent used over a range of known molecular weights sufficient to establish a reliable calibration curve.

The viscosity, η, of a series of dilute solutions (usually from 10^{-3} to 10^{-2} g/ml) of the sample is compared to that of the solvent, η_0, by the following equation:

$$\frac{\eta_{sp}}{c} = \frac{\eta - \eta_0}{\eta_0 c} \tag{18}$$

where the specific viscosity, η_{sp}, divided by concentration is termed the *viscosity number*. Because of interactions between molecules, this result is dependent on concentration, and a graph of η_{sp}/c versus c is often a straight line which permits extrapolation to infinite dilution. The intercept is termed the *limiting viscosity number* or in older terminology, the *intrinsic viscosity*, with the symbol $[\eta]$. There also may be dependence of this value on the rate of shear used in the viscometer, and an extrapolation to zero shear may be necessary. This usually occurs at very high molecular weights or if the polymer is stiffened by charge or steric effects.

The relation of the limiting viscosity number to molecular weights is expressed by the following equation:

$$[\eta] = KM^a \tag{19}$$

in which K and a are constants for a given polymer–solvent system, and the molecular weight is a viscosity average. This equation is commonly used as a linear graph of log $[\eta]$ versus log M_v, which enables rapid conversion of measured viscosities to molecular weights by use of the best average line. Many relations of this kind have been published, and some of these are tabulated (34). It is important in using these results to duplicate the conditions originally used, that is, the same solvent, temperature, and in some cases the same type of viscometer, to provide the same shear rate. In addition, the source of each such relationship should be checked to determine that the calibrating samples had been suitably characterized. It is preferable that weight-average molecular weights be used in preference to number-average values in such calibrations, since they are much closer to the type of average governing the viscosity property. This is especially important when unfractionated samples are used for calibration, since there can be little assurance that the molecular weight distribution of all unknowns will be comparable to the standards. When fractionated standards are used, all average molecular weights will be closer together, but it has been found that many fractions still have a fairly broad distribution, so that viscosity–molecular weight relations determined from number-average molecular weights often differ substantially from those standardized with fractions whose weight-average molecular weight was used.

The statements above obviously apply to the preparation of calibration curves and the selection of methods for determining the molecular weight of fractions. This suggests that light-scattering or Archibald ultracentrifugal methods should be employed for molecular weight measurements in these cases.

More detailed discussions of the theory and practice of viscometry of polymer solutions are given by Onyon (35), Frith and Tuckett (36), and Flory (37). See also Rheological measurements, fluid.

END-GROUP ANALYSIS

This chemical method involves determination of the total number of reactive end groups present in a polymer sample, and calculation of the molecular weight by dividing the sample weight by the number of moles of groups found or half this quantity if both terminal groups are determined. Applications are most often made in the molecular weight range up to 25,000, usually limited by the decrease in concentration of reactive groups; however, in some cases results have been reported up to 400,000 molecular weight. Various applications of the method are discussed by Price (38).

Linear condensation polymers are most suitable for this analysis because of the certainty that all chains are terminated by reactive groups. The polymers examined include many polyamides and polyesters, and some applications have been made to cellulose derivatives. The presence of branched chains, introduced by a trifunctional reactant, provides more than two end groups per molecule and thus makes the method useful for determining the extent of branching, but does not ordinarily permit molecular weight determinations. Vinyl polymers are generally unsuitable for this method, because the free radical mechanisms of their formation may leave active end groups on one or both ends; also the number of such groups is very small because of the much higher molecular weights (above 50,000) usually produced.

The end groups which are most easily determinable are free carboxyl groups present in polyesters and polyamides, free hydroxyl groups in polyesters, and free amino groups in polyamides. It is usually necessary to determine both end groups present in these polymers, since there are several reasons not to assume equal amounts of each type of group, even when the quantities of monomers had been initially equal. Carboxyl and amino groups can be titrated directly as acids or bases when dissolved in selected nonaqueous solvents, though several alternative methods are available. Hydroxyl groups are determined by acetylation or conversion to urethanes, with determination of the amount of reagent consumed, or of some easily detectable constituent of the derivative. Again, several other procedures have been applied to particular polymers.

The quantities of end groups are very small, and so microtitration techniques must be used. The polymer sample must be fully soluble in a solvent which is unreactive during the titration. These conditions restrict the choice of solvents, especially for polymers which are insoluble in many common solvents.

X-RAY DIFFRACTION METHOD

This is a powerful tool for determining the structure, including the size and weight, of molecules which can be crystallized. However, the method is complicated and slower than many techniques which provide molecular weights of accuracy sufficient for many purposes, and so is usually employed only when the detailed analysis of the arrangement of atoms in crystals is the goal. The theory and applications of x-ray crystallography are reviewed by Lipscomb (39) and by Klug and Alexander (40). See also X-RAY METHODS.

The sample to be examined must have a high degree of crystalline order, and is preferably a single crystal at least 0.1 mm in size; such samples are prepared fairly readily from many inorganic and nonpolymeric organic compounds. Alternatively, crystalline powders of certain crystal types may provide suitable results. Polymers are usually not suitable for detailed structure analysis by this method, because most polymers are not highly crystalline and so produce poorly defined patterns.

X rays are diffracted from ordered planes of atoms in crystalline compounds, and form diffraction patterns by several measurement methods which can be used to calculate the lattice spacings, and thus the size of the unit cell in the crystal. This unit cell is the smallest volume unit which retains all geometrical features of the crystalline class, and it usually contains a small integral number of molecules. A rough estimate of the molecular weight of a compound is needed from an independent method in order to obtain this integral number. Finally, the resultant molecular volume is multiplied by the exact bulk density of the crystal and by the Avogadro number to yield the molecular weight.

This method has been often used for the identification of compounds, and for determination of the structure of simple inorganic and organic compounds, as well as very complex molecules, such as proteins, when they provide well-ordered crystals. The molecular weights which are obtained during the structure determination are of high accuracy, being accurate to one atom.

SURFACE PRESSURE OF MONOLAYERS

When polymeric molecules can be spread on the surface of a liquid or at an interface between two liquids, they form expanded films under some conditions at very low surface pressures which obey a two-dimensional analog of the gas law. This behavior has been used to determine the molecular weights of a number of proteins and some synthetic polymers in the molecular weight range from about 2000 to 100,000.

The molecules must be sufficiently surface active to spread and yet must not dissolve upon compression of the surface film; they also must not associate, that is, they must behave as single units. Many proteins and some polymers have been successfully studied on the surface of a saturated aqueous solution of ammonium sulfate, which appears to overcome association and solubility problems, though pure water and benzene have also been used.

The method requires that measurements be made of the very small pressures developed at various stages of compression of a very dilute film, with all measurements usually below 0.3 dyne/cm. Compression is usually accomplished in a Langmuir trough and measurements of pressure are made with a vertical plate (Wilhelmy) balance, or with a horizontal float balance modified to give much greater sensitivity than is required for usual monolayer studies. The surface pressures are divided by the surface concentration, in g/cm^2, and the P/c values are extrapolated to infinite dilution in much the same way as is done in osmometry in order to avoid molecular interactions. The molecular weight derived from the gas law type of calculations is expected to be a number-average quantity.

The advantages of the method are the fairly high speed of measurement when the necessary experience has been gained, and the use of small samples (often less than 1 mg). It suffers from the necessity to produce and maintain clean liquid surfaces, and the need for some experimentation with every new substance to establish conditions for good spreading. It also must be noted that while some workers have reported very satisfactory results for various proteins, polypeptides, gelatin, cellulose acetate, polystyrene, poly(methyl methacrylate), and a polyester, others have not obtained satisfactory results with some of these under similar conditions. Review articles on the subject should be consulted for detail (41,42).

MASS SPECTROMETRY METHOD

This method is most used for the detection, identification, and quantitative analysis of volatile compounds in mixtures, and to elucidate the structure of organic compounds. It also enables the determination of molecular weights up to about 700 with great accuracy, and this has recently been used as a powerful tool in this structure analysis. It is unlikely that the method would normally be used exclusively for the determination of molecular weights, because methods which are simpler and use much less expensive equipment currently give satisfactory results in the range of applicability. The method is listed here because molecular weights are customarily observed during structure analysis by this technique and could be obtained independently if the advantages of very small sample requirements (often less than 1 mg) and high accuracy are important. Reviews covering the theory, instruments and applications are given by Stewart (43) and Biemann (44). See also MASS SPECTROMETRY.

The following are the basic features of the method. Samples are vaporized and introduced into an electron beam which ionizes and fragments the molecules. These charged particles are accelerated in an electric field, and then are spread out by a magnetic field so that each individual mass is isolated and detected in a scan of the ratio of mass to charge (m/e) for all components. The intensity of the sharp peaks obtained is related to the concentration of each fragment. Unless some reaction occurs between fragments, the peak found at the highest m/e value when low electron excitation is used corresponds to the weight of the singly charged ion, and thus to the molecular weight of the sample. For a small proportion of the compounds tested, the molecular ion is missing or only present as a small concentration, because of its instability. In these cases, the molecular weight can only be determined from identification of the fragments and reconstruction of the original formula.

Conventional mass spectrometers have a resolution of about 1 mass unit, which is quite satisfactory for normal molecular weight measurements. Recently, high resolution spectrometers have been built with the ability to distinguish peaks differing by 0.002 mass units or less, and still able to cover the mass range as high as 900 units. These have been utilized already to great advantage in structure determinations where various possible formulas are met which have apparently equal molecular weights and therefore produce peaks which superimpose in ordinary instruments. These can now be distinguished at high resolution on the basis of differences in the fractional masses contributed by the various constituent atoms.

BIBLIOGRAPHY

General References

1. G. L. Beyer, "Determination of Particle Size and Molecular Weight," in A. Weissberger, ed., *Physical Methods of Organic Chemistry,* Part I (Vol. 1 of *Technique of Organic Chemistry*), Interscience Publishers, Inc., New York, 1959, pp. 191–257.
2. R. U. Bonnar, M. Dimbat, and F. H. Stross, *Number-Average Molecular Weights,* Interscience Publishers, Inc., New York, 1958, pp. 288–293.
3. P. M. Doty and H. M. Spurlin, "Determination of Molecular Weight and Molecular Weight Distribution," in E. Ott et al., eds., *Cellulose and Cellulose Derivatives,* Part 3, Interscience Publishers, Inc., New York, 1955, pp. 1173–1188.
4. D. F. Rushman, "Other Methods for the Determination of Number-Average Molecular Weights," in P. W. Allen, ed., *Techniques of Polymer Characterization,* Butterworth & Co., Ltd., London, 1959, pp. 113–130.

Specific References

5. R. W. Hall, "Fractionation of High Polymers," in Ref. 4, pp. 19–57.
6. L. H. Cragg and H. Hammerschlag, *Chem. Rev.* **39,** 79 (1946).
7. J. B. Nichols and E. D. Bailey, "Determinations in the Ultracentrifuge," in Ref. 1, Part II, p. 1029.
8. G. V. Schulz, *Z. Physik. Chemie B* **43,** 25 (1939).
9. P. J. Flory, *Principles of Polymer Chemistry,* Cornell University Press, Ithaca, N.Y., 1953, p. 321.
10. H. W. McCormick, F. M. Brower, and L. Kin, *J. Polymer Sci.* **39,** 87 (1959).
11. T. G Fox and S. Loschaek, *J. Polymer Sci.* **15,** 371, 391 (1955).
12. H. Fujita and K. Ninomiya, *J. Polymer Sci.* **24,** 233 (1957).
13. N. Bauer and S. Z. Lewin, "Determination of Density," in Ref. 1, Part I, pp. 131–190.
14. Ref. 2, pp. 17–58.
15. E. L. Skau, J. C. Arthur, Jr., and H. Wakeham, "Determination of Melting and Freezing Temperatures," in Ref. 1, Part I, pp. 287–356.
16. W. Swietoslawski and J. R. Anderson, "Boiling and Condensation Temperatures," in Ref. 1, Part I, pp. 395–398.
17. R. H. Wagner and L. D. Moore, Jr., "Determination of Osmotic Pressure," in Ref. 1, Part I, pp. 816–894.
18. F. B. Rolfson and H. Coll, *Anal. Chem.* **36,** 888 (1964).
19. R. E. Steele, W. E. Walker, D. E. Burge, and H. C. Ehrmantraut, *Paper, Pittsburgh Conference on Analytical Chemistry and Applied Spectroscopy, March 1963;* reprint available from Mechrolab, Inc., Mountain View, Calif.
20. A. J. Staverman, *Rec. Trav. Chim.* **70,** 344 (1951).
21. A. J. Staverman, D. T. F. Pals, and C. A. Kruissink, *J. Polymer Sci.* **23,** 57 (1957).
22. D. B. Bruss and F. H. Stross, *J. Polymer Sci.* **55,** 381 (1961).
23. J. B. Donnet and B. Roth, *Bull. Soc. Chim. France* **49,** 1257 (1954).
24. K. A. Stacey, *Light Scattering in Physical Chemistry,* Butterworth & Co., Ltd., London, 1956.
25. F. W. Peaker, "Light-Scattering Techniques," in Ref. 4, pp. 131–170.
26. D. McIntyre, *ASTM Spec. Tech. Publ. No. 247* (1948).

27. G. Oster, "Light Scattering," in Ref. 1, Part I, pp. 2107–2146.
28. Ref. 7, pp. 1007–1038.
29. E. M. Frith and R. F. Tuckett, *Linear Polymers*, Longmans, Green & Co., Inc., New York, 1951, pp. 237–249.
30. C. Tanford, *Physical Chemistry of Macromolecules*, John Wiley & Sons, Inc., New York, 1961, pp. 364–390.
31. S. G. Weissberg, S. Rothman, and M. Wales, "Molecular Weights and Sizes," in G. M. Kline, ed., *Analytical Chemistry of Polymers*, Part II, Interscience Publishers, a division of John Wiley & Sons, Inc., New York, 1962, pp. 1–77.
32. H. Fujita, H. Inagaki, T. Kotaka, and H. Utiyama, *J. Phys. Chem.* **66,** 4 (1962).
33. N. E. Weston and F. W. Billmeyer, Jr., *J. Phys. Chem.* **67,** 2728 (1963).
34. G. M. Burnett, *Mechanism of Polymer Reactions*, Interscience Publishers, Inc., New York, 1954.
35. P. F. Onyon, "Viscometry," in Ref. 4, pp. 171–206.
36. Ref. 29, pp. 220–237.
37. Ref. 9, pp. 308–314, 611–622.
38. G. F. Price, "Techniques of End-Group Analysis," in Ref. 4, pp. 207–230.
39. W. N. Lipscomb, "X-ray Crystallography," in Ref. 1, Part II, pp. 1641–1738.
40. H. P. Klug and L. E. Alexander, *X-ray Diffraction Procedures*, John Wiley & Sons, Inc., New York, 1954.
41. H. B. Bull, *J. Biol. Chem.* **185,** 27 (1950).
42. E. Mishuk and F. Eirich, in H. Sobotka, ed., *Monomolecular Layers*, American Association for the Advancement of Science, Washington, D.C., 1954.
43. D. W. Stewart, "Mass Spectrometry," in Ref. 1, Part IV, pp. 3449–3539.
44. K. Biemann, "Applications of Mass Spectrometry," in K. W. Bentley, ed., *Elucidation of Structures by Physical and Chemical Methods*, Part I, Interscience Publishers, a division of John Wiley & Sons, Inc., New York, 1963, pp. 261–316.

George L. Beyer
Eastman Kodak Company

CHEMICAL ANALYSIS

The chemical characterization of polymers is not basically different from the analysis of organic compounds of low molecular weight. However, the absence of functional groups of sufficient reactivity, the generally low solubility and chemical inertness, and the complexity of most polymeric systems necessitate modifications of existing procedures so that the component or functional group being determined can be brought into intimate contact with the reagent. Depending upon the analysis being performed, contact with the reagent is brought about by degradation or ashing, by extraction of the polymer, or by solution techniques.

Recent advances in our knowledge of the chemical and physical structures of polymeric substances have been due in a large measure to new or improved instrumental techniques. These methods are (a) often nondestructive so that the sample can be recovered for other studies, (b) rapid, (c) highly sensitive, (d) accurate, (e) direct, ie, films or dispersions can be handled and therefore limited solubility is no problem, (f) capable of supplying data on the physical behavior as well as the chemical structure not obtainable by purely chemical techniques, and (g) supply

Table 1. Techniques Used in Analysis for Chemical and Physical Structure

Chemical structure	Physical structure
chromatography	infrared spectroscopy
column	x-ray diffraction
paper	electron diffraction
thin-layer	optical microscopy
electrophoresis	optical rotation
electrochemical tests	nuclear magnetic resonance
absorption spectroscopy	differential thermal analysis
infrared	rheology
ultraviolet	cryoscopy
Raman	ebulliometry
optical rotatory dispersion	osmometry
fluorescence, phosphorescence	light scattering
nuclear magnetic resonance	ultracentrifugation
electron paramagnetic resonance	fractionation
emission spectroscopy	miscellaneous physical tests
x-ray spectroscopy	
electron probe	
pyrolysis	
infrared	
mass spectrometry	
gas chromatography	
thermogravimetry	
radiochemical tests	
neutron activation	

other useful information which may assist in characterizing the polymer. Examples of techniques used in the analyses for chemical and physical structure are listed in Table 1.

Separate articles deal with many of the physical methods such as ultraviolet-absorption spectroscopy (qv), infrared-absorption spectroscopy (qv), nuclear magnetic resonance (qv), differential thermal analysis (qv), x-ray diffraction (qv), isotopic labeling (qv) and molecular-weight determination (qv). This article will discuss mainly the more conventional methods such as characterization tests, end-group analysis, systematic procedures, and similar topics. Analysis of specific types of polymers is treated in the articles covering these materials.

It should be emphasized that complete characterization of a newly developed or completely unknown polymer is not an easy task and may require the successful application of many chemical and physicochemical techniques. Besides their limited solubility, many of these materials do not have sharp melting points. Commercial plastics may contain plasticizers, solvents, stabilizers, pigments, fillers, dyes, etc, which often must be removed before identification can be made. Presence of substances of low molecular weight and of catalyst fragments among the macromolecules, variations in the degree of branching and crystallinity, differences in monomer ratios in copolymers, stereospecificity, and grafting will change the physical properties and chemical behavior, and thus increase the difficulties of identification.

The literature on methods of analysis of polymers is scattered in many technical journals covering the fields of analytical chemistry and polymer science. A few books have been published since 1958 that deal mainly with this subject matter (1–4). General identification schemes for organic compounds (5–8) and quantitative methods for the analysis of functional groups (9–10a) are also applicable to polymer analysis.

A comprehensive systematic identification procedure for polymers was published by Shaw (11) in 1944. Obviously, it does not include the newer polymeric materials such as polycarbonates, polyurethans, polypropylenes, etc. This procedure is based on the following steps: (a) purification of the polymer by removal of fillers, vehicles, or solvents, usually by methods based on differences in solubility; (b) division of the resin into groups by determination of the elementary composition, saponification number, and acetyl number; and (c) distinguishing polymers in a given group by characteristics such as physical properties (density, refractive index, etc), burning characteristics, solubility, and chemical classification tests. A summary of this and other identification schemes is given in Ref. 12. A typical short scheme for analysis based on solubility, elemental analysis, and a few specific tests has been proposed by Gruben and Leiner (13).

The biennial reviews on analysis of coatings (14) and natural and synthetic rubbers (15) should be consulted for the more recent developments. The reviews on fundamental developments in analysis (16) provide up-to-date background information on such diversified techniques as chromatography, microchemistry, chemical and electron microscopy, polarography, spectroscopy (including ultraviolet, infrared, light-absorption, mass, magnetic-resonance, and Raman techniques), differential thermal analysis, titrations in nonaqueous solvents, volumetric and gravimetric analytical methods, and x-ray diffraction.

Various techniques employed for the characterization of polymers to obtain information as to the size of the polymer molecule, molecular-weight distribution, degree of branching, copolymer segment length, etc have been described by Allen

(17). Instrumental methods for the analysis of polymeric materials have been reviewed by Mitchell and Lord (18). Kupfer (19) has reviewed the analysis of synthetic resins. The comprehensive treatises for the analysis of resin-based coating materials (3,20–22), natural and synthetic fibers (23–26), textile finishes (27), adhesives (28), plastics (1,29), rubbers (30–33), and surface-active agents (34) should be consulted when these specific materials are under investigation.

Preparation of the Sample

A representative sample, usually reduced to a powder, must be selected for the analysis. The material can be crushed in a mortar, but great care should be taken that the material does not become hot during the pulverization, or changes in the chemical composition may occur. The various constituents of the polymeric system should then be isolated in as pure a form as possible before attempting the ultimate analysis.

The principle used repeatedly in the analysis of plastic materials is the solution of the polymer in one solvent followed by precipitation by addition of a nonsolvent. The subsequent examination of the recovered polymer or of the additive remaining in solution can then be accomplished.

The most commonly encountered problem is the separation of filler and plasticizer from the base plastic. The plasticizer can be removed by means of a suitable solvent, such as ether or some other low-boiling liquid. Every effort should be made to select a solvent that will remove the plasticizer only. Liquids that react with the polymer, partially dissolve it, or are strongly absorbed must be avoided.

Thermoplastic resin and plasticizer may be separated from inorganic fillers by dissolving the mixture in a solvent, preferably using an extraction apparatus. The undissolved residue in the extraction thimble should be tested for the complete removal of the polymer. Successive fractionation employing other solvents or solvent mixtures to obtain complete separation may be necessary. The polymer in the extract may be recovered by precipitation (by the slow addition of a nonsolvent that is a solvent for the plasticizer) or by distilling off the extraction liquid. The proper solvents are selected on the basis of some prior knowledge of the material or by trial and error. If a quantitative determination of the purified polymer is required, the material can be dried, preferably in a high vacuum, and weighed.

Separation of the filler by extraction methods is not feasible for the thermosetting insoluble plastics. In this case, modification of the polymer, as by hydrolysis or esterification, may be helpful. Sometimes it is not possible to remove an organic filler that has properties nearly identical to those of the plastic. Microscopic examination of the specimen will usually detect the presence of such a constituent. Inorganic constituents are determined by the customary chemical or spectroscopic analytical procedures.

A method for the separation of polymers from solvents and monomers by a freezing technique is described by Lewis and Mayo (35). The polymer is dissolved in eight to ten times its weight in an appropriate volatile solvent, eg benzene, and the solution is frozen quickly. The benzene, in this case, is sublimed from the polymer without melting the benzene or sintering the polymer. The polymer is left as a soft, fluffy, very porous solid. After the benzene has been largely removed at 0°C, the product is allowed to stand at room temperature in a high vacuum for a few hours before heating in order to reduce the tendency of remaining small amounts of solvent to cause sintering. Final drying should be conducted at 70 to 100°C in a high vacuum.

Since the whole procedure is carried out at reduced pressure, no difficulties arise from oxidation or moisture adsorption. Some chlorine-containing compounds lose hydrogen chloride under these conditions. Solvents other than benzene which are conveniently sublimed can also be employed.

Determination of Purity

One of the most difficult problems in polymer chemistry is the determination of the purity of the sample. Direct analysis by chemical or physical procedures for probable contaminants is most commonly employed. Determinations of volatile matter such as moisture or solvents or of noncombustible constituents (usually inorganic) in polymers can be readily conducted. Functional groups other than those which should be present can be detected or quantitatively estimated by various procedures, involving chemical or instrumental techniques.

Trace concentrations of contaminants in the ppm range may be detrimental to the stability of the polymer, may affect its physical behavior, or may prohibit its use in food packaging. Methods employed for the preparation of the sample prior to the actual determination include dry and wet ashing, extraction, and solution techniques. Extraction of metals is usually not satisfactory because of the inability of the extractant to penetrate the polymer completely. Solution techniques are valuable when they can be applied because of their simplicity. However, satisfactory solvents cannot always be found. Because of these limitations wet- or dry-ashing procedures are generally employed. Wet-ashing techniques can be used for all the elements studied. Only with mercury is recovery low; however, large amounts of metals in the reagents may produce excessively high blanks. Low blanks are obtained by dry-ashing procedures, but they are more time consuming and losses of volatile metals can be serious. Thus, neither technique is a panacea and choice of the proper method depends on the problem at hand.

Traces of monomer in polymers can be detected by a variety of techniques such as end-group analyses or physicochemical procedures. Quantitative determination of monomer content in the polymer can be conducted by the freeze-drying technique of Lewis and Mayo (35): the volatile products are trapped and the monomer content is determined by an appropriate method. Monomer content in acrylic and some allyl or vinyl polymers can be estimated by halogen addition to the carbon double bond, using aqueous bromide–bromate reagent which, on acidification, releases bromine (36).

Assay of the main constituents by analysis for elementary composition, measurement of physical or optical properties, and behavior on degradation also give an indication of purity, provided the structure of the major component is accurately known and the precision of the particular measurement is high enough to detect possible impurities. For crystalline materials of narrow molecular-weight spectra, purity determinations based on freezing point, thermogravimetric or differential thermal analysis, or solubility curves may be useful. A discussion of the various techniques is given by Stenger, Crummett, and Cobler (37).

Physical Tests

Preliminary Examination. The physical appearance and qualitative aspects of some physical characteristics are useful guides for identification (see also Characterization of polymers). The physical state, color, transparency, and odor should be noted. It should be determined whether the material is amorphous or crystalline. Such criteria as sharp melting point, sharp x-ray diffraction pattern, insolubility in at least some common solvents, and high toughness rather than high degree of elasticity suggest that the polymer is at least partially crystalline. Surface imperfections or inhomogeneities can often be detected readily under a microscope. The polymer should be stretched, if possible. Materials such as poly(vinyl acetate), polyacrylates, elastomers, and plasticized thermoplastic resins are quite extensible. Hardness can be judged by attempting to mar the sample with the thumbnail or a coin. A comparison with a known polymer is often helpful. Polystyrene gives a characteristic "metallic" ring when dropped; no other plastic behaves in the same manner.

Fibers or yarns should be broken and the strength and extensibility noted. Good strength with little extensibility suggests highly oriented fibers or, perhaps, glass fibers. Glass fibers are indicated by their rapid disintegration on rubbing. Good strength combined with moderate extensibility and good elasticity suggests silk or one of the synthetic fibers, such as nylon or poly(ethylene terephthalate).

Another preliminary test consists of heating the dry sample in a test tube in an oil bath. Thermosetting resins remain hard if already set, and harden, after intermediate softening or melting, if in the A-stage (qv). Thermoplastic resins soften or melt at different temperatures and retain their characteristic solubility properties after cooling, provided that decomposition has not taken place.

The preliminary examination should be followed by an ignition test and determination of the melting, softening, or glass temperature, density, index of refraction, absorption spectra, and fluorescence.

Burning Characteristics. A sample of 0.05–0.2 g, preferably a specimen of small cross section, is placed on a spatula and is brought to the edge of a small Bunsen flame to determine its flammability. Ease of ignition, self-extinguishing characteristics, and any odor given off are noted. The sample should be heated slowly. If the flame is too large, the decomposition will be too fast to observe the various phenomena. It should be observed whether the sample blackens, melts sharply with or without decomposition, chars, or burns. The color of the flame, the formation of any sublimate, and the acidic or basic nature of any vapors formed should be noted. Polymers that decompose on heating into aromatic hydrocarbons burn with a yellow, sooty flame; those from aliphatic hydrocarbons burn with a flame that is much less sooty. With increasing oxygen content of the decomposition product the flame becomes more blue.

The acidic or basic reaction of the decomposition product can be detected by placing a strip of wet litmus paper on the top of the test tube containing the sample and heating. If an acid reaction is obtained, add a piece of Congo paper, which in the presence of mineral acid turns blue.

Finally, the sample should be ignited strongly. The characteristics of any ash formed, whether colored, bulky, hygroscopic, etc, should be recorded. The residue may be analyzed either by microchemical or by spectroscopic methods.

To detect the odors produced by the polymers, place a small specimen on a steel spatula blade, hold the blade in a small Bunsen flame for a few seconds, and withdraw. If the polymer ignites, extinguish the flame immediately. Note the odor of the rising vapors. The heating may be continued but the blade should be withdrawn every few seconds and the vapors arising from the hot plastic should be tested after each withdrawal. Odors evolved during successive stages may help to identify the presence of modifying resins, plasticizers, oils, or other additives. No valid conclusions can be reached from smelling the plastic while it burns; too much vapor will produce a deadening effect upon the olfactory organs. In many cases it will be necessary to describe the odors as characteristic since no better terms are known. The identification of odors is a matter of experience. Known specimens should also be decomposed and the results compared with those noted for the unknown material. Generally, methods of identification based on odor alone are too subjective for a satisfactory scientific analysis.

A summary of the behavior of the commonly used plastics and elastomers when subjected to the ignition test is given by Brauer and Horowitz (12). A systematic scheme of analysis, based on placing the sample at the edge of a flame for 10 sec, has been suggested by Nechamkin (38).

Transition Points. The melting point, softening point, and glass-transition temperature are used in characterizing polymers.

Melting Point. Melting points are of limited use in polymer identification: amorphous polymers usually soften over a considerable temperature range, hence show no sharp melting point; other polymers decompose completely before their melting point is reached. A comparatively small number of polymers, such as polyesters, polyamides, polyethylene, poly(vinylidene chloride), and acrylic ester resins with long-chain alcohol groups, exhibit fairly sharp melting points; since these polymers are usually incompletely crystallized, heating causes a softening of the amorphous portions before actual melting of the crystalline part occurs. Many crystalline polymers are susceptible to some degree of decomposition when heated in air, which contributes to the disagreements in melting points reported in the literature.

The melting point is most readily determined by placing some of the finely powdered material in a capillary, which is then heated in an oil bath. A Maquenne block is very suitable for measuring melting points above 200°C.

Melting points determined by the capillary tube method and the Maquenne block are not always in agreement. More refined melting point apparatus or hot stages for microscopes that permit the determination of melting points with microquantities of material are commercially available. An automatic melting-point apparatus especially suitable for macromolecular substances has been developed by Überreieer and Orthmann (39). A procedure for estimating the melting point of a resin by measuring the temperature at which the molten resin rises to the top of a pool of mercury has been described (40).

Identification of synthetic fibers by microfusion methods has been described by Grabar and Haessly (41). Melting points are determined on a hot stage of a polarizing microscope at magnifications of 100–200 diameters using both ordinary transmitted and polarized light. The following behavior and corresponding temperatures are noted: abrupt longitudinal shrinkage, discoloration caused by decomposition, changes in polarization colors, transverse swelling which usually precedes or marks melting, the beginning of melting, and the end of melting marked by disappearance of the last traces of birefringence. A supplementary observation can be made by placing a larger sample between a slide and cover glass and heating on a hot plate. If the fiber melts, slight pressure on the cover glass will cause the melt to flow into a thin film. If the slide is immediately placed on a cold microscope stage, observation will show whether the melt crystallizes upon cooling or whether it supercools to the glassy state. For example, nylon characteristically crystallizes into a mass of spherulites, whereas

Dacron supercools to an isotropic film which recrystallizes only when reheated to 100°C.

Many crystalline polymers decompose below their melting points. However, the melting point can be depressed below the decomposition temperature by addition of a suitable second component. The eutectic melting point with this second component, as observed on a microscope hot stage, becomes as useful for identification purposes as the melting point of the original material. A suitable second component for use as reference compound with synthetic fibers is *p*-nitrophenol. Fibers that neither melt nor show eutectic melting on addition of this compound exhibit characteristic, easily distinguishable, and reasonably reproducible behavior patterns, such as abrupt longitudinal shrinkage, transverse swelling, and partial or complete solubility of the fiber in the molten *p*-nitrophenol. *p*-Nitrophenol starts to sublime above 80°C and sometimes condenses on the cover glass in droplets. Therefore, indications of eutectic melting above 80° must be looked for in the fiber rather than in the *p*-nitrophenol.

Melting-point measurements have been reviewed by Skau, Arthur, and Wakeham (42) and Herbrandson and Nachod (43).

Softening Point. Softening or brittle-point tests of polymers have, in most cases, a purely empirical significance and cannot be considered as strictly reproducible and scientifically well-established procedures (see also CHARACTERIZATION OF POLYMERS). They are very useful and important in the absence of better approaches as long as their empirical character is clearly realized. Most plastics readily distort within a fixed temperature range. The temperature at which the deformation begins is the softening point. The physical significance of this property has been described by Alfrey (44).

The softening point is a characteristic property of amorphous polymers and, in conjunction with other data, provides valuable information for identification. Because of overlapping among various types of polymers and variations due to addition of plasticizers and modifying agents, the results of such measurements should be accepted only after additional confirming tests have been conducted.

A rapid method for estimating the softening point is as follows: Place a portion of polymer film, approximately 0.25 in. wide and 2 in. long, on a brass block. Place a 5- or 10-g weight on the specimen. Insert a thermometer and slowly heat the block. At intervals of 5°C remove the weight and quickly pull the film away from the brass block. After each trial, cut away the part of the strip that has been heated. The temperature at which the polymer first sticks to the block or shows other evidence of transition to the liquid state is the softening point. Reproducibility of the procedure is ±10°C.

A number of more refined tests have been described, such as the ball and ring method and the Vicat test. Since these tests are highly arbitrary, the results will depend on such factors as time, method of loading, and fiber stress. They have only limited significance in polymer identification.

A number of other test methods for softening point are described by Teeple (45) and Gardner and Sward (20). Extensive tables of softening points of polymers, resins, and synthetic fibers are given in Ref. 12.

Glass-Transition Temperature. The brittle-point temperature cannot be specified uniquely, but depends on the manner of testing used in its determination. The glass-transition temperature (second-order transition temperature), on the other hand, is manifested as a change in slope of the curve obtained by plotting any of the primary thermodynamic properties against temperature (see also CHARACTERIZATION OF POLYMERS). This change in slope falls in the same temperature range as that in which the mechanical softening point occurs.

For polymer analysis, rapid methods of determination of glass-transition temperature are suitable. It is sufficient to measure the respective physical property at a few temperatures above and below the glass-transition temperature. Methods based on dilatometric (46), volumetric (47), and refractive-index procedures (48) have been described.

The glass-transition temperature is affected by the degree of polymerization; it increases rather rapidly in the region of low molecular weight and then levels off. The glass-transition temperature depends on the degree of tacticity; for polystyrene it decreases linearly with increasing isotacticity and the results are independent of whether the atactic units are extractable or incorporated in the polymer chain. The presence of small amounts of monomer, solvents, or plasticizers in the polymer lowers the glass-transition temperature.

A useful test for fiber identification, reported by Preston (49), is the temperature at which contraction occurs. In the case of oriented fibers, this appears to correspond with the glass-transition temperature.

Density. Density is the mass of a unit volume of material. It varies inversely with temperature and with chain length. The presence of atoms of large atomic mass in the polymer usually increases the density: thus, if hydrogen atoms in polyethylene are replaced by chlorine atoms, the density is increased. The degree of crystallinity in the polymer also affects the density, but the maximum observed range for individual polymers is quite small and does not alter appreciably the order of polymer densities. Tabulated data (12) should not be considered as absolute values, but primarily as a general guide to the relative position of the various materials. Knowledge of the density in conjunction with other specific data, particularly those furnished by refractive-index measurements, is extremely useful.

All density measurements of solids are subject to a number of errors caused by the presence of inhomogeneities such as surface imperfections, trapped air bubbles, and presence of monomer. Many polymers contain air inclusions. These will increase buoyancy and give an apparent density much lower than that characteristic of the material itself. Therefore, observations of density must be interpreted with additional information about the presence of hollow spaces in the polymer, ie, density measurements should be combined with microscopic observations.

The density may be determined by any standard method if great care is taken to exclude all voids or air bubbles. It is best to condition the sample at 23 ± 2°C for at least 24 hr. Densities of sheets, rods, or molded articles can be measured by ASTM test D 792-60T (50) using a hydrostatic weighing technique. For molding powders, pellets, or flakes, a pycnometer may be advantageously employed.

The density of very small specimens may be readily determined by immersion in salt solutions of known density or in a density-gradient tube. The latter method is well adapted for rapid, precise work in routine analysis for small, as well as large, solid particles and fibers (51,52). Tubes can be prepared by filling each half of the column with either two pure liquids or two different mixtures of the liquids. Many liquids can be found that will make a satisfactory column. For practical use, however, the field will be narrowed by the need for using liquids that do not interact chemically, possess fairly low viscosity and volatility, are additive by volume, and are inert to the materials to be immersed in them. Despite these restrictions it is usually possible to find pure or mixed liquids covering a wide range of specific gravities as well as permitting a flexible choice of sensitivity.

Other methods for measuring density are discussed by Bauer and Lewin (53).

Refractive Index. The refractive index of a polymeric material is useful in identifying a specimen by comparison with a known specimen or with data from the literature (12). Refractive-index values assist in eliminating certain polymers from consideration in the identification of an unknown and serve as a check of the identity of reaction or pyrolysis products of polymers. Detailed discussion of the theory and instruments employed in refractive-index measurements are given by Bauer, Fajans, and Lewin (54), Tilton and Taylor (55), and Forziati (56).

The molar refraction of a compound may be considered as the sum of the refractions of the constituent atoms, groups, and linkages, and can aid in identifying a polymeric substance. A probable structural unit for the unknown polymer is assumed and the formula weight and the measured density and refractive index are substituted in the Lorentz-Lorenz equation to calculate the "unit refractivity." The result is compared with tabulated values or with the sum of individual atomic or group refractivities (56).

The refractive index of a polymer is usually independent of the direction in which it is measured. If the molecules are oriented by some process such as stretching or drawing, the refractive index in the direction of stretch or drawing will be different from that measured perpendicularly to it. The magnitude of this difference is a measure of the birefringence of the substance and is a sensitive indicator of structural dissymmetry, whether it is introduced by strain or by a natural orientative process as in cotton or wool fibers.

Absorption Spectra. A discussion of ultraviolet and infrared-absorption spectroscopy as applied to polymers will be given elsewhere in this encyclopedia. Excellent reviews of this field are available for polymers (18,57–61) and for coating materials (62,63).

Infrared-absorption spectra are extremely useful for ascertaining the presence or absence of such groups as CH_3, CH_2, C_6H_5, OH, C≡N, NH_2, SH, C═O, and larger groups such as olefinic substituents, carboxyl, ester, amide, and imide groups. Several tables of correlating spectra and functional groups have been published and Bellamy's book (64) is very useful if identification and assignment of empirical bands are required. Besides the identification of functional groups, infrared data allow direct comparison of the spectrum of an unknown with spectra of known materials. These "fingerprints" show the many absorption bands due to skeletal vibrations that are characteristic of the molecule as a whole. By empirical comparison of the absorption bands it is often possible to establish unequivocally the identity of the unknown polymer. A large library of spectral data is a prerequisite for successful identification. Spectra for reference in identifying unknowns are given by ASTM Committee D-13 (24), Barnes and co-workers (65), Hummel (66), Kagarise and Weinberger (67), Nyquist (68), Sadtler Research Laboratories (69,69a), and the Federation of Societies of Paint Technology (70). Hummel (4,66) discusses spectroscopic methods of polymer analysis with particular emphasis on infrared. The same author has outlined a systematic scheme of analysis of plastics and lacquers based on the characteristic infrared-absorption bands of polymeric compounds (71). Another systematic scheme of analysis based on the infrared spectrum of an unknown polymer is given by Tryon and Horowitz (61).

Infrared-absorption spectroscopy (qv) has found increasing usefulness in the determination of such structural features of polymers as type of unsaturation (for instance degree of *cis*- and *trans*-1,4 repeating units), side-chain characterization, tacticity, crystallinity, and determination of molecular orientation in polymer films. For instance, by measuring the absorbance ratios of bands characteristic for each phase from suitably prepared standards and plotting a working curve the atactic–isotactic content of polypropylenes can be measured with reasonable accuracy. Compensated spectra obtained with double-beam spectrophotometers have been used to determine branching in polyethylene (72) and additives in polymers.

Spectra of polymer films obtained by attenuated total reflectance techniques are quite similar to transmission spectra. The technique is independent of sample thickness, and the sample does not have to be clear. This method is suitable for the identification of organic surface coatings (73–75), curing rates, thermal and photostability studies, and characterization of thermoset resins on metallic containers.

Ultraviolet-absorption spectroscopy (qv) has found its main application in the detection, identification, and quantitative analysis of additives such as accelerators, antioxidants, or ultraviolet absorbers (60). A number of methods for the determination of styrene in styrene copolymers as well as the determination of phenolic groups in lignin have been proposed.

Spectrophotometric methods should be employed for polymer analysis, if proper instruments are available, before the generally more time-consuming chemical tests are conducted.

Magnetic Resonance Spectroscopy. Nuclear magnetic resonance (NMR) (qv) has become increasingly important for both physical and chemical structure studies on polymers primarily associated with proton or ^{19}F resonances. Wide-line studies provide information on transitions, crystallinity, rates of crystallization, plasticization and swelling, branching and crosslinking, and bond directions and distances. High-resolution investigations can be used profitably for analyses involving group and molecule identifications, hydrogen bonding, proton-exchange reactions, and stereoisomerism.

For a long time it was felt that high-resolution NMR could not profitably be applied to the study of polymers. Bovey and Tiers (76), working with chloroform solutions of methyl methacrylate, were able to identify isotactic, syndiotactic, and heterotactic sequences and to correlate their relative abundances with the method of preparation. Chen (77) determined *cis*- and *trans*-1,4 contents of polyisoprenes. Takeda, Tanaka, and Nagao (78) applied NMR to natural rubbers, cis and trans isomers of polybutadienes, and *cis*-polyisoprene. Naylor and Lasoski (79) studied some fluorine-containing polymers and found that most of the generalizations previously deduced from small molecules apply also to polymers. Quantitative estimates from peak areas seem to indicate that the frequency of head-to-head linkages in poly(vinylidene fluoride) is about 10%.

A great weakness of high-resolution NMR, finding suitable solvents, as applied to polymer work has been largely eliminated for noncrosslinked materials by recent developments of variable-temperature probes. Many polymers that are only sparingly soluble in suitable solvents at room temperature are sufficiently soluble in the same solvents at higher temperatures to give excellent NMR data. Recent reviews on this subject include those of Wall and Florin (80), Slichter (81), and Roberts (82).

Electron-spin resonance (ESR) (qv) contributes mainly to the elucidation of chemical processes, such as kinetics and mechanisms of polymerization and degradation phenomena. With ESR one can detect free radicals and measure them at concentrations as low as 10^{-9} mole/l. Details of this technique are given by Wall and Florin (80).

Fluorescence. The fluorescence of polymeric materials is not sufficiently characteristic to establish the presence or identity of any polymer, but it may aid in distinguishing between some materials when considered together with other properties. A discussion of the application and limitations of fluorescence in polymer analysis is given by Forziati (83). The presence of even trace amounts of catalysts, accelerators, inhibitors, stabilizers, ultraviolet absorbers, or plasticizers in the polymer will affect significantly the color of the fluorescence. According to Thinius (84), it is often possible to recognize polymers with certainty after dyeing them with one or more suitable dyes. Characteristic fluorescent colors are obtained by which the polymers can be distinguished. Millson (85) has reported the phosphorescence of textile fibers and many other substances.

Thermal Degradation Techniques

When heated under carefully specified conditions, a polymeric product usually decomposes to a specific number of products in a definite ratio. Reliable and reproducible measurements are obtained only through careful control in sampling and pyrolysis and complete separation and high sensitivity in detection and identification. The low order of diffusion of degradation products through the viscous polymer mass must be considered in the sampling step. As pyrolysis temperature is increased, viscosity is lowered, but secondary reactions between pyrolysis products are enhanced. Thin films generally give best results.

Pyrolytic Behavior. Polymers may be depolymerized or broken down by dry distillation, heating with alkali, either dry or in aqueous or solvent media, or heating with strong acid. Dry distillation is particularly useful with acrylic resins, phenolic resins, and polystyrene. Decomposition with alkali is valuable for cellulose derivatives, phenolic resins, alkyd resins, and various types of esters. Odor on carbonate fusion is a useful classification test.

Products of destructive distillation, either alone or in the presence of sodium carbonate, sodium hydroxide, etc, can be collected by passing the evolved products through water. Solids or water-insoluble liquids that separate can be removed and subjected to further analysis. Water-soluble materials, such as acids, bases, or low-molecular-weight esters or carbonyl compounds, can be detected by the various classification tests or by infrared techniques. Volatile acids and bases may be trapped by passing the products through dilute basic or acidic solutions, respectively.

Mark and Raff (86) have described an identification scheme based on destructive distillation: they carry out the dry distillation with approximately 100-mg samples in a test tube; the gaseous products are entrapped in another tube that is surrounded by an ice bath.

Some polymers yield monomeric decomposition products (methyl methacrylate, styrene, α-methylstyrene, tetrafluoroethylene); phthalate resins, on heating, give off phthalic acid; cellulose nitrate gives off oxides of nitrogen; chlorinated polymers, hydrochloric acid; and polysulfide polymers, hydrogen sulfide.

Many textiles fibers can be identified by determining the pH of the distillate resulting from pyrolysis of the fibers (87).

Various vinyl and cellulose derivatives may be readily identified by dropping a small piece of the material into a test tube containing cold 25% sulfuric acid, warming slightly, and noting the odor. Cellulose acetate gives off the odor of acetic acid; cellulose acetate butyrate, the odor of butyric acid; and ethylcellulose, the odor of ethylene.

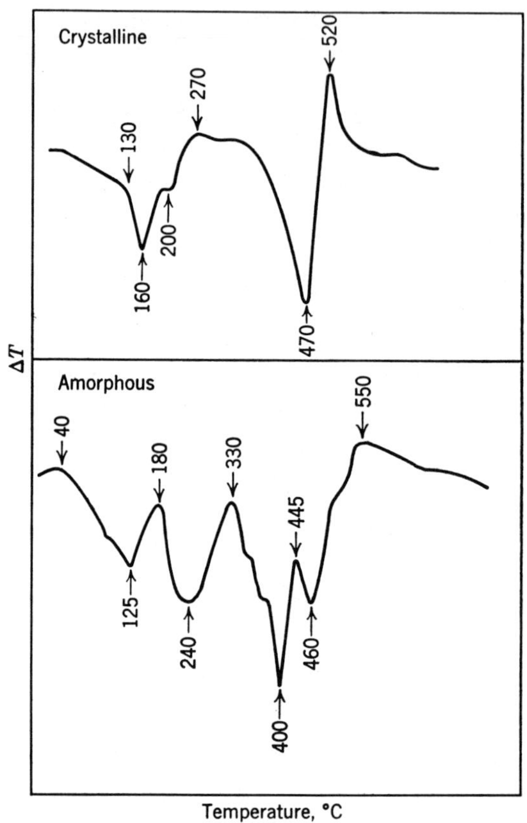

Fig. 1. High-temperature thermograms of crystalline and amorphous polypropylene (89a).

A detailed discussion of the mechanism and kinetics of pyrolysis is given by Wall (88). Vassallo (89) has summarized pyrolysis techniques and Kupfer (19) has tabulated degradation products of polymers. Recent books have thoroughly reviewed the thermal degradation of organic polymers (195) and thermal methods of analysis (196,197).

Thermogravimetric Analysis (qv). Thermogravimetry facilitates polymer identification either through knowledge gained of degradation behavior or through identification of pyrolysis products. The weight loss is continuously measured versus time or temperature. This information provides a rapid means for following the degradation of polymers over a wide temperature range. Critical factors are sample thickness, temperature, and the atmosphere in which the sample is pyrolyzed. The chief analytical value is the comparison of the characteristic shapes and points of inflections of thermograms obtained under identical experimental conditions; by this means polymers can be distinguished and classified according to their heat stability. More detailed discussion of thermogravimetry is given by Wall (88) and Vassallo (89).

Differential Thermal Analysis (qv). Differential thermal analysis (DTA) techniques are highly useful for studying phase transitions and chemical reactions involving heat. The sample is heated in an inert atmosphere and distinction is made between exothermic and endothermic activity. The temperatures at the peaks, areas of the exotherms and endotherms, number of peaks in the thermogram (Fig. 1), maximum rates of change in differential temperature, and temperatures at which these maxima occur are unique and characteristic for each polymer and can be used for its identification. Although it is often not possible to interpret all of the peaks, differences in molecular configurations have a decided influence upon the shape of the thermogram. The technique has been applied for the characterization and study of the thermal stability of a variety of polymers and for determining the degree of crosslinking and cure of resins, as well as for irradiation effects. DTA can detect physical mixtures of polymers that melt sufficiently far apart; the thermogram peak areas are proportional to the amount of crystalline material present. Thus mixtures of polyethylene with isotactic polypropylene give two peaks in the thermogram, the areas of which are proportional to composition. Schwenker and Beck (90) have applied DTA to a number of polymers employed as textile fibers, including polyacrylonitrile, poly(ethylene terephthalate) (Dacron), and nylon. Differences were observed between drawn and undrawn poly(ethylene terephthalate). Differentiation between phase transitions and decompositions is possible and solid–liquid transitions of polyolefins and polystyrene have been determined. From the heats and entropies of fusion obtained from the thermograms the degree of crystallinity of the polymer can be estimated.

For details of DTA techniques the reviews by Kissinger and Newman (91), Ke (92), and Murphy (93,94) should be consulted.

Gas Chromatographic Analysis of Pyrolysis Products. Pyrolytic techniques used in conjunction with mass or infrared spectroscopy or gas chromatography have received increasing attention in recent years. These procedures are most useful for detecting minor differences in similar polymers. The sample is heated on a hot filament, in a furnace, or in a small boat.

Harms (95), Kruse and Wallace (96), and Hummel (4,97) have shown that characteristic spectra of the pyrolysis products are useful for the identification of the original polymer. The technique is especially important for the qualitative and quantitative analysis of insoluble, crosslinked, or heavily filled materials such as a vulcanized mixture of natural rubber and styrene–butadiene rubber. An empirical relationship between adjacent peaks, one of which is characteristic of the natural rubber pyrolysis products and the other of the products of the styrene–butadiene copolymer, is used to determine the composition.

Gas chromatograms of the pyrolysis products of polymers are used not only to identify polymers (98–101), but also for the quantitative determination of the monomer ratio of copolymers (100,102–105) from the height ratio of peaks characteristic of each component. A calibration curve of peak-height ratio versus composition obtained on pyrolysis of samples of known composition is constructed, and from it the composition of polymers of unknown monomer ratio can be determined to within 1–3%. Determination of functional groups such as alkoxyl groups in polymers and copolymers can be accomplished by their conversion to the corresponding iodides, which are analyzed quantitatively by gas chromatography (103).

A procedure suitable for determining esters such as 2-ethylhexyl, decyl, and lauryl acrylates in copolymers consists in saponifying the esters at elevated temperatures and chromatographing the alcohols thus formed (100). Polyhydric alcohols (106) and carboxylic acids in alkyd and polyester coating resins (104) have been identified by programmed-temperature gas chromatography. The chromatograms obtained at different pyrolysis temperatures will vary markedly because secondary reactions of the degradation products will take place more rapidly at higher temperatures (107–112).

Ettre and Varadi (110) describe a specially designed pyrolysis chamber that allows well-controlled thermal degradation studies over a wide temperature range; a modification of this apparatus has become commercially available. The chromatographic fractions of the more complex degradation products can be analyzed further by infrared-absorption spectroscopy (113).

The effluents from gas chromatography can also be subjected to qualitative functional-group analysis (114), which, in conjunction with retention-volume data from the corresponding chromatogram, gives positive identification of the components of the pyrolysis mixture.

Owing to the different amounts of degradation products formed on pyrolysis, it is possible by gas chromatography to distinguish between random or block copolymers and mechanical mixtures of the same composition as well as to detect traces of copolymers or crosslinking agents in resins. A review of pyrolysis–gas chromatographic techniques for the analysis of polymers has been given by Brauer (114a).

Quantitative determination of residual monomer in resins can be achieved by dissolving the resin and chromatographic analysis of the resulting solution (100,115). It is usually preferable to precipitate the bulk of the polymer by the addition of a nonsolvent prior to the injection into the chromatographic column to prevent clogging. Rapid distillation followed by gas chromatographic analysis has also been suggested to determine the residual monomer content in emulsions (116). Nelsen, Eggertsen, and Holst (117) claim that as little as 0.05% of the residual monomers and other volatile organic compounds can be measured by their procedure, in which the effluent from a gas chromatography column is oxidized to carbon dioxide and water over hot copper oxide. A hydrogen flame detector which is insensitive to water is especially suitable for the determination of unreacted monomer in aqueous emulsions of copolymers (118).

Gas chromatographic techniques have been adapted for the analysis of rubbers (119,120), plasticizers (121–123), and resins and solvents in coatings (124–128). For a more detailed discussion see Gas Chromatography under CHROMATOGRAPHY.

Chemical Tests

Qualitative and Quantitative Elementary Analysis. Most schemes of polymer analysis classify polymers according to their elementary composition (12).

Ignition tests are usually sufficient for the detection of carbon. The detection of hydrogen in organic compounds can be based on the production of hydrogen sulfide when the compound is heated with sulfur (129). Davidson (130) has suggested a test for oxygenated organic compounds using ferric thiocyanate; alcohols, ethers, esters, aldehydes, ketones, and amides give a positive test. Sulfur, nitrogen, and halogens can be readily determined by converting them by sodium fusion to water-soluble ionic compounds which can be analyzed by the usual techniques of qualitative analysis (6,12). The presence of phosphorus is detected by decomposing the polymer with boiling concentrated nitric acid and concentrated sulfuric acid, followed by addition of ammonium molybdate. A test for silicon is described by Feigl and Krumholz (131). For details of these tests see Ref. 12.

Quantitative analyses for individual elements are recommended only if no conclusive identification has been made after completion of the classification and specific tests, or when a quantitative determination of polymer mixtures or the determination of the monomer ratio of copolymers of known constituents is desirable. The analyses can generally be performed by the commonly accepted methods for microquantitative elementary analysis described in books of Milton and Waters (132) and Steyermark (133). Details of individual procedures are beyond the scope of this article. Brauer and Horowitz have summarized the most commonly used techniques (12).

Besides carbon and hydrogen determinations, those for nitrogen, sulfur, or the halogens are most useful for polymer analysis. Carbon and hydrogen are usually determined by burning the compound at red heat in an atmosphere of dry oxygen in which the carbon and hydrogen are converted into carbon dioxide and water, respectively. The water is absorbed by magnesium perchlorate and the carbon dioxide is taken up by sodium hydroxide on asbestos. Oxygen is usually arrived at by difference. Sulfur can be converted to sulfuric acid by a dry or wet combustion technique or by oxidation in a Parr bomb. The Kjeldahl method for the determination of nitrogen is applicable to nearly all types of polymers.

The halogens can be determined by a variety of techniques. For chlorine and bromine the dry combustion method, in which the sample is burned in oxygen with a piece of filter paper, is very useful; the combustion products are absorbed and the halogen is determined volumetrically. In the fusion technique the polymer is fused with a mixture of potassium nitrate, sodium peroxide, and cane sugar, and the halide is precipitated with silver nitrate.

In the ASTM method (134) for the determination of total chlorine in poly(vinyl chloride) polymers and copolymers used for surface coatings, fuming sulfuric acid and potassium persulfate are used as the oxidizing agents. The liberated halogen is absorbed and reduced to halide in an alkaline solution of sodium arsenite. Special methods for the determination of chlorine in poly(vinyl chloride) are discussed by Thinius (135).

Unlike the determination of the other members of the halogen family, quantitative determination of fluorine in organic compounds can seldom proceed to the final stage immediately after the decomposition process, because of the presence of inorganic compounds of nitrogen, sulfur, chlorine, bromine, iodine, or phosphorus produced upon oxidation or reduction from the organic sample containing these elements. Separation of these interfering ions may be accomplished by precipitation with a suitable reagent, by formation of thermally stable compounds, or by volatilization. The separation may also be based on steam distillation of fluosilicic acid, leaving foreign matter behind in the reaction mixture. After separation of interfering substances, the fluoride (present as hydrogen fluoride, sodium fluoride, potassium fluoride, or silicon tetrafluoride) can be determined gravimetrically as lead chlorofluoride, volumetrically with sodium hydroxide or thorium nitrate, colorimetrically with ammonium molybdate, or potentiometrically with calcium ions.

The simultaneous determination of carbon and fluorine in organic compounds containing a high percentage of fluorine has been reported (136), as has a rapid microdetermination of fluorine using the Schöniger combustion technique (137). Haslam and Whettem (138) determine fluorine in polytetrafluoroethylene by fusing the sample in a Parr bomb. Determination of chlorine and fluorine in fluorocarbon polymers by fusion with potassium carbonate has also been described (139).

Phosphorus in organic compounds can be determined by destroying the sample with suitable oxidizing agents (potassium nitrate–potassium hydroxide mixture, digestion in a Kjeldahl flask with 30% hydrogen peroxide–concentrated sulfuric acid or concentrated nitric acid–sulfuric acid mixture, or combustion in a Schöniger flask). Phosphorus is oxidized to phosphoric acid, which is determined gravimetrically as ether-dried ammonium phosphomolybdate.

There is no completely satisfactory routine method for determining silicon in organosilicon compounds that can be applied to all samples. The most convenient procedure is digestion of the material with fuming sulfuric acid. After ignition to 800°C the residue can be dissolved and the silicon determined acidimetrically. Volatile materials are best handled by fusion with sodium peroxide and sugar in a nickel Parr bomb and volumetric analysis of the fusion mixture for silicon.

Solubility. Solubility not only depends on the recurring units present in the polymer, but is also greatly affected by the degree of polymerization, branching, crosslinking, and steric configuration, tacticity, and crystallinity of the material. It is often possible to find solvents and nonsolvent systems that are characteristic for each polymer. Sometimes a polymer becomes soluble in a mixture of nonsolvents.

In observing the solubility behavior of a polymer, it should be noted whether (a) it swells before going into solution, (b) it is partially soluble, (c) the solution becomes viscous, (d) there is any change of color in the solution, and (e) the solution has become turbid or opalescent. If there is any doubt whether any of the material has dissolved, the solvent should be evaporated on a watch glass to determine if any residue remains.

The solubility characteristics of polymers and resins in organic solvents and the chemical resistance of various polymers have been tabulated (12). Many plastics are insoluble in common solvents. Phenolic, urea–formaldehyde, melamine–formaldehyde, and aniline–formaldehyde resins in the fully cured state, lignin, nylon, polyethylene, and polytetrafluoroethylene are not readily soluble. Phenolic plastics are slowly attacked by molten 1- or 2-naphthol or resorcinol. Nylon is soluble in cresol and 60% hydrochloric acid (by volume); polyethylene in xylene at 75°C. Elastomers swell to varying degrees in different solvents; these swelling characteristics may be used to assist in identifying them.

With a few exceptions, such as poly(acrylic acid), poly(vinyl alcohol), poly(vinyl methyl ether), methylcellulose, and starch, polymers are insoluble in water.

Since the solubility of commercial resins may vary according to grade, solubility should not be considered as an absolute criterion for identification. Generally, materials of higher molecular weight or greater degree of crystallinity have decreased solubility. However, solubility is not always related to crystallinity or to stereoregularity. Crosslinks greatly diminish solubility and swelling. Introduction of comonomers, even in small amounts, will change the solubility behavior. If such modifications exist, comparison with the published data for the unmodified polymer may be misleading. Studies of the relation of solubility to composition of plastics and the solubility of various plastics in twenty-one common solvents have been made by Reinhart and Kline (140).

The behavior of plastics in solvents is not affected by fillers, but when the plasticizer content is high an increase in solubility due to cosolvent effects can often be observed. The plastic may dissolve in solvents that ordinarily only swell it. Combinations of solvent mixtures have been used by Gordijenko and Schenck (141) to identify a variety of resins, elastomers, and plasticizers. Thinius (142) describes an identification procedure that uses characteristic solvents, nonsolvents, nonsolvent pairs acting as solvents, and precipitating agents. A systematic scheme of analysis based mainly on solubility behavior (12) is useful only for homopolymers because mixtures and copolymers exhibit deviations in solubility behavior.

Nitsche and Toeldte (143) distinguish polymers by their precipitability in different groups of solvents under definite conditions. The technique is simple and does not require special apparatus. The precipitation number is defined as the number of milliliters of precipitating agent needed to produce turbidity. The type of precipitation and whether the precipitate dissolves in an excess of the solvent are also recorded.

The precipitation temperature of a polymer solution at a constant concentration has been utilized to characterize polymeric vinyl ethers (144). For polymers such as poly(vinyl isobutyl ether), it depends essentially on the steric structure of the polymer molecule and not on the degree of polymerization. Selection of the proper solvent is very important for the measurement of the precipitation temperature. Preliminary tests must be conducted to find a solvent in which the precipitation temperature of the polymer is affected mainly by its steric structure, and is nearly independent of the degree of polymerization.

Functional-Group Analysis. The detection and determination of functional groups can be accomplished by ultraviolet or infrared spectroscopic methods, color tests, or gravimetric and volumetric techniques.

Spectrophotometric procedures have become the most important tool for the analysis of many functional groups. Ultraviolet spectra determined in suitable solvents (alcohol, dioxane, water, halogenated hydrocarbon, dimethylformamide) quickly detect the presence of conjugated double bonds or aromatic groups. The wavelengths of the absorption maxima serve as qualitative identification, and the heights of the absorption peaks for the quantitative estimation of functional groups. A larger number of conjugated double bonds in the molecule shifts the absorption peak to a higher wavelength and increases the extinction.

The absorption spectra in the infrared region from 2–15 μ yield valuable information on the presence of functional groups in the polymer molecule; see, eg, Bellamy (64). The use of color tests for the identification of functional groups in polymers is considered below, p. 656.

The basis for the quantitative analysis of functional groups by means of chemical methods is the reaction of a specific group with a known reagent, followed by a quantitative determination of either a reaction product or an excess of the reagent. Reactions that consume or liberate an acid, base, oxidant, or reductant are particularly desirable, since quantitative measurements can be readily made of such substances. Reactions in which gases are consumed or liberated are also very useful. If water is a product or reactant it can frequently be measured by means of the Karl Fischer reagent (a solution of iodine, sulfur dioxide, and pyridine in methanol or in methyl Cellosolve). Determination of ester, carboxylic acid, hydroxyl, and vinylene groups

Table 2. Analytical Methods for Functional Groups in Polymers

Polymer	Functional group	Method used
di- and polyolefins	$>C=C<$	halogen addition
alkyd resins, modified cellulose, modified rosin, phenolic resins, polyoxyalkylene ethers, poly(vinyl alcohol)	—OH	acetic anhydride in pyridine
polyoxyalkylene ethers	—OH	acetic anhydride in ethyl acetate with *p*-toluenesulfonic acid catalyst
epoxy resins	—OH	lithium aluminum hydride
unsaturated polyesters	—OH and —COOH	phenyl isocyanate in chlorobenzene (titrate with diisobutylamine)
acidic polymers (resin acids, poly(acrylic acid), poly(methacrylic acid), etc)	—COOH	titration with alkali
ester polymers (alkyd resins, cellulose esters, polyesters, poly(acrylic esters), poly(methacrylic esters), poly(vinyl esters), resin esters, etc)	—COOR	saponification with potassium hydroxide in alcohol or diethylene glycol
carboxymethylcellulose	$—COO^-$	conversion to sl sol copper or uranium salt
oxidized cellulose	$>C=O$	hydroxylamine hydrochloride in pyridine and titration of hydrochloric acid liberated
thiourea resins	free formaldehyde + methylol groups	potassium cyanide–sodium hydroxide excess (potassium cyanide is determined iodometrically)
phenolic resins	$—CH_2OH$	phenol in hydrochloric acid
polycaprolactam	methylene ($—CH_2—$) group in disulfide and methylene sulfide crosslinks	hydrolyze with concd sulfuric acid; determine formaldehyde formed colorimetrically with chromotropic acid
epoxy resins	$>C—C<$ (bridged by O)	hydrochloric acid in pyridine or in dioxane, hydrogen bromide in glacial acetic acid
cellulose and poly(vinyl ethers)	ROR′	hydrogen iodide (Zeisel method)
cellulose nitrate	$RONO_2$	saponification, reduction, and determination of ammonia
all polymers	water	Karl Fischer reagent

is described below under saponification, acid, acetyl, and iodine numbers, respectively. Siggia (9) describes the determination of all functional groups that are likely to be encountered in the analysis of polymeric molecules. Some gravimetric and volumetric procedures that are useful in the estimation of functional groups in polymers are given in Table 2.

Saponification and Acid Numbers. The *saponification number* is the number of milligrams of potassium hydroxide that reacts with the free acids and the esters in

1 g of material. The *acid number* is the number of milligrams of potassium hydroxide required to neutralize 1 g of material. The *neutralization equivalent* is the number of grams of sample that combines with 56.1 g of potassium hydroxide and is, therefore, equal to or a small submultiple fraction of the molecular weight of the acidic material, depending on the number of carboxyl groups per mole. *Ester number* (saponification number minus acid number) is sometimes used as a measure of ester content. The molecular weight of an ester can be calculated from the saponification equivalent if no free acid is present or from the ester equivalent by multiplying by the number of ester linkages in the molecule. Saponification numbers are especially significant if the polymer contains no elements other than carbon, hydrogen, and oxygen. Saponification and acid numbers of various polymers have been tabulated (12). The values are approximate since not all of them were determined under identical conditions.

Saponification Number. Many schemes of analysis classify polymers according to their saponification equivalents. Furthermore, examination of saponification products is another valuable tool in identifying polymers. For classification tests the saponification must always be carried out under identical conditions. If examination of the products is contemplated, as complete a saponification as possible is required. Since some polymers have very low saponification rates, more drastic conditions such as longer saponification times and stronger alkaline solutions than those used for classifying the materials are desirable.

The saponification rate depends on the position of reactive groups in the molecule. If a carboxyl group is bound directly to a carbon of the polymer chain (as in methyl acrylate), the ester is more difficult to saponify than if it is on a side chain. Introduction of an α-methyl group, as in methyl methacrylate, produces steric hindrance which makes it very difficult to saponify even the monomer. Steric hindrance effects are also responsible for the fact that rosin acid ester linkages, once formed, are very resistant to acid or base hydrolysis. Esters involving hydroxyl groups bound directly to carbon atoms of the polymer chain, as in cellulose acetate and poly(vinyl acetate), are easily saponified. The presence of acid groups in copolymers of methacrylic acid esters enhances markedly the ease of hydrolysis.

Differences in the stereochemical arrangement in the immediate neighborhood of ester groups change the rate of hydrolysis. For example, different steric forms of poly(methyl methacrylate) also have different chemical reactivities. Amorphous and syndiotactic polymers hydrolyze relatively slowly compared to the isotactic form. The ease of saponification can, therefore, be helpful in identification of steric arrangement and can be used for separation of physical mixtures of such polymers.

Hall and Shaefer (145) have reviewed the saponification of easily and difficultly saponifiable esters and have suggested procedures for determining saponification numbers.

Additional information may be gained by determining the saponification curve. Fixed quantities of material are taken and are saponified in a suitable solvent with the same quantities of standard alkali. Only the saponification time is varied. The curve obtained is compared with saponification curves of known polymers.

For classification tests the sample is dissolved in an alcohol–benzene mixture and is heated under reflux with alcoholic potassium hydroxide. The excess base is titrated with standard hydrochloric acid solution. Dark-colored materials can be titrated electrometrically (146).

Many difficultly saponifiable esters, such as methyl esters of modified rosin acids, do not yield correct saponification numbers on refluxing with alcoholic potassium hydroxide. The method of Redemann and Lucas (147) employs diethylene glycol as a solvent; this permits the use of a higher reaction temperature and often results in complete saponification. Most high-boiling alcohols can be distilled from the reaction mixture. Saponification numbers of esters that cannot be saponified quantitatively by this technique can be determined by using a reagent consisting of potassium hydroxide in a mixture of diethylene glycol and ethyl phenyl ether.

Sometimes the stability of the ester linkage can be used as a separation procedure. Conner (148) describes a procedure by which mixtures of resin acid and fatty acid esters can be analyzed by selective saponification.

A large saponification number (above 200) usually indicates the presence of polybasic acids. It may be desirable to separate and identify the individual acids, particularly if the presence of alkyd resins is suspected. Crystals that form during saponification in alcohol may be separated by filtering. On acidification of the salt in aqueous solution, the polybasic acid precipitates. The purified acid may be examined microscopically and its melting point, refractive index, neutralization equivalent, iodine number, and ultraviolet and infrared-absorption spectra can be determined. Derivatives such as the anilides, toluidides, or the *p*-nitrobenzyl or *p*-phenacyl esters may also be prepared for identification.

Acid Number. Determination of the acid number is valuable for natural resins, some modified resins, and certain of the alkyds. These polymers generally possess appreciable portions of neutralizable acidic groups, which can be detected even in the presence of large quantities of neutral resinous materials. For resins that contain free acid but no saponifiable groups, the acid number is often considerably lower than the saponification number. Differences in the two values may be caused by the slow neutralization of the acid when the resin is titrated directly with an alkali solution at room temperature. There is also the possibility that small amounts of aldehydes are present which react with alkali during saponification. Widely varying acid numbers have been reported, which are generally due to the differences in the procedures employed by the various investigators.

Acid numbers of polymers are usually determined by dissolving the sample in neutral dioxane or alcohol and titrating the free acid, using phenolphthalein as indicator (12). For dark-colored resins, this method is unsatisfactory because the end point is difficult to detect. Coburn (149) suggests titration in a two-phase medium, an alcohol–benzene layer in which the colored material remains, and a lower aqueous layer which is saturated with sodium chloride to obtain sharper separation between the two phases. The lower layer contains the titration indicator. Addition of a large quantity of sodium chloride is very useful in determining the acid value of phenolic resins, since the sodium chloride suppresses the ionization of the phenolates which should not be included as carboxylic acids.

Acetyl Number. The acetyl numbers are very useful in the examination of natural and synthetic polymers that contain hydroxyl groups. The material is acetylated with acetic anhydride and then titrated with potassium hydroxide. The acetyl number is the number of milligrams of potassium hydroxide that will react with the acetic anhydride required to acetylate the hydroxyl groups of 1 g of material. The method is applicable only to nearly anhydrous resins. Water in the resin exhausts the reagent by reacting with the acetic anhydride to form acetic acid. Details of the procedure are given elsewhere (12,150).

Aldehydes and primary and secondary amines will interfere with the determination. In the presence of these groups, phthalic anhydride should be used as an esterification agent. The phthalic anhydride reacts more slowly than the acetic anhydride and a larger excess is needed. Phthalic anhydride will not react with phenols, as does acetic anhydride, so that alcohols can be determined in the presence of phenols. Amines, mercaptans, higher fatty acids, and easily saponifiable esters interfere. Tertiary alcohols are both esterified and dehydrated and cannot be determined by this procedure. Details of the procedure are given by Siggia (9) and Kline (151).

Iodine Number. The iodine number is the number of grams of iodine absorbed by 100 g of the sample under standard conditions. It is, therefore, a measure of unsaturation, and indicates the possible presence of natural resins, rosin or modified rosin, certain alkyd resins, natural rubber, and many synthetic elastomers. A large number of procedures have been suggested for the determination of iodine values (152). Most use iodine monochloride or iodine monobromide as the active agent since without any catalyst iodine adds only with difficulty to ethylenic groups at room temperature.

All procedures are subject to certain limitations such as reduced absorption of iodine by conjugated double bonds, steric hindrance, and effects due to ester, carboxyl, or electronegative groups near the carbon–carbon double bond. Addition to one double bond of a conjugated diene and to two double bonds of a conjugated triene goes rather rapidly, but the saturation of the remaining double bond is extremely slow, resulting in low iodine numbers. Low values and recurring end points in the back-titration of iodine may be the result of the liberation of halogen from the addition product by the re-forming of the original double bond (eq. 1). The reaction rate is

$$\text{—CHX—CHX} \rightarrow \text{—CH=CH—} + X_2 \qquad (1)$$

influenced by the group adjacent to the double bond. The effectiveness in increasing the reaction rate is of the order $—COCH_3 > —C_6H_5 > —COOC_2H_5 > —COOH$.

Polymers present special difficulties since the addition reaction is not always rapid and side reactions (substitution and "splitting out") between the iodine monochloride and the polymer may occur. In the "splitting out" reaction, the halogen first adds to the double bond and subsequently a molecule of hydrogen iodide splits out. This latter reaction occurs readily when iodine monochloride is added to olefins and polymers that are branched in the neighborhood of the double bond because the addition product undoubtedly possesses steric strains. The presence of double bonds in end groups or of unreacted monomer in the polymer owing to incomplete polymerization will also give high iodine values. Iodine values of complex mixtures depend on the reaction conditions, especially on the type of solvent used, the reaction time and temperature, and the light intensity. A suitable solvent should give a highly stable solution of iodine monochloride and should be a good solvent for the polymer–iodine monochloride addition product. Each of these factors may exert different effects according to the type of double bond present. However, by rigorous control of reaction conditions, reproducible results satisfactory for specification purposes may be obtained. It should be realized that reproducible iodine values do not necessarily imply that the amount of halogen consumed is a correct measure of the olefinic double bond content of the material examined. Results should be interpreted with the greatest care and generally are useful only in conjunction with results of other measurements, such as absorption spectra.

The Wijs method, which uses iodine monochloride as reagent, has gained the widest acceptance (12); other well-known procedures include those of Rosenmund-Kuhnhenn (153), Hübl (154), and Hanus (155). The subject has been reviewed by Bradley and Kropa (156), Stafford and Williams (157), and Polgár and Jungnickel (152).

Unsaturation can also be determined by reaction with excess mercuric acetate (158) or by the more reliable but cumbersome bromination (159,160) or hydrogenation techniques. Unger (159) showed that the bromine number of mono- and diolefins is affected by the position of the double bond, extent of branching, and position of branching. Hydrogenation requires elaborate equipment and is more difficult to carry out. Hydrogenation procedures, however, have the advantage that the danger of substitution reactions is removed and interference by carbonyl or other groups conjugated with the double bond is eliminated (9). The diene or maleic anhydride value, which is defined as the quantity of maleic anhydride absorbed by 100 g of the substance under analysis, and is expressed in terms of grams of iodine, is particularly useful in identifying compounds containing conjugated double bonds such as are encountered in drying oils (161). An excellent review of determination of olefinic unsaturation is given by Polgár and Jungnickel (152).

End-Group Analysis. The ends of polymer chains sometimes consist of groups different from the monomer units that make up the body of the polymer molecule. Although analysis of these end groups is primarily useful for the determination of molecular weights and for studies of the kinetics of polymerization and depolymerization, it is also an important method for characterizing polymers.

The theoretical background for the determination of molecular weights of condensation or addition polymers by end-group analysis can be found elsewhere (162). Estimation of the total end-group content by analysis and of the number-average molecular weight by a method such as osmometry permits the calculation of the average number of end groups per molecule, which is a measure of the degree of branching if each branch terminates with the kind of end group in question.

For linear polymers determination of end groups gives the number-average molecular weight. The technique has the advantage of not requiring calibration against another method. However, successful application of end-group analysis requires a knowledge of the nature of the terminal groups, the number of groups per molecule, and methods capable of accurate estimation of the group in question.

Condensation polymers, by their very nature, have reactive functional end groups, and this type of polymer, when prepared from difunctional monomers for which there is a simple ratio between number of end groups and number of molecules, is therefore most amenable to end-group techniques. The end groups are often acidic or basic, as exemplified by the carboxyl groups of polyesters or the amine groups of polyamides. Such groups can be readily estimated by titration. Furthermore, as a consequence of the nature of the condensation reaction, these polymers usually possess molecular weights below 20,000, which is the molecular-weight range for which end-group methods are most effective.

Chemical procedures for counting end groups are usually considered inadequate for polymers of molecular weights in excess of 20,000 to 30,000. The adoption of more sensitive instrumental techniques such as infrared spectroscopy, gas chromatography, or mass spectrometry to estimate reaction products of the terminal groups may, however, raise this limit considerably. Methods using radioactive-labeled end groups have also been developed and increase the sensitivity to a considerable extent. These latter methods, however, are limited by the availability of certain isotopes and by the special equipment required for measurement of radioactivity (163).

An unusual method for analyzing an insoluble polymer has been described in the determination of unreacted epoxy groups in epoxy resins (164). These resins are insoluble in the hydrochloric acid–dioxane reagent used to cleave the oxirane ring. However, with very finely divided polymer, swelling occurs and intimate contact is obtained between the sample and the reagent. The reaction proceeds rapidly and no difficulty is experienced in determining the unreacted hydrochloric acid.

Mercapto groups in polycaprolactam fibers having disulfide and alkylene sulfide crosslinks have been determined by swelling the sample with methanol and titrating the suspended strips with alcoholic silver nitrate solution (165). Similar techniques could conceivably be applied to other insoluble polymers; however, sample size would undoubtedly be limited because of the high viscosity of the swollen polymer–solvent mixture.

End groups in polymers can be detected by two methods involving reactions with dyes (166). The dye-partition test (167) is based on the production, through salt formation, of a color, eg, when a benzene solution of an anionic polymer is shaken with an aqueous solution of a cationic dye such as pinacyanol at a controlled pH. The test is suitable for detecting carboxyl, sulfate, and other strong acid groups, hydroxyl, amino, quaternary ammonium, and halogen groups. The more sensitive dye-interaction test depends on the color change produced in benzene extracts of some dye solutions at suitable pH by acids, bases, or salts (168). The dyes are normally insoluble in benzene and are not acid–base indicators in aqueous solution. This test is capable of estimating fairly accurately carboxyl end groups in a polymer chain (169).

Colorimetry, infrared-absorption spectroscopy, and gas chromatography are among the analytical techniques of value for determining certain classes of end groups. The amino end groups of certain polyamides can be reacted with dinitrofluorobenzene and the colored dinitrobenzene derivative can be determined colorimetrically (170). Hydroxyl and carboxyl end groups of poly(ethylene terephthalate) may be estimated by the infrared-absorption spectra of these groups after calibration by a special method involving deuteration (171).

Identification of carboxylic acids in alkyd and polyester coatings has been accomplished by programmed-temperature gas chromatography (104). The method involves transesterification of the resin with lithium methoxide and subsequent chromatographic separation of the methyl esters formed.

Besides the direct analysis of the terminal group (such as titration of carboxyl groups in polyesters or polyamides with base, titration of amino groups in polyamides with acids, or determination of hydroxyl groups by reacting them with a titratable reagent), indirect methods may be used in which the end group is first converted to one more suitable for analysis. Thus, to check direct titration methods by an indirect procedure, Staudinger and Schmidt (172) converted the carboxyl group of a polyester to a carboxymethyl group by using diazomethane in benzene. The product was analyzed for the methoxy group content by the Zeisel method and the results found to be in good agreement with the direct titration. Another example of indirect analysis is the determination of the acetyl end group often present as a molecular-weight stabilizer in polyamides; in this case, the acetyl group is estimated by completely hydrolyzing the polymer to regenerate the monomer. The acetic acid in the mixture is isolated by distillation and estimated by titration. Obviously, techniques involving hydrolysis and subsequent isolation of the product are more laborious than direct estimation.

The analysis of an end group can sometimes be facilitated by modifying it with another group containing a readily analyzable element. The procedure is especially useful for addition polymers, but may also be applied to condensation polymers. An example is the acetylation of hydroxyl groups of a polyester with bromoacetyl bromide, followed by estimation of the bromine content of the polymer. Similarly, the hydroxyl groups of a polyester may be determined by reacting them with succinic anhydride to convert to titratable carboxyl groups.

One of the earliest and most widely used applications of end-group analysis has been in the field of cellulose (qv) and its derivatives. The procedures are based on the assumption that there is one reducing and one nonreducing end group per molecule. Although numerous techniques have been suggested, no end-group method can be regarded as giving a reliable absolute estimate of the chain length of polysaccharides. Some procedures are of definite value in following processes involving chain scission, such as occur in acid degradation. The results of determinations such as the well-known copper number method, a procedure involving the reduction of cupric ion to cuprous ion by the reducing group, should be interpreted with the greatest of care (173). A number of reviews of various methods for the analysis of polysaccharides have been published (174,175).

For proteins the most widely used end-group determination is that of Van Slyke for amino nitrogen; it involves treatment of protein with nitrous acid and subsequent measurement of the nitrogen evolved.

Price (163) describes details of selected techniques for the end-group analysis of some condensation polymers that are important industrially. The end groups involved are carboxyl end groups of polyester and polyamides, amino and acetyl end groups of polyamides, hydroxyl end groups of polyesters, and reducing end groups of cellulose.

The high molecular weight of addition polymers makes the chemical methods of end-group analysis prohibitively difficult. Furthermore, the number and kind of end groups present cannot always be predicted with accuracy. The main value of determining end groups of addition polymers has been in elucidating the mechanism of polymerization. In such cases it is necessary to find the molecular weight by other means and to use the information obtained from end-group determinations to determine the number of end groups per molecule.

Sometimes it is possible to synthesize a polymer with an analyzable terminal group, such as a fragment of a polymerization catalyst, transfer agent, or inhibitor which is incorporated into the chain during the initiation, transfer, or termination step. When the polymerization kinetics are well known, analysis may then be made, for instance, for the initiator fragments containing identifiable groups. Since most commonly used initiators such as benzoyl peroxide or azobisisobutyronitrile cannot be determined by simple methods, it is necessary to label these catalysts with easily analyzable functional groups, elements, or radioactive groups. Initiation with compounds such as bromo- and chlorobenzoyl peroxides, bromobenzenediazonium hydroxide, and *N*-nitroso-*p*-bromoacetanilide, and subsequent determination of the halogen content of the resultant polymer have been suggested by a number of investigators (176–178).

Sully (179) determined end groups in polystyrene polymerized in the presence of sodium sulfite. The monothionic acid end group is converted to a salt by addition of base and back-titrated with hydrochloric acid in benzene. The end point is determined by viscosity changes that occur. Evans (180) has prepared vinyl polymers with hydrogen peroxide–ferrous ions and analyzed for hydroxyl groups.

In free-radical polymerizations performed without the use of a catalyst, the polymer is expected to terminate by a mechanism involving the formation of double bonds. Lee, Kolthoff, and Mairs (181) used iodine monochloride to determine double bonds in rubber. Infrared-absorption spectroscopy has also been employed to measure unsaturation in polyethylenes and poly(α-olefins.). Thus, Natta and co-workers (182) determined vinyl groups in polypropylene from infrared-absorption data. The carboxylic acid end group in polychlorotrifluoroethylene prepared by suspension polymerization with peroxydisulfate was determined by bands at 5.90 and 4.41 μ (183); to eliminate the interference of hydrogen bonds, the sample was first converted into the sodium salt.

By making use of radioactive-labeled initiators the sensitivity of many methods is greatly increased. This technique has been used in conjunction with the number-average molecular weight determined osmotically to provide information as to the nature of the termination reaction (184).

Transfer agents containing elements that can be easily analyzed are often used to control the molecular weight range. Thus, Allen (163) prepared poly(methyl methacrylates) of low molecular weight with *t*-butyl mercaptan as transfer agent under conditions such that each polymer molecule was effectively initiated by a radical derived from the transfer agent; the molecular weight was then obtained by sulfur analysis. Carbon tetrachloride or other halomethanes are frequently used in chain-transfer studies. For instance, ethylene has been polymerized with benzoyl peroxide in the presence of carbon tetrachloride, followed by determination of the chlorine and trichloromethyl groups in the molecule (185).

End-group analysis can also be applied to studies of inhibition. Thus, Price and Durham (186) investigated the effects of nitrobenzene, 2,4-dinitrochlorobenzene, and nitromethane on the polymerization of styrene by analyzing the polymers for nitrogen or chlorine. Methods have been devised to follow the action of chloranil, the chlorinated analog of the commonly used hydroquinone inhibitor, during polymerization reactions (187). Hydroquinone in polymers can also be determined by iodometric, colorimetric, and polarographic methods (188).

Some of the difficulties and limitations of end-group analysis for the characterization of high polymers will be evident from the foregoing discussion. The low concentration of end groups, even in condensation polymers of comparatively low molecular weight, necessitates microchemical methods of analysis. In particular, the presence of small amounts of impurities in the solvent or the polymer can cause appreciable errors and every precaution should be taken to minimize such errors.

Color Tests. Color tests utilize the reactions of polymers or constituents of polymers with certain reagents to form colored products. Color reactions are still of greatest usefulness for the detection of structural characteristics and of functional groups even in laboratories that have modern instruments at their disposal. Advantages of color tests are sensitivity, simplicity and ease of manipulation, economy of material, savings in time and space, and use of readily available inexpensive equipment, which makes these procedures especially suited for smaller laboratories.

A great number of selective and characteristic color reactions of polymers or their constituents have been suggested. The causes of the color formation and the mechanisms of many of these reactions are not fully understood. Most of these tests depend on maintaining very definite experimental conditions; even a slight change of variables, such as the quantity or concentration of the material to be analyzed or the reagents used, temperature, or reaction time, may lead to unsatisfactory results. The sensitivity of many procedures is greatly reduced in the presence of additives. To avoid erroneous results it is nearly always necessary to examine the behavior of even an accepted reaction on as many polymer mixtures of known composition as possible.

Dye stains also provide a simple and rapid means of identification. Those of widest application consist of combinations of dyes, each chosen for its specificity for certain fibers or polymers. A number of identification stains have been developed and marketed by dye manufacturers and chemical supply houses. They are intended primarily for application to bulk fiber or to fabrics but have been used effectively on other materials. Each has characteristics that make it most satisfactory for certain determinations. Previous dyeing and other treatments may cause some variation in the results obtained so that known fibers with the same history should be used as controls. Some of the commercial identification stains are supplied with a separate bleach or stripping agent.

Table 3 lists some of the more important color reactions. A comprehensive treatise of color tests for polymers, including experimental details, is given by Brauer and Newman (189). Many other useful color tests for solvents and other materials associated with polymers are given by Feigl (190). Paulson (191) has summarized the color reactions of common surface-coating resins. A large number of these qualitative reactions can be readily adopted for quantitative determinations of the concentration of individual polymers in complex mixtures. This type of colorimetric analysis is especially useful when a sample of limited size is available.

Besides the general identification schemes and specific tests for chemical elements or groups described in the previous sections, a number of tests for the identification of specific polymers have been developed. Many of these procedures are concerned with the analysis of polymers in various products such as adhesives, fibers, leather, paints, varnishes, lacquers, or paper products. Such methods may be extremely useful for special problems and the analyst should consult the appropriate paper or book.

Determination of Monomer Ratio of Copolymers

The ratio of monomers present in copolymers has a distinct bearing upon the properties of these polymers. Often careful control of this factor is essential to the production of a satisfactory copolymer for a particular end use.

Elementary analyses frequently can be applied with a minimum of difficulty because these procedures usually involve destruction of the polymer and solubility problems are not encountered. Examples of such methods are the analysis of vinyl chloride copolymers via Parr bomb or Schöniger flask chloride determinations, and the determination of the monomer ratio of acrylonitrile copolymers by the Kjeldahl nitrogen method.

When attempts are made to apply chemical methods other than elemental analyses to the determination of monomer ratio, solubility problems are frequently overwhelm-

Table 3. Selected Color Tests for Detection of Polymers or Functional Groups in Polymers or Polymer Fragments[a]

Polymer, functional group, or fragment	Reagent or method	Color reaction
acetate	lanthanum nitrate–iodine ammonium hydroxide	varies
acrylic acid or esters	reflux pyrolyzate with phenylhydrazine; treat with formic acid and hydrogen peroxide	dark green or blue
methacrylate resins	react pyrolyzate with concd nitric acid and zinc powder	blue, extractable with trichloromethane
hydroxyl	vanadium oxinate	red
ethanol	nitroprusside–morpholine	blue
glycerol	periodic acid	red
aldehydes in acetals	azobenzenephenylhydrazinesulfonic acid[b]–sulfuric acid	red to purple
	fuchsin–sulfurous acid	violet
formaldehyde	carbazole–sulfuric acid	blue
	chromotropic acid[c]–sulfuric acid	violet or purple
formaldehyde	chloroglucinol–sodium hydroxide	red
amine and imine	phenol–sodium hypochlorite	blue
aniline	ferric chloride–*p*-phenylenediamine	green
	diazotize with solution of R-salt (disodium salt of 2-naphthol-3,6-disulfonic acid)	red
polyamides	treat pyrolyzate with *p*-dimethylaminobenzaldehyde–hydrochloric acid	red
nylon-6,6	treat pyrolyzate with *o*-nitrobenzaldehyde–sodium hydroxide	mauve to black
urea	phenylhydrazine–ammonium hydroxide, nickel sulfate	red
thiourea	mercuric iodide–sodium hydroxide	yellow
carbohydrates	1-naphthol–sulfuric acid	yellow
and cellulose	aniline acetate	pink to red
	anthrone–sulfuric acid	blue-green
carbonyl	2,4-dinitrophenylhydrazine–sulfuric acid	yellow
carboxylic acid esters	hydroxylamine hydrochloride–ferric chloride	violet to brown-red
coumarone–indene resins	bromine–glacial acetic acid in trichloromethane	red
epoxy resins	sodium nitroprusside–morpholine	blue
	nitric acid–sulfuric acid, then sodium hydroxide	red-orange
	mercuric sulfate	red-orange
	formaldehyde–sulfuric acid	blue
furfural resin	aniline acetate	red
isocyanate resins	sodium nitrite	yellow to red
natural resin	acetic anhydride–sulfuric acid	red-violet
	dichloroacetic acid	varies
	phosphomolybdic acid–ammonium hydroxide	varies
nitrate	diphenylamine–sulfuric acid	blue
nitrile	copper acetate–benzidine acetate	blue
	potassium mercuric chloride, Congo paper	blue
phenolic resin	2,6-dibromoquinone chloroimide–sodium hydroxide	blue
	mercuric nitrate	red
	phthalic anhydride	red
	couple with *p*-nitroaniline diazonium salt	red
	uranium nitrate–ammonium hydroxide	yellow to red

(*continued*)

Table 3 (*continued*)

Polymer, functional group, or fragment	Reagent or method	Color reaction
phthalate	phenol	red
	thymol	blue
	resorcinol	greenish yellow fluorescence
polycarbonate	treat pyrolyzate with *p*-dimethylaminobenzaldehyde	green-blue
protein	copper sulfate–potassium hydroxide	violet
	mercury nitrate	red
	ninhydrin	blue or mauve
	ammonium molybdate–nitric acid	white
sulfhydryl	sodium nitroprusside–ammonium hydroxide	amber
disulfide	sodium nitroprusside–potassium cyanide	pink
vinyl polymers	chloroacetic acid	varies
	dichloroacetic acid	varies
halogen-containing vinyl polymers	pyridine and methanolic potassium hydroxide	brown to black
poly(vinylidene chloride)	morpholine	black
poly(vinyl alcohol)	iodine–borax	green
polyvinylpyrrolidone	iodine	red to yellow
polyvinylcarbazole	sulfuric acid–nitric acid	green
polystyrene	convert to phenol with nitric acid and react with 2,6-dichloroquinone-4-chloroimine	blue

[a] For details see Ref. 189.

[b] Formula: C_6H_5—N=N—C_6H_4—NH—$NHSO_3H$

[c] 1,8-Dihydroxynaphthalene-3,6-disulfonic acid.

ing. Sometimes a special solvent mixture can be found in which the reaction can be successfully carried out. Thus, Critchfield and Johnson (192) were able to saponify an ethylene–ethyl acrylate copolymer using a solvent of 2% water, 13% triethylene glycol, 30% hexanol, and 55% xylene; at refluxing temperature the solvent dissolves both the polymer and potassium hydroxide. The ratio of the monomer units in the copolymer found by this saponification technique was later used to calibrate a simpler infrared method. This is a good example of how chemical techniques can be used to standardize physical methods that are usually simpler to use but often difficult to calibrate.

Barall, Porter, and Johnson (112), and Strassburger and co-workers (102) demonstrated the usefulness of the pyrolysis–chromatography combination in determining copolymer ratios on a quantitative basis. For instance, for ethylene–ethyl acrylate and ethylene–vinyl acetate copolymers an absolute accuracy of 2% was obtained in correlation with neutron-activation oxygen analysis of both polymers.

Many copolymer systems lend themselves to analysis by infrared or ultraviolet spectrophotometric techniques. A major problem with the spectrophotometric approach, however, is the need for standard copolymers of known composition for calibration purposes. Absorptivities based on the absorbance of individual homopolymers are not always identical with those obtained on copolymers. Consequently, each new system must, at least, be checked in this regard. For instance, the preparation of copolymers of vinyl acetate and acrylonitrile using vinyl acetate-1-^{14}C has enabled primary standards to be established for evaluating wet chemical and infrared spectrophotometric methods (193). Employment of carbon-14 radioisotopes minimized errors in calibration and improved the accuracy of the infrared spectrophotometric method. In addition, it has substantiated the completeness of the alkaline hydrolysis procedure.

In another example of the spectrophotometric technique Brown, Tryon, and Mandel (194) rapidly analyzed ethylene–propylene copolymers by using a calibration curve to interpret, in terms of propylene content, infrared spectra of specimens pyrolyzed at 450°C. The strong absorption bands at 909 and 889 cm^{-1}, which vary in intensity according to the propylene concentration, were used to calculate the propylene content.

Determination of Trace Concentrations of Polymers

The determination of trace concentrations of polymers is motivated largely by the increased use of polymers in contact with food. In many cases the direct determination of the polymer is practically impossible, and migration studies in "simulated-food" solvents are carried out so that predictions can be made as to the level that would exist in the food.

An example of such an analysis is the determination of the amount of polyethylene that would migrate from a packaging film into vegetable oil (192). The oil extract of the film is treated with a mixture of hexane, ethanol, and 2-propanol which causes a portion of the dissolved polyethylene to precipitate. The turbidity of the solution is then measured and referred to a calibration curve prepared by extracting the film with *n*-hexane at various temperatures. Portions of these extracts are evaporated to determine the concentration of dissolved polyethylene; corresponding portions are added to vegetable oil and precipitated with the alcohol mixture and the turbidity is determined. Since only the higher-molecular-weight species of the dissolved polyethylene are precipitated by the alcohol, the amount of nonprecipitating polymer must be the same in the standards and the samples. This factor is controlled by using the same film–solvent ratio in the standards and sample and by preparing the standards at various temperatures.

With the increasing use of polymers in food-packaging materials, clothing, and prosthetic appliances it is likely that increased emphasis will be placed on analyses of this type.

Bibliography

1. G. M. Kline, ed., *Analytical Chemistry of Polymers,* Interscience Publishers, a division of John Wiley & Sons, Inc., New York, Part I, *Analysis of Monomers and Polymeric Materials. Plastics–Resins–Rubbers–Fibers,* 1959; Part II, *Analysis of Molecular Structure and Chemical Groups,* 1962; Part III, *Identification Procedures and Chemical Analysis,* 1962.
2. B. Ke, ed., *Newer Methods of Polymer Characterization,* Vol. 6 in *Polymer Reviews* Series, Interscience Publishers, a division of John Wiley & Sons, Inc., New York, 1964.
3. C. P. A. Kappelmeier, ed., *Chemical Analysis of Resin-Based Coating Materials,* Interscience Publishers, Inc., New York, 1959.
4. D. Hummel, *Kunststoff-, Lack- und Gummi-Analyse. Chemische und infrarotspektroskopische Methoden,* Vol. I, Carl Hanser Verlag, Munich, 1958.
5. N. D. Cheronis, J. B. Entrikin, and E. M. Hodnett, *Semimicro Qualitative Organic Analysis: The Systematic Identification of Organic Compounds,* 3rd ed., John Wiley & Sons, Inc., New York, 1964.
6. R. L. Shriner, R. C. Fuson, and D. Y. Curtin, *Systematic Identification of Organic Compounds,* 5th ed., John Wiley & Sons, Inc., New York, 1964.

7. W. T. Smith and R. L. Shriner, *Examination of New Organic Compounds,* John Wiley & Sons, Inc., New York, 1956.
8. F. Wild, *Characterization of Organic Compounds,* 2nd ed., Cambridge University Press, London, 1958.
9. S. Siggia, *Quantitative Organic Analysis via Functional Groups,* 3rd ed., John Wiley & Sons, Inc., New York, 1963.
10. F. E. Critchfield, *Organic Functional Group Analysis,* The Macmillan Co., New York, 1963.

10a. N. D. Cheronis and T. S. Ma, *Organic Functional Group Analysis by Micro and Semimicro Methods,* Interscience Publishers, a division of John Wiley & Sons, Inc., New York, 1964.

11. T. P. G. Shaw, *Ind. Eng. Chem., Anal. Ed.* **16,** 541 (1944).
12. G. M. Brauer and E. Horowitz, "Systematic Procedures," in Part III of Ref. 1, pp. 1–140.
13. A. Gruben and H. H. Leiner, *Mod. Plastics* **37** (11), 88 (July 1960).
14. M. H. Swann, M. L. Adams, and G. G. Esposito, "1963 Review of Application of Analysis: Coatings," *Anal. Chem.* **35,** 35R–39R (1963).
15. E. J. Parks and F. J. Linnig, "1963 Review of Application of Analysis: Natural and Synthetic Rubbers," *Anal. Chem.* **35,** 160R–178R (1963).
16. "Analytical Reviews: 1964 Review of Fundamental Developments in Analysis," *Anal. Chem.* **36,** 1R–472R (1964).
17. P. W. Allen, ed., *Techniques of Polymer Characterization,* Butterworths Publications Ltd., London, 1959.
18. J. Mitchell, Jr., and S. S. Lord, Jr., *Rubber Chem. Technol.* **34,** 1553–1573 (1961).
19. W. Kupfer, *Z. Anal. Chem.* **192,** 219–248 (1962).
20. H. A. Gardner and G. G. Sward, *Physical and Chemical Examination of Paints, Varnishes, Lacquers and Colors,* 12th ed., Gardner Laboratory, Bethesda, Md., 1962.
21. O. D. Shreve, "Synthetic Organic Coatings," in *Organic Analysis,* Vol. III, Interscience Publishers, Inc., New York, 1956, p. 443.
22. O. D. Shreve, in L. Meites, ed., *Handbook of Analytical Chemistry,* McGraw-Hill Book Co., Inc., New York, 1963, pp. 13–131 to 13–159.
23. C. E. M. Jones, ed., *Identification of Textile Materials,* 4th ed., The Textile Institute, Manchester, England, 1958.
24. "ASTM Designation D 276-62T: Tentative Methods for Identification of Fibers in Textiles," *ASTM Standards on Textile Materials (with Related Information),* 34th ed., sponsored by ASTM Committee D-13 on Textile Materials, American Society for Testing and Materials, Philadelphia, 1963, pp. 111–147.
25. "Analysis of Textiles: Quantitative, Tentative Test Method 20A - 1959T," *Technical Manual of the American Association of Textile Chemists and Colorists,* Vol. XXXIX, 1963 ed., AATCC, Durham, N.C., 1963, p. B-8.
26. "Fibers in Textiles: Identification, Tentative Test Method 20–1962T," p. B-17 in Ref. 25.
27. "Finishes in Textiles: Identification, Tentative Test Method 94–1962T," p. B-37 in Ref. 25.
28. W. H. Neuss, ed., *Testing of Adhesives,* No. 26 in TAPPI Monograph Series, Technical Association of the Pulp and Paper Industry, New York, 1963.
29. C. A. Lucchesi and J. D. McGinness, in L. Meites, ed., *Handbook of Analytical Chemistry,* McGraw-Hill Book Co., Inc., New York, 1963, pp. 13–210 to 13–226.
30. H. E. Frey, *Methoden zur chemischen Analyse von Gummimischungen,* Springer Verlag, Berlin, 1960.
31. J. C. J. Poulton, in W. J. C. Naunton, ed., *The Applied Science of Rubber,* Edward Arnold and Co., London, 1960, pp. 874–991.
32. M. Tryon and E. Horowitz, pp. 13–233 to 13–256 in Ref. 29.
33. W. C. Wake, *Analysis of Rubber and Rubberlike Polymers,* Maclaren and Sons, Ltd., London, 1958.
34. M. J. Rosen and H. A. Goldsmith, *Systematic Analysis of Surface-Active Agents,* Interscience Publishers, Inc., New York, 1960.
35. F. M. Lewis and F. R. Mayo, *Ind. Eng. Chem., Anal. Ed.* **17,** 134 (1945).
36. C. E. Albertson and I. R. McGregor, *Anal. Chem.* **22,** 806 (1950).
37. V. A. Stenger, W. B. Crummett, and J. G. Cobler, p. 587 in Ref. 3.
38. H. Nechamkin, *Ind. Eng. Chem., Anal. Ed.* **15,** 40 (1943).
39. K. Überreiter and H. J. Orthmann, *Kunststoffe* **48,** 525 (1948).
40. Ref. 1, Part I, p. 155.

41. D. G. Grabar and R. Haessly, *Anal. Chem.* **28,** 1586 (1956).
42. E. L. Skau, J. C. Arthur, Jr., and H. Wakeham, in *Physical Methods of Organic Chemistry,* Vol. 1 of A. Weissberger, ed., *Technique of Organic Chemistry,* 3rd ed., Part I, Interscience Publishers, Inc., New York, 1959, pp. 287–356.
43. H. F. Herbrandson and F. C. Nachod, in E. A. Braude and F. C. Nachod, eds., *Determination of Organic Structures by Physical Methods,* Academic Press, Inc., New York, 1955, p. 3.
44. T. Alfrey, *Mechanical Behavior of High Polymers,* Vol. 6 in *High Polymers* Series, Interscience Publishers, Inc., New York, 1948, p. 77.
45. J. H. Teeple, in R. Houwink, ed., *Elastomers and Plastomers,* Vol. III, Elsevier Publishing Co., New York, 1948, p. 52.
46. N. P. Bekkedahl, *J. Res. Natl. Bur. Std.* **43,** 145 (1949).
47. R. F. Clash and L. M. Rynkiewicz, *Ind. Eng. Chem.* **36,** 279 (1944).
48. R. H. Wiley and G. M. Brauer, *J. Polymer Sci.* **3,** 455, 647, 705 (1948).
49. J. M. Preston, *J. Textile Inst.* **40,** T767 (1949).
50. "Test for Specific Gravity of Plastics (Tentative)," D 792-60T, *Am. Soc. Testing Mater., ASTM Std.,* 1964, Part 27, p. 325.
51. J. M. Preston and M. V. Nimkar, *J. Textile Inst.* **41,** T446 (1950).
52. C. R. Stock and E. R. Scofield, *ASTM (Am. Soc. Testing Materials) Bull. No. 175,* July 1951, p. 78.
53. N. Bauer and S. Z. Lewin, pp. 131–190 in Ref. 42.
54. N. Bauer, K. Fajans, and S. Z. Lewin, pp. 1140–1281 in Ref. 42, Part II.
55. L. W. Tilton and J. K. Taylor, in W. G. Berl, ed., *Physical Methods in Chemical Analysis,* 2nd ed., Academic Press, Inc., New York, 1960, pp. 411–462.
56. A. F. Forziati, "Optical Methods," Part II, Ref. 1, pp. 109–157.
57. A. Elliott, "The Infra-Red Spectra of Polymers," in H. W. Thompson, ed., *Advances in Spectroscopy,* Vol. I, Interscience Publishers, Inc., New York, 1959, pp. 214–287.
58. S. Krimm, *Fortschr. Hochpolymer.-Forsch.* **2,** 51–172 (1960).
59. J. P. Luongo, *Anal. Chem.* **33,** 1816 (1961).
60. M. Tryon and E. Horowitz, "Ultraviolet Spectrophotometry," Part II of Ref. 1, pp. 269–289.
61. M. Tryon and E. Horowitz, "Infrared Spectrophotometry," Part II of Ref. 1, pp. 291–333.
62. N. Brenner, *Offic. Dig. Federation Soc. Paint Technol.* **33** (432), 51 (1961).
63. J. G. Frazer, N. M. Peacock, and A. W. Pross, "Infrared Spectrophotometry in the Analysis of Coating Materials," pp. 407–444 in Ref. 3.
64. L. J. Bellamy, *The Infrared Spectra of Complex Molecules,* 2nd ed., John Wiley & Sons, Inc., New York, 1958.
65. R. B. Barnes, R. C. Gore, R. W. Stafford, and V. Z. Williams, *Anal. Chem.* **20,** 402 (1948).
66. D. Hummel, *Kunststoff- Lack- und Gummi-Analyse. Chemische und Infrarotspektroskopische Methoden,* Vol. II, Carl Hanser Verlag, Munich, 1958.
67. R. E. Kagarise and L. A. Weinberger, *Naval Research Laboratory Report 4369,* U.S. Naval Research Laboratory, Washington, D.C., 1954.
68. R. A. Nyquist, *Infrared Spectra of Plastics and Resins,* The Dow Chemical Co., Midland, Mich.
69. *Monomers and Polymers,* Vol. IV, *Spectra of Monomers and Polymers with Classification Indices,* Sadtler Research Laboratories, Philadelphia, 1962.

69b. *Infrared Spectra of Commercial Products. Pyrolyzates,* Sadtler Research Laboratories, Philadelphia, 1964.

70. *Offic. Dig. Federation Soc. Paint Technol.* **32** (427), 517 (1960).
71. D. Hummel, *Kunststoffe* **46,** 442 (1956).
72. D. L. Wood and J. P. Luongo, *Mod. Plastics* **38,** 132 (March 1961).
73. J. P. Deley, R. J. Gigi, and A. J. Liotti, *Tappi* **46** (2), 188A (Feb. 1963).
74. H. Dannenberg, J. W. Forbes, and A. Jones, *Anal. Chem.* **32,** 365 (1960).
75. R. L. Harris and G. R. Svoboda, *Anal. Chem.* **34,** 1655 (1962).
76. F. A. Bovey and G. V. D. Tiers, *J. Polymer Sci.* **44,** 173 (1960).
77. H. Y. Chen, *Anal. Chem.* **34,** 1793 (1962).
78. M. Takeda, K. Tanaka, and R. Nagao, *J. Polymer Sci.* **57,** 517 (1962).
79. R. E. Naylor and S. W. Lasoski, Jr., *J. Polymer Sci.* **44,** 1 (1960).
80. L. A. Wall and R. E. Florin, "Magnetic Resonance Spectroscopy," Part II of Ref. 1, pp. 487–559.
81. W. P. Slichter, *Rubber Chem. Technol.* **34,** 1574–1600 (1961).

82. J. D. Roberts, *Introduction to the Analysis of Spin-Spin Splitting in High Resolution Nuclear Magnetic Resonance Spectra,* W. A. Benjamin, Inc., New York, 1961.
83. A. F. Forziati, "Fluorescence," Part II of Ref. 1, pp. 335–357.
84. K. Thinius, *Farbe Lack* **56,** 3 (1950).
85. H. E. Millson, *The Phosphorescence of Textile Fibers and Other Substances,* Calco Technical Bull. No. 753, Calco Chemical Division, American Cyanamid Co., Bound Brook, N.J., 1944.
86. H. Mark and R. Raff, *High Polymeric Reactions: Their Theory and Practice,* Interscience Publishers, Inc., New York, 1941, p. 40.
87. J. N. Banks and P. S. Bembrick, *Textile J. Australia* **28,** 1522 (1954).
88. L. A. Wall, "Pyrolysis," Part II of Ref. 1, pp. 181–248.
89. D. A. Vassallo, *Anal. Chem.* **33,** 1823 (1961).
89a. N. A. Nechitaĭlo, I. M. Tolchinskiĭ, and P. I. Sanin, *Plasticheskie Massy* **1960** (11), 54.
90. R. F. Schwenker, Jr. and L. R. Beck, Jr., *Textile Res. J.* **30,** 624 (1960).
91. H. B. Kissinger and S. B. Newman, "Differential Thermal Analysis," Part II of Ref. 1, pp. 159–179.
92. B. Ke, "Application of Differential Thermal Analysis to High Polymers," in *Organic Analysis,* Vol. IV, Interscience Publishers, Inc., New York, 1960, pp. 361–394.
93. C. B. Murphy, *Mod. Plastics* **37** (12), 125 (Aug. 1960).
94. C. B. Murphy, *Anal. Chem.* **34,** 298R (1962).
95. D. L. Harms, *Anal. Chem.* **25,** 1140 (1953).
96. P. F. Kruse, Jr. and W. B. Wallace, *Anal. Chem.* **25,** 1156 (1953).
97. D. Hummel, *Kunststoffe* **46,** 442 (1956).
98. J. Haslam and A. R. Jeffs, *J. Appl. Chem.* (*London*) **7,** 24 (1957).
99. D. F. Nelson, J. L. Yee, and P. L. Kirk, *Microchem. J.* **6,** 225 (1962).
100. J. G. Cobler and E. P. Samsel, *SPE Trans.* **2,** 145 (1962).
101. W. H. Parriss and P. D. Holland, *Brit. Plastics* **33,** 372 (1960).
102. J. Strassburger, G. M. Brauer, M. Tryon, and A. F. Forziati, *Anal. Chem.* **32,** 454 (1960).
103. D. L. Miller, E. P. Samsel, and J. G. Cobler, *Anal. Chem.* **33,** 677 (1961).
104. G. G. Esposito and M. H. Swann, *Anal. Chem.* **34,** 1048 (1962).
105. E. W. Neumann and H. G. Nadeau, *Anal. Chem.* **35,** 1454 (1963).
106. G. G. Esposito and M. H. Swann, *Anal. Chem.* **33,** 1854 (1961).
107. F. Lehman and G. M. Brauer, *Anal. Chem.* **33,** 673 (1961).
108. A. Barlow, R. S. Lehrle, and J. C. Robb, *Polymer* **2,** 27 (1961).
109. K. Ettre and P. F. Varadi, *Anal. Chem.* **34,** 752 (1962).
110. K. Ettre and P. F. Varadi, *Anal. Chem.* **35,** 69 (1963).
111. C. W. Stanley and W. R. Peterson, *SPE Trans.* **2,** 298 (1962).
112. E. M. Barrall, III, R. S. Porter, and J. F. Johnson, *Anal. Chem.* **35,** 73 (1963).
113. J. Haslam, A. R. Jeffs, and H. A. Willis, *Analyst* **86, 44** (1961).
114. J. T. Walsh and C. Merritt, Jr., *Anal. Chem.* **32,** 1378 (1960).
114a. G. M. Brauer, *Polymer Preprints, Am. Chem. Soc., Div. Polymer Chem., Chicago, Sept. 1964.*
115. E. P. Ragelis and R. J. Grajan, *J. Assoc. Offic. Agr. Chemists* **45,** 918 (1962).
116. O. Tweet and W. K. Miller, *Anal. Chem.* **35,** 852 (1963).
117. F. M. Nelsen, F. T. Eggertsen, and J. J. Holst, *Anal. Chem.* **33,** 1150 (1961).
118. P. Shapras and G. C. Claver, *Anal. Chem.* **34,** 433 (1962).
119. J. Voigt, *Kunststoffe* **51,** 18, 314 (1961).
120. H. Feuerberg and H. Weigel, *Kautschuk Gummi* **15,** WT 276 (1962).
121. C. D. Cook, E. J. Elgood, G. C. Shaw, and D. H. Solomon, *Anal. Chem.* **34,** 1177 (1962).
122. G. G. Esposito, *Anal. Chem.* **35,** 1439 (1963).
123. J. Zulaica and G. Guiochon, *Anal. Chem.* **35,** 1724 (1963).
124. J. Haslam and A. R. Jeffs, *Analyst* **83,** 455 (1958).
125. J. Haslam, A. R. Jeffs, and H. A. Willis, *J. Oil Colour Chemists' Assoc.* **45,** 325 (1962).
126. G. G. Esposito and M. H. Swann, *Offic. Dig. Federation Soc. Paint Technol.* **33,** 1122, 1460 (1961).
127. L. A. Adcock, *Paint Technol.* **26** (5), 20 (1962).
128. Baltimore Society for Paint Technology, *Offic. Dig. Federation Soc. Paint Technol.,* **32** (430), 1517 (1960).
129. F. Feigl and E. Jungreis, *Microchim. Acta* **1958,** 812.
130. D. Davidson, *Ind. Eng. Chem., Anal. Ed.* **12,** 40 (1940).

131. F. Feigl and P. Krumholz, *Mikrochemie* **1929,** 82.
132. R. F. Milton and W. A. Waters, *Methods of Quantitative Microchemical Analysis,* 2nd ed., St. Martin's Press, London, 1955.
133. A. Steyermark, *Quantitative Organic Microanalysis,* The Blakiston Co., Inc., Philadelphia, 1957.
134. "Test for Total Chlorine in Poly(vinyl chloride) Polymers and Copolymers Used for Surface Coatings," D 1156-52, *Am. Soc. Testing Mater., ASTM Std.,* 1964, Part 20, p. 526.
135. K. Thinius, *Analytische Chemie der Plaste,* Springer Verlag, Berlin, 1952.
136. H. E. Freier, B. W. Nippoldt, P. B. Olson, and D. G. Weiblen, *Anal. Chem.* **27,** 146 (1955).
137. W. Schöniger, *Mikrochim. Acta* **1955,** 123.
138. J. Haslam and S. M. A. Whettem, *J. Appl. Chem.* **2,** 339 (1952).
139. R. Kojima and H. Ueno, *J. Chem. Soc. Japan, Ind. Chem. Sect.* **61,** 270 (1958).
140. F. W. Reinhart and G. M. Kline, *Ind. Eng. Chem.* **31,** 1522 (1939).
141. A. Gordijenko and H. J. Schenck, *Kunststoffe* **37,** 123 (1947); **39,** 29 (1949).
142. K. Thinius, *Deut. Farben-Z.* **5,** 73 (1951).
143. R. Nitsche and W. Toeldte, *Kunststoffe* **40,** 29 (1950).
144. S. Okamura, T. Higashimura, and I. Sakurada, *J. Polymer Sci.* **39,** 507 (1959).
145. R. T. Hall and W. E. Shaefer, "Determination of Esters," in *Organic Analysis,* Vol. II, Interscience Publishers, Inc., New York, 1954, p. 19.
146. R. W. Stafford and E. F. Williams, in J. J. Mattiello, ed., *Protective and Decorative Coatings,* Vol. V, John Wiley & Sons, Inc., New York, 1946, p. 122.
147. C. E. Redemann and H. S. Lucas, *Ind. Eng. Chem., Anal. Ed.* **9,** 521 (1937).
148. A. Z. Conner, "Rosin and Rosin Derivatives," pp. 70–74 in Ref. 3.
149. H. H. Coburn, *Ind. Eng. Chem., Anal. Ed.* **2,** 181 (1930).
150. Part I, in Ref. 1, pp. 206, 263, 552, 613.
151. Part I, in Ref. 1, pp. 208, 552.
152. A. Polgár and J. L. Jungnickel, in *Organic Analysis,* Vol. III, Interscience Publishers, Inc., New York, 1956, p. 203.
153. G. H. Benham and L. Klee, *J. Am. Oil Chemists' Soc.* **27,** 127 (1950).
154. Part I, Ref. 1, p. 236.
155. Part I, Ref. 1, p. 550.
156. T. F. Bradley and E. L. Kropa, p. 202 in Ref. 146.
157. Ref. 146, p. 134.
158. J. B. Johnson and J. P. Fletcher, *Anal. Chem.* **31,** 1563 (1959).
159. E. H. Unger, *Anal. Chem.* **30,** 375 (1958).
160. J. C. S. Wood, *Anal. Chem.* **30,** 372 (1958).
161. Part I, Ref. 1, p. 612.
162. M. Hellmann and L. A. Wall, "End Group Analysis," in Part III, pp. 397–411 in Ref. 1.
163. G. F. Price, "Techniques of End Group Analysis," pp. 207–230 in Ref. 17.
164. H. Dannenburg and W. R. Harp, Jr., *Anal. Chem.* **28,** 86 (1956).
165. S. D. Bruck and S. M. Bailey, *J. Res. Natl. Bur. Std.* **A66,** 185 (1962).
166. S. R. Palit, *Pure Appl. Chem.* **4,** 459 (1962).
167. S. R. Palit and M. K. Saha, *J. Polymer Sci.* **58,** 1233 (1962).
168. S. R. Palit, *Anal. Chem.* **33,** 1441 (1961); *Chem. Ind. (London)* **1960,** 1531.
169. S. R. Palit and P. Ghosh, *J. Polymer Sci.* **58,** 1225 (1962).
170. H. Zahn and P. Rathgeber, *Melliand Textilber.* **34,** 749 (1953).
171. I. M. Ward, *Trans. Faraday Soc.* **53,** 1406 (1957).
172. H. Staudinger and H. Schmidt, *J. Prakt. Chem.* **155,** 153 (1940).
173. A. R. Martin, L. Smith, R. L. Whistler, and M. Harris, *J. Res. Natl. Bur. Std.* **27,** 449 (1941).
174. F. Smith and R. Montgomery, in D. Glick, ed., *Methods of Biochemical Analysis,* Vol. III, Interscience Publishers, Inc., New York, 1957, pp. 153–212.
175. A. M. Sookne and M. Harris, "End Groups," in E. Ott and H. Spurlin, eds., *Cellulose and Cellulose Derivatives,* 2nd ed., Part I, Interscience Publishers, Inc., New York, 1954.
176. P. D. Bartlett and S. G. Cohen, *J. Am. Chem. Soc.* **65,** 543 (1943).
177. C. C. Price and B. E. Tate, *J. Am. Chem. Soc.* **65,** 517 (1943).
178. A. T. Blomquist, J. R. Johnson, and J. H. Sykes, *J. Am. Chem. Soc.* **65,** 2446 (1946).
179. B. D. Sully, *Nature* **159,** 882 (1947).
180. M. G. Evans, *J. Chem. Soc.* **1947,** 272.

181. T. S. Lee, I. M. Kolthoff, and M. A. Mairs, *J. Polymer Sci.* **3,** 66 (1948).
182. G. Natta, P. Pino, E. Mantica, F. Danusso, G. Mazzanti, and M. Peraldo, *Chim. Ind.* (*Milan*) **38,** 124 (1956).
183. H. Muramatsu, M. Iwasaki, and R. Kojima, *J. Chem. Soc. Japan, Ind. Chem. Sect.* **64,** 1096 (1961).
184. J. C. Bevington, H. W. Melville, and R. P. Taylor, *J. Polymer Sci.* **12,** 449 (1954); **14,** 463 (1954).
185. R. M. Joyce, W. E. Hanford, and J. Harmon, *J. Am. Chem. Soc.* **70,** 2529 (1948).
186. C. C. Price and D. A. Durham, *J. Am. Chem. Soc.* **64,** 2508 (1942).
187. C. C. Price, *J. Am. Chem. Soc.* **65,** 2380 (1943).
188. E. L. Stanley, "Acrylic Plastics," Part I of Ref. 1, p. 5.
189. G. M. Brauer and S. B. Newman, "Color Tests," Part III of Ref. 1, pp. 141–259.
190. F. Feigl, *Spot Tests in Organic Chemistry,* 6th ed., Elsevier Publishing Co., New York, 1960.
191. P. Paulson, *J. Oil Colour Chemists' Assoc.* **36,** 127 (1953).
192. F. E. Critchfield and D. P. Johnson, *Anal. Chem.* **33,** 1834 (1961).
193. M. E. Gibson and R. H. Heidner, *Anal. Chem.* **33,** 1825 (1961).
194. J. E. Brown, M. Tryon, and J. Mandel, *Anal. Chem.* **35,** 2172 (1963).
195. S. L. Madorsky, *Thermal Degradation of Organic Polymers,* Interscience Publishers, a division of John Wiley & Sons, Inc., New York, 1964.
196. B. Ke, ed., "Thermal Analysis of High Polymers," *J. Polymer Sci.* Part C, Polymer Symposia No. 6, 1964.
197. W. W. Wendlandt, *Thermal Methods of Analysis,* John Wiley & Sons, Inc., New York, 1964.

G. M. Brauer and G. M. Kline
National Bureau of Standards

INFRARED–ABSORPTION SPECTROSCOPY

A rapid development of infrared-absorption spectroscopy began at the end of World War II when the first commercial instrument provided with reliable electronic amplifiers and radiation detectors was constructed in the United States. Studies of polymers were among the first applications of the method (1). At the present time, infrared spectroscopy is probably the method most extensively used for the investigation of polymer structure and the analysis of functional groups.

The basis of this method rests upon the interaction of infrared electromagnetic radiation with mass. This interaction results in absorption of certain wavelengths of radiation, the energy of which corresponds to the energy of transition between various vibrational–rotational states, or only rotational states, of the molecules or groups of atoms within the molecule. In the spectrum produced, the absorption intensity is recorded as a function of wave number or wavelength (see, for example, Fig. 8); specific groups of atoms in the molecule give rise to characteristic absorption bands whose wave numbers fall within a definite range regardless of the composition of the remainder of the molecule. This constancy of absorption wave numbers makes possible the determination of functional groups present in the substance being analyzed. From the exact values of the wave numbers at which absorption is observed, conclusions may be drawn about the influence of adjacent groups in the molecule or in neighboring molecules on the vibration of the group in question. It is thus possible to create a picture of the molecular configuration and to determine details of the probable structure. Because of the great variety of possible vibrations in most molecules, it is practically impossible to find two compounds which would have identical rotational–vibrational spectra. For this reason, the infrared spectrum is often spoken of as the "fingerprint" of a molecule. Since the intensity of absorption is a measure of the concentration of a group, infrared spectra may be used in quantitative analysis. In addition to structural, qualitative, and quantitative analyses, infrared spectroscopy is also useful in the determination of number-average molecular weight and the degree of branching and in studies of the course of chemical reactions of polymer molecules. Another application is the determination of the degree of regularity of arrangement of the macromolecules, ie, degree of crystallinity and the degree of orientation of stretched films and fibers.

An important factor which favors extension of infrared spectroscopy in polymer chemistry is that both soluble and insoluble materials can be analyzed. The sample can be prepared as a film or a powder, and thick specimens can be analyzed by the method of attenuated total reflectance.

The combination of data on vibrational–rotational states from the infrared spectra with data obtained by Raman spectroscopy (qv) is very important for theoretical studies not only of low-molecular-weight compounds but also of polymers (2). For common analytical and structural purposes, however, the combination of infrared spectroscopy with nuclear magnetic resonance (qv) or with mass spectrometry (qv) is more frequent. These three physical–chemical methods, because of their convenience and speed, are the methods predominately used at the present time in structural and analytical studies of high polymers. In pyrolytic analyses of polymers the infrared spectral method is usually combined with gas chromatography (qv under CHROMATOGRAPHY). See also CHARACTERIZATION OF POLYMERS; CHEMICAL ANALYSIS.

Theoretical Background

As mentioned above, infrared spectra originate in the interaction of electromagnetic radiation, ranging from 1 to 500 μ in wavelength (ie, from wave numbers of 10,000 to 20 cm^{-1}), with mass. (The wave number ν, the number of waves per centimeter, is equal to $10^4/\lambda$, where λ is the wavelength in microns.) The relationship between wave number and wavelength or frequency ν_f is given by the well-known expressions in equations 1a and 1b, where c is velocity of light.

$$\nu_f = c\nu \tag{1a}$$

$$\nu_f\lambda = c \tag{1b}$$

Number and Position of Absorption Bands. The interaction of radiation with mass is governed by quantum conditions, so that radiation of wave number ν is absorbed only when equation 2 holds; E_1 and E_2 are discrete values of energy corresponding to two states and h is Planck's constant.

$$\nu = \frac{E_2 - E_1}{hc} \tag{2}$$

The energy of infrared radiation cannot cause transitions between electron energy levels. It is capable only of bringing about changes in *vibrational* and *rotational* states of a molecule; these are closely related to the molecular structure. The establishment of these relations is a problem whose exact solution must be approached quantum mechanically using the Schrödinger wave equation. The solution shows that approximation as a simple harmonic oscillator is possible, and that the whole problem is reducible to a classical analysis of the vibrating system.

For a system of N atoms, the solution leads to $3N$ wave number values that satisfy the set of equations of motion of individual atoms. Of these $3N$ wave numbers, three representing translations and three representing rotations of the molecule as a whole are zero; the total number of normal vibrations is then $3N - 6$. In the case of a linear molecule which rotates only around axes perpendicular to the linear axis of the molecule, the total number of normal vibrations is $3N - 5$. Table 1 shows the vibrations of the $—CH_2—$ group and the common notation for the various modes of vibration. The first three vibration modes correspond to the $—CH_2—$ group as an isolated entity $[(3 \times 3) - 6]$; the other three result from the fact that the $—CH_2—$ group is actually part of a linear chain. In addition to fundamental vibrations, overtone and combination vibrations appear in the spectrum. Their intensity is lower than that of the fundamental vibrations.

Not every vibration is necessarily detected in the infrared spectrum. Only those vibrations that are accompanied by a change in dipole moment result in infrared absorption. However, if the electrical polarizability is changed during the normal vibration, this vibration appears in the Raman spectrum. The number of absorption bands found in the spectrum decreases also with increasing symmetry of a molecule, because the vibrations become identical.

Table 1. Vibration Modes of the —CH_2— Group

Mode of vibration[a]	Direction of transition moment	Assignment
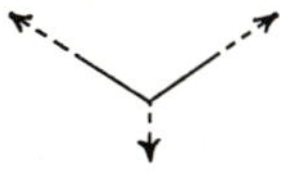	↑	ν_s, symmetric stretching
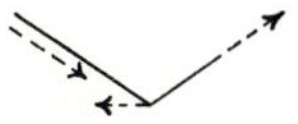	→	ν_a, asymmetric stretching
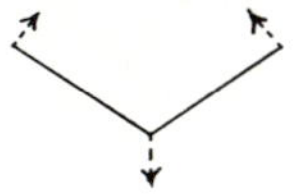	↑	δ, deformation, symmetric bending
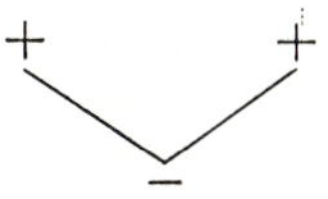 + + −	⊙	ω or γ, wagging
 + −	none	t, twisting
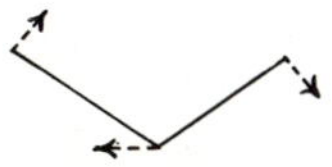	→	r, rocking

[a] + and − indicate motion perpendicular to the plane of the paper.

Based on theoretical considerations, it is possible to derive the values of individual vibration frequencies. They depend on the masses of the atoms and on their geometric positions with respect to one another; these latter are determined by the bond lengths and angles and by the force constants of the bonds. Magnitudes of all of these cannot be determined simultaneously from the spectrum. For example, if the wave number of the vibration is to be determined, either some of these values must be estimated or another method must be used for determining them, such as x-ray diffraction (qv) in the case of crystals.

Polymer molecules contain large numbers of atoms, but their infrared spectra are relatively simple. The reason lies in the structure of polymers. Each linear or branched macromolecule contains end groups and a sequence of repeating base units. Regular repetition of a series of identical units means that the number of spectroscopically active vibrational modes is relatively small. In analyses of the vibrations of the end groups, it can be assumed, as a first approximation, that the interaction between the characteristic vibration of the group in question and the rest of the macromolecule is negligible. This concept of *group frequencies* is very useful for practical interpretation of infrared spectra. The concentration of the end groups is relatively small and so their absorption bands are weak. The main contribution to the absorption spectrum usually comes from vibrations of atoms or groups in the repeating units.

Figure 1 shows a part of a linear planar zig-zag polyethylene chain. Because of its regular configuration, the problem of theoretical analysis of the spectrum is similar to the analysis of a crystalline compound. The linear macromolecule can, as a first approximation, be characterized by a one-dimensional space group. For a more rigorous treatment, it is necessary to use a three-dimensional space group model of the crystalline lattice. According to the one-dimensional method, the repeating unit (R,

Fig. 1. Planar model of zig-zag polyethylene chain with repeating unit R (3).

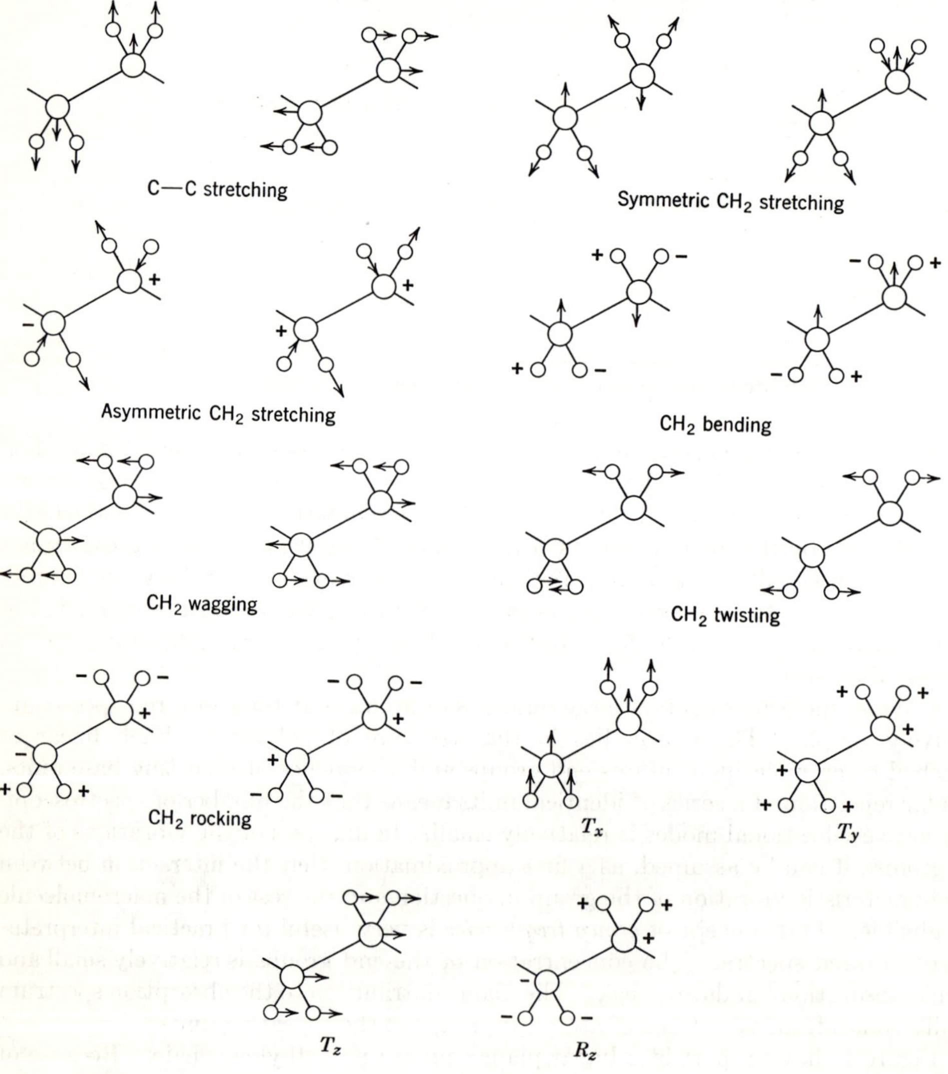

Fig. 2. Vibrations of a polyethylene chain, with first approximation by one-dimensional space groups (+ and − indicate motion perpendicular to the plane of the paper) (3).

Fig. 3. One-dimensional model of a polymer chain represented by point masses and springs.

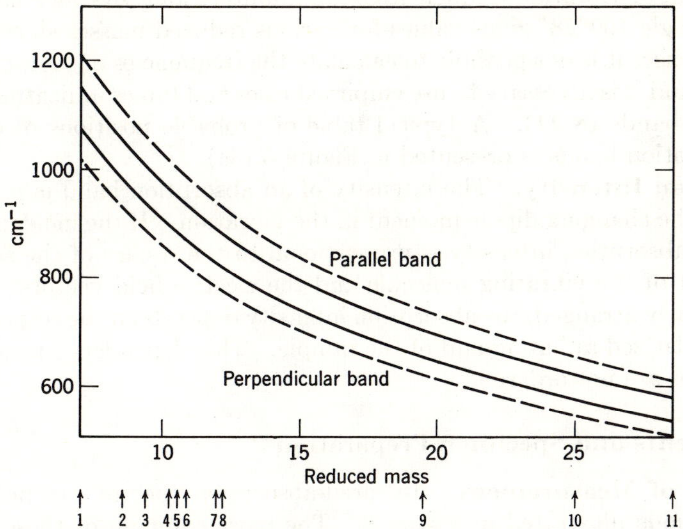

Fig. 4. Vibration frequencies of infinitely long zig-zag chains; stretching force constant (k_s) = 4.20 × 10^5 dynes/cm and bending force constant (k_b) = 0.37 × 10^5 dynes/cm. The dashed lines correspond to a 10% variation of the values for the C—C stretching force constants. Repeating units are: (1) CH_2CH_2, (2) CH_2CHCH_3, (3) CH_2CHOH, (4) CH_2CHCN, (5) $CH_2C(CH_3)_2$, (6) CH_2CHCl, (7) CH_2CCl_2, (8) $CH_2CHC_6H_5$, (9) $CHFCF_2$, (10) CF_2CF_2, and (11) CF_2CFCl (3).

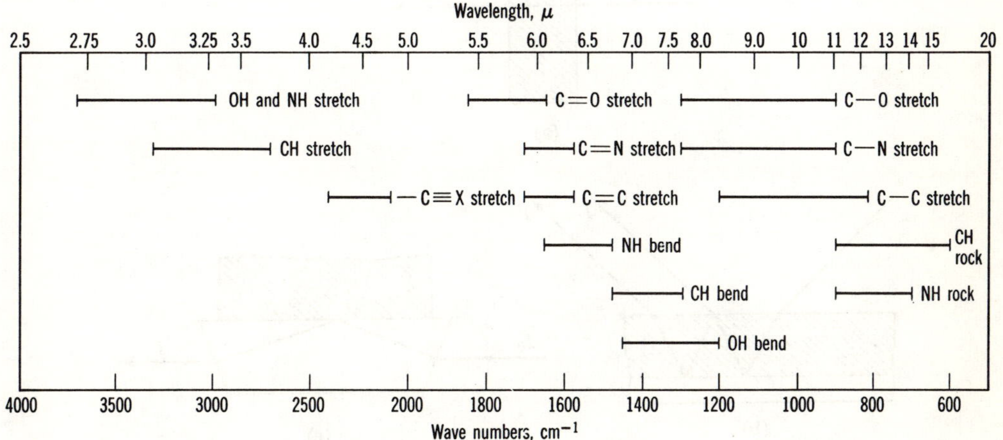

Fig. 5. Characteristic infrared-absorption bands (3a). Courtesy *Journal of the Optical Society of America.*

Fig. 1) consists of two —CH_2— groups, ie, six atoms with eight elements of symmetry. This unit has 18 vibrations, as shown in Figure 2.

A more exact model of the three-dimensional unit cell of crystalline polyethylene contains four —CH_2— groups; 36 vibrations of a unit cell are possible, of which 15 are infrared active and 18 are Raman-active.

Frequencies of atoms or groups of atoms in a linear polymer chain can be derived from the classical model of atoms represented by point masses connected with chemical bonds represented by springs (Fig. 3).

Calculation of vibration frequencies of an infinitely long zig-zag chain with tetrahedral bond angle 109°28′ gives values for various reduced masses shown in Figure 4 (3). In most cases it is not possible to calculate the frequencies of vibrations for every atomic group and it is necessary to use empirical rules and tables indicating the position of absorption bands (8–11). A typical table of probable positions of characteristic infrared-absorption bands is presented in Figure 5 (3a).

Absorption Intensity. The intensity of an absorption band is proportional to the square of the changing dipole moment in the vibration. If the incident radiation is *polarized,* the absorption intensity is proportional to the square of the scalar product of the moment of the vibrating molecule and the electric field vectors. If the molecules are regularly arranged, the absorption intensity depends on the respective orientations of the polarized radiation and of the sample. This dependence is not observable if the molecules are not ordered.

Instruments and Specimen Preparation

Methods of Measurement. Infrared-absorption spectra can be obtained by the three methods illustrated in Figure 6. The most common of these is the transmission method in which the fraction of the incident light that is absorbed (or transmitted) is measured. The *attenuated total reflectance method* (ATR) is especially

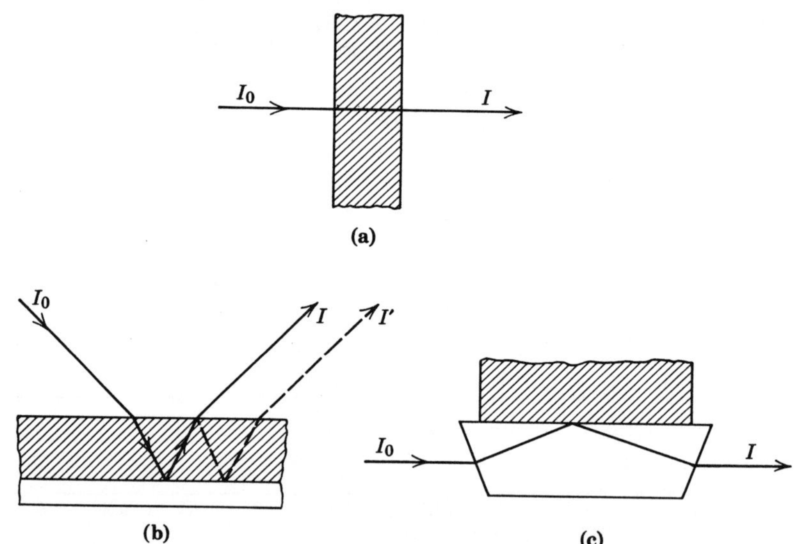

Fig. 6. Schematic representation of (**a**) transmission method, (**b**) reflection method, and (**c**) attenuated total reflectance method.

applicable in cases in which the sample cannot be prepared in the required thickness and is thus not suitable for analysis by the transmission method. In other cases, it is most convenient to use the reflection method.

The principle of the transmission and reflection methods lies in the absorption of radiation on passage through the sample. In the ATR method, the angle of the incident radiation at the boundary between the sample and the optical material of the instrument is wider than critical. Part of the radiation penetrates beneath the reflecting surface of the material being analyzed. Internal reflection occurs and, at appropriate ratios of the refractive indexes of both materials, the reflected light returns to the reflecting surface, modified by absorption in the sample at some wave numbers. The ATR spectrum thus obtained resembles the transmission spectrum. Since in the ATR method the beam penetrates only a few microns into the sample, the thickness of the sample beyond the depth of penetration is of no significance and the spectral intensity is independent of sample thickness; absorption spectra of coatings, for example, may thus be obtained.

Instruments. Spectroscopic instruments can be divided into three categories: special devices with very high resolving power; devices which cover a broad range of wave numbers, which are usually automated and possess a great variety of registration possibilities; and simple devices with rapid recording.

The instruments of the first type are designed for structural studies of gaseous substances and are not used with polymers. For liquids and solids, where splitting of absorption bands to bands of different rotation states does not occur, their high resolving power cannot be utilized. The resolving capacity of the instrument used, however, should be high enough to differentiate absorption bands of the same type of functional group, but associated with different chain structures. Simple devices, on the other hand, such as the Perkin-Elmer Infracord, the Beckman IR 5A or Unicam SP200 are sufficient for ordinary purposes, eg, for identification by direct comparison of the spectrum of the analyzed compound with various reference spectra. With the extension of spectral studies to the range of C–H stretching vibrations at about 3 μ (3000 cm^{-1}) and to the long-wavelength range of 15–40 μ (700–250 cm^{-1}), devices of the second type, with sufficient resolving power in the 3000-cm^{-1} region and covering the spectral range as far as 200–250 cm^{-1}, came to the fore, and their use is now widespread in polymer chemistry. The applicability of the 800–200 cm^{-1} region for distinguishing the stereoregular forms of polymers will probably result in even more frequent use of this type of instrument. Wholly organic polymers such as polyethylene, poly(ethylene terephthalate), polyamides, etc, have low absorption in the region below 200 cm^{-1} (4), so that this range is not suitable for their analysis. On the other hand, this spectral range will be useful for polymers with metal–carbon bonds. The choice of this type instrument is supported by the fact that utilization of measurements of absolute values of absorption intensities both for structural elucidation and analyses is expected in the near future. The type of instrument described, eg, the Perkin-Elmer 621 R, usually requires a specialized operator.

The construction principles of most infrared spectrometers do not differ a great deal from one another. The newer instruments are mostly double-beam type. They consist of a source of radiation, monochromator, and radiation detector with recording equipment. A simplified block scheme is shown in Figure 7.

Common sources of radiation are Globar, ie, a rod of silicon carbide fired to 1000–1200°C, and the Nernst lamp, the filament of which is composed of rare earth metal oxides, operating at about 2000°C. The emitted energy of both sources is maximum in the near infrared region, 1–2 μ (5000–10,000 cm^{-1}), then decreases rapidly so that at 25 μ (400 cm^{-1}) it reaches only 0.01% of its maximum value. The problem of measurement at these values is one reason why the older commercial instruments did not cover the long-wavelength region. Another reason is the necessity of employing

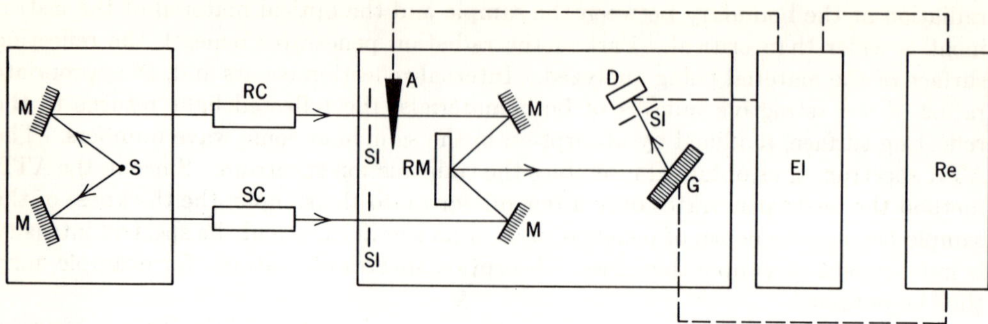

Fig. 7. Schematic diagram of a double-beam spectrometer; S = source of radiation, M = mirror, SC = sample cell, RC = reference cell, A = attenuator, RM = rotating mirror, Sl = slit, G = grating or prism, D = detector, El = electronic amplifying system, and Re = recorder; (———), optical path; (- - - - -), electrical or mechanical connection.

less common materials for monochromators and cells than when the shorter wavelengths are used.

A beam emitted from the source is split into two paths, ie, the sample and the reference beams. The former passes through the sample to be analyzed, whereas the reference beam passes through the reference cell containing solvent but no sample. The monochromator disperses both beams according to wavelength. Prisms and diffraction gratings are used for this purpose. Only a few materials satisfy the requirements for the optical properties of the prisms. The most common optical materials are lithium fluoride, sodium chloride, potassium bromide, KRS-5 (thallium bromide–iodide), and cesium bromide and iodide (5). Originally, diffraction gratings were used only in special instruments, but at present they are in many commercial devices. Several prisms or gratings are necessary if the whole spectral range from 2 μ (5000 cm^{-1}) to 50 μ (200 cm^{-1}) is to be covered. The sample beam and the reference beam pass alternately through the monochromator, once or several times. A set of slits isolates a certain wavelength, which then falls on the radiation detector.

The intensities of the beam which has passed through the sample (I) and that of the reference beam (I_0) are compared in the photometric part of the instrument. The recording assembly presents either the ratio of both intensities, ie, the transmission ($T = I/I_0$, in %), the absorption (100–T, in %), or the logarithm of the ratio I_0/I (the absorbance) as a function of wave number ν (in cm^{-1}) or of wavelength λ (in μ). The recording speed is limited by the ability of the detector and the recorder to respond to changes in radiation intensity without delay. Thermocouples, bolometers, or pneumatic Golay cells are employed as radiation detectors.

Oriented polymers can be studied on instruments with *polarized infrared radiation.* This is an important method now frequently used. Polarization of the infrared radiation is achieved by transmission polarizers consisting of silver chloride or selenium plates (6).

Other special equipment for infrared spectrometers designed for polymer studies are the attenuated total reflection unit, described above, and the infrared microscope, which, attached to the spectrometer, permits the use of extremely small samples.

Applicability of the instrument for polymer chemistry must be considered from various aspects: the spectral range covered by the device, the resolving power, the accuracy of the wave number and of the absorption intensity measurements, and the speed of recording. The resolving power and the spectral range have been discussed above. For polymer studies the whole spectral range available is utilized. It is therefore necessary for the recording speed to be high enough for the analysis to be completed in a reasonable time. High accuracy in the determination of absorption intensity is an inevitable condition for quantitative analysis. It is also important that the instrument be capable of recording spectra after the intensity of the reference beam is as low as 5%. This is important in differential measurements.

Specimen Preparation. Samples of polymers for infrared analysis by transmission are usually prepared in the form of solutions, but sometimes as films or, if insoluble, in potassium bromide pellets or in nujol mulls. The use of solutions is limited by the large number of absorption bands of most solvents and by the necessity of combining at least two solvents if the whole infrared spectrum is to be recorded. The ranges of applicability of individual solvents are presented in many books on practical infrared spectroscopy (see the section on Literature).

When using the film technique, a complication may arise by orientation during sample preparation. This can result in a change of absorption intensities, since the optical system of the instrument causes partial polarization of radiation.

Pellets of potassium bromide are advantageous for insoluble samples. Even in this case, a number of factors such as crystallinity and homogeneity of the sample can influence the recorded spectrum. Instead of potassium bromide, other suitable materials such as thallium bromide, silver chloride, or polyethylene can be used as the basis of sample pellets. Mulls in nujol are also suitable for insoluble samples. The presence of absorption bands of nujols (CH_2 vibrations) makes the analysis of the vibrations of methylene groups impossible.

Polymer specimens can also be prepared as very thin platelets by cutting with a microtome, in the form of fibers if special microscopic equipment is available, by compressing with polyethylene or another suitable film, by swelling, etc.

In the reflection methods based on passage of the infrared beam through the sample, reflection from the carrier plate, and reversal through the sample (Fig. 6), the polymer is deposited as a thin film on a suitable reflecting base, such as a polished metal plate.

For analysis by attenuated total reflectance, which is more commonly used than reflection methods, a sample of any thickness is pressed onto the proper optical material, usually silver chloride or KRS-5 (thallium bromide–iodide, 42% TlBr and 58% TlI).

Possible changes in the spectra of different or identical samples caused by different methods of sample preparation must be taken into account in comparative studies. Several studies have been published in which the applicability of these various methods of sample preparation is considered (7).

Elucidation of Structure

As has been discussed in the section on theory, repeating units as well as end groups may be analyzed with great accuracy. The application of empirical methods is inevitable in the elucidation of the structure of polymers. In polymers with more or less known chemical structures, infrared examination can supply a great deal of other information about the physical state of the polymer, ie, about crystallinity, orientation, etc; these aspects will be discussed below in the section on study of the Physical Nature of Polymers.

In this section we shall limit our discussion to the use of empirical methods in the analysis of the chemical structure of polymers. The same correlation tables can be employed here as for low-molecular-weight compounds (8–11). These correlation tables present the range of wave numbers within which the vibrations of functional groups can vary, as well as some discussion of the influence of adjacent groups on these vibrations.

Some groups, such as —OH, —C=O, —C≡N, and $—CH_3$, can be identified with reasonable certainty from the occurrence of one typical absorption band. The presence of other groups can be revealed only through the occurrence of several characteristic absorption bands. For example, the vinyl group, $—CH=CH_2$, always appears as the C—H stretching vibration in the $H_2C=$ group, the C=C stretching vibration, as well as the $H_2C=$ out-of-plane vibration and its overtone. Identification of unknown polymers by various methods, including infrared spectroscopy, is discussed in CHEMICAL ANALYSIS; ELASTOMERS, SYNTHETIC; FIBERS, IDENTIFICATION; and RESINOGRAPHY.

A more difficult subject for infrared analysis than mere identification of functional groups is the determination of the microstructure of polymers, for example, determination of the relative content of cis and trans configurations of double bonds, or the type of stereoregularity, such as isotactic and syndiotactic structures, etc. Stereoregular polymers are characterized by so-called "stereospecific absorption bands" (also referred to as "stereoregular" bands) that indicate the particular type of stereoregularity. Atactic polymers have bands of more structural forms than do the stereoregular polymers. The stereoregular bands, according to Zerbi and co-workers (12), are associated with well-defined configurations which can be transformed to others only by chemical reactions (eg, isomerization); the same number of these bands occur in the spectrum of the polymer in all possible phases.

The most difficult form of analysis of stereoregular polymers is the determination of steric configuration of the main polymer chain. A necessary condition for the appearance of a vibration of the "infinite chain" is that all equivalent atoms oscillate in phase (13). The regularity of the skeleton of a macromolecule is broken by melting or dissolving. Absorption bands called "regular bands" are associated with the well-defined configuration of certain groups in the repeating unit which can be transformed from one configuration to another only by physical methods (12). In the analysis of a polymer skeleton for which the configuration of the main chain has been assumed or established by other analytical methods, vibration modes are assigned to the absorption bands found in the spectrum of the polymer. Analysis of these spectra is similar to the process of analysis of the physical nature of the polymer, as will be discussed below (p. 638).

The final aim of infrared analysis is the determination of all functional groups and their mutual positions, and the assignment of vibration modes to all absorption bands. In this respect, the most thoroughly studied polymer has been polyethylene (3,14).

In questionable cases, the analysis may be carried out after chemical modification of the polymer. A typical example is hydrogenation or bromination of a polymer with carbon–carbon double bonds. Synthesis of polymers with isotopic substitution is also very useful. Most polymers contain hydrogen which can be substituted by deuterium. In this case vibrations of CH, OH, or NH groups will be replaced by vibrations of CD, OD, or ND groups. The ratio of wave numbers of vibrations of the groups with hydrogen to those of the same groups with deuterium is $\nu_H/\nu_D = 1.3–1.4$, the exact value depending upon the type of vibration. The result is a shift of absorption bands and better separation of overlapping bands so that additional conclusions about the type of vibration may often be made (see also ISOTOPIC LABELING).

Qualitative Analysis

Qualitative analysis is, by its character, analogous to the structural studies. The aim of the latter is to disclose the structure by characterization of the functional groups present and the mode of their attachment to the polymer backbone. The aim of qualitative analysis of mixtures is to determine the presence of individual components; eg, in the case of copolymers, it is to determine the presence of individual monomer units. The spectrum of a mixture is additively composed of the spectra of the individual components. All absorption bands of the spectrum should be ascribed to individual components. No band should be in surplus; no band should be missing.

It is extremely difficult to distinguish between a mixture of homopolymers and a true copolymer. Theoretically, only barely perceptible differences in the spectra of both extreme cases should be expected. One method of distinguishing between a copolymer and a mixture of homopolymers is by extraction of the product and analysis of the resulting soluble and insoluble fractions. A more reliable criterion is based on fractionation carried out in such a way that separation of different structures predominates over separation according to molecular weight; the individual fractions are then analyzed. In spite of these complications, it can be concluded that the qualitative analysis both of copolymers and of polymer mixtures is in most cases successful. The greater the difference between structures of the monomer units, the easier is the analysis. Figure 8 presents the infrared spectra of butadiene–styrene copolymer, butadiene–acrylonitrile copolymer, and butadiene–styrene–acrylonitrile terpolymer. In the first and the third spectrum the characteristic absorption bands of styrene units at 700 cm^{-1}, 760 cm^{-1}, and 1500 cm^{-1} are present. In the second and the third spectrum the absorption bands at 2250 cm^{-1} show the presence of acrylonitrile units. From the bands typical of phenyl or nitrile groups it can be seen what kind of rubber is the subject of the investigation.

Another extreme case of difficult qualitative analysis is a polymer with *cis*-1,4-isoprene units (**1**), *trans*-1,4-isoprene units (**2**), 3,4-isoprene units (**3**), and cyclic units (**4**).

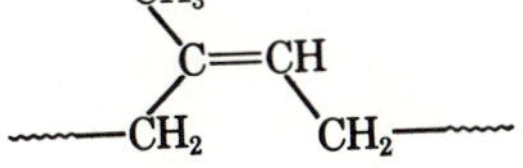

CH_3, C=CH, ~CH_2, CH_2~

(1)

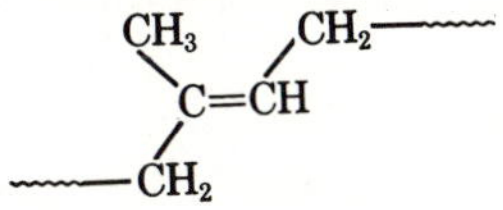

CH_3, CH_2~, C=CH, ~CH_2

(2)

~CH_2—CH~, C—CH_3, ‖, CH_2

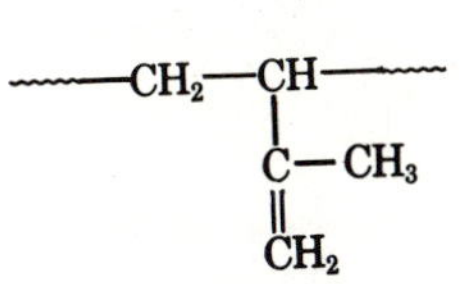

(3)

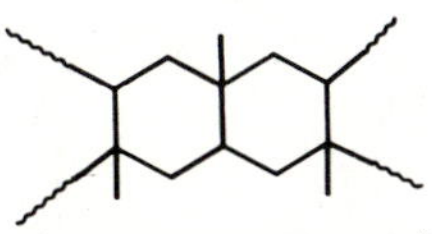

(probable structure)

(4)

Spectra of stereoregular polymers composed of each one of these forms are presented in Figure 9. The spectrum of the *cis*-1,4 polymer is characterized by the absorption bands at 572 cm^{-1}, 840 cm^{-1}, and 1130 cm^{-1}, which are typical for this structural form. On the other hand, the spectrum of 3,4-polyisoprene does not contain these absorption bands, but it is characterized by the bands at 566 cm^{-1}, 890 cm^{-1}, and 1780 cm^{-1}. A minor difference in configuration of the cis and trans forms causes much less difference in the spectra than between *cis*-1,4- and 3,4-polyisoprene (the spectrum of *trans*-1,4-polyisoprene is not presented in Fig. 9). In addition to the bands mentioned, both structures have the absorption bands at 1380 cm^{-1} and 1460 cm^{-1}. These two bands are present also in the spectrum of cyclopolyisoprene, which is otherwise very simple; the absorptivities are so small that it is impossible to give direct evidence of the presence of cyclic units in the polymer containing 1,4 and 3,4 units. However, quantitative infrared and nuclear magnetic resonance spectra confirm the presence of the cyclic structure in some polymers (15).

The difficulties of qualitative analysis become obvious with the first visualization of all possible structures of all isomers of a polymer. It can be reasonably expected that the analysis of hydrocarbon polymers will be difficult because similarity of the spectra is to be expected. On the other hand, polymers containing various groups with atoms such as O, S, N, etc, are analyzed relatively easily because they may be analyzed by functional groups (51).

The most reliable evidence for establishing the identity of two compounds is that from *differential spectrophotometry*. In this method the sample being analyzed is in-

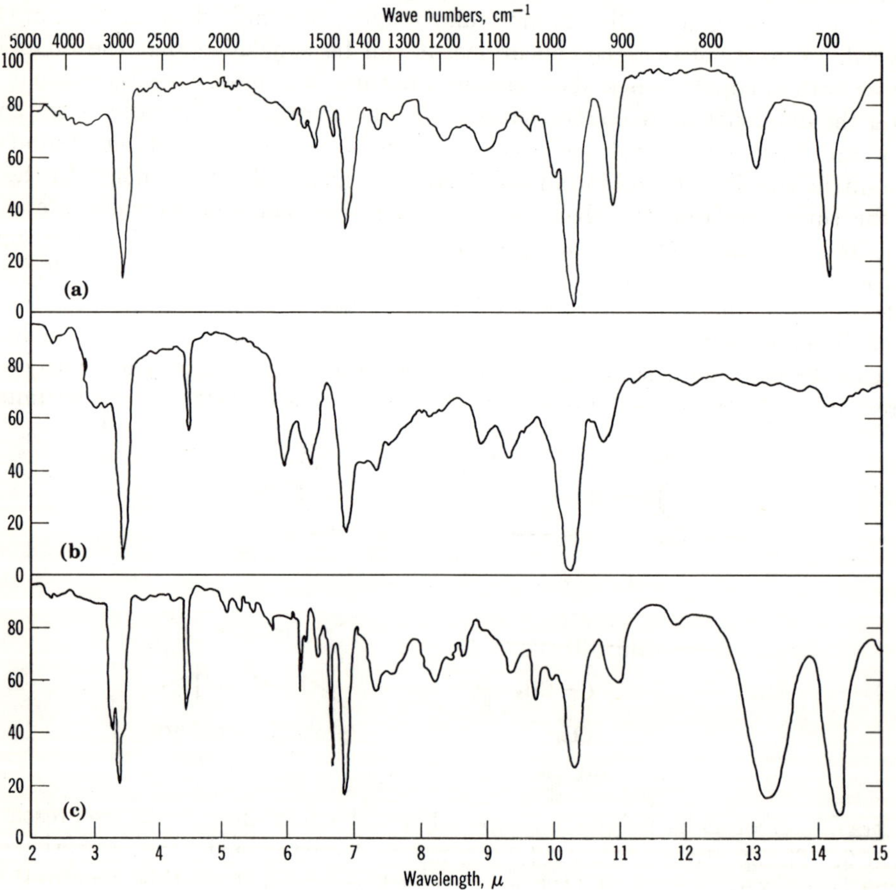

Fig. 8. Infrared spectra of (**a**) butadiene–styrene copolymer, (**b**) butadiene–acrylonitrile copolymer, and (**c**) butadiene–styrene–acrylonitrile terpolymer.

serted into the path of the sample beam and a known compound into the path of the reference beam. If the samples are identical, the recorded curve is free of absorption bands. A slightly different content of the same group in the two samples produces a spectrum containing only the bands of this group. However, when the wave numbers are slightly shifted, the difference spectrum is very complicated. Figure 9 gives the difference spectra of natural rubber and synthetic *cis*-1,4-polyisoprene, as well as of natural rubber and cyclopolyisoprene. In the former case, the identity of the two structures is evident; in the later case, the difference in frequencies of CH_3 and CH_2 groups at 1380 cm^{-1} and 1460 cm^{-1} is very distinct although it could be overlooked in an examination of the separate spectra.

Many polymers easily undergo oxidation, which is indicated by the appearance of the absorption band of the C=O group near 1720 cm^{-1}. The occurrence of this group affects absorption bands of other groups and complicates interpretation of the spectra. For example, series of bands can originate as a result of conjugation of other groups, such as C=C, with the carbonyl group.

Qualitative analysis of commercial products is frequently carried out by simple comparison of spectra, preferably after extraction of various additives, with the published spectrum of the suspected compound. Collections of spectra are important tools for this purpose (see the section on Literature).

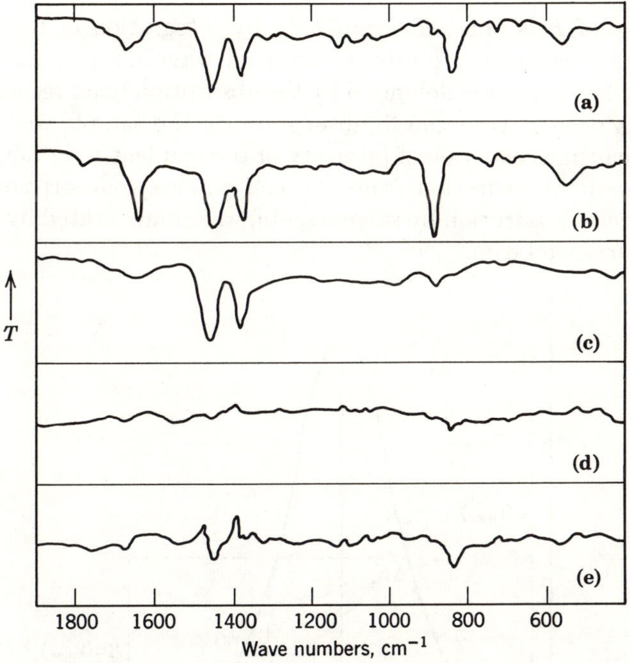

Fig. 9. Infrared-absorption spectra of (**a**) *cis*-1,4-polyisoprene, (**b**) 3,4-polyisoprene, (**c**) cyclopolyisoprene, (**d**) differential spectrum of natural rubber–synthetic *cis*-1,4-polyisoprene, and (**e**) differential spectrum of natural rubber–cyclopolyisoprene.

Quantitative Analysis

Quantitative infrared analysis, as well as other optical spectral methods, is based on Lambert-Beer's law establishing the relationship between concentration of the compound (or the functional group) and the amount of radiation absorbed (eq. 3).

$$\log I_0/I = \sum_{i=1}^{n} a_i c_i d \tag{3}$$

Here I_0 is intensity of incident radiation, I is intensity of radiation after passage through the sample, c_i is concentration of component i (usually in g/l), d is thickness of the sample (cell) (usually in cm), and a_i is a constant, the so-called absorptivity (or, less desirably, the "extinction coefficient"). The expression $\log I_0/I = A$ is called the absorbance (sometimes "optical density").

Since the resolution by the monochromator is not perfect, distortion of the theoretical shape of the absorption band takes place. In some cases it is therefore useful to characterize the absorption band not only by its intensity at a wave number of maximum absorption but also by certain other descriptions (Fig. 10). The concept of the *half-width* of the absorption band, ie, the width at half of the absorption maximum, is frequently used for this purpose. The integrated absorption, B, defined by equation 4, where $a(\nu)$ is absorptivity at wave number ν, is another common description.

$$B = 2.303 \int_{-\infty}^{+\infty} a(\nu) d\nu \tag{4}$$

The value of B is obtainable not only by integration of the curve in the plot characterizing the dependence of absorptivity on wave number ($a = f(\nu)$) (Fig. 10), but also directly from the area delimited by the absorption band recorded ($I/I_0 = f(\nu)$).

In the derivation of equation 3, absorption by the sample was considered to be the only factor causing reduction of intensity of the incident radiation. The reduction of intensity caused by reflection from the cell windows, absorption by the optical material of the cells, absorption by solvents, etc, was compensated by the reference cell placed in the reference beam.

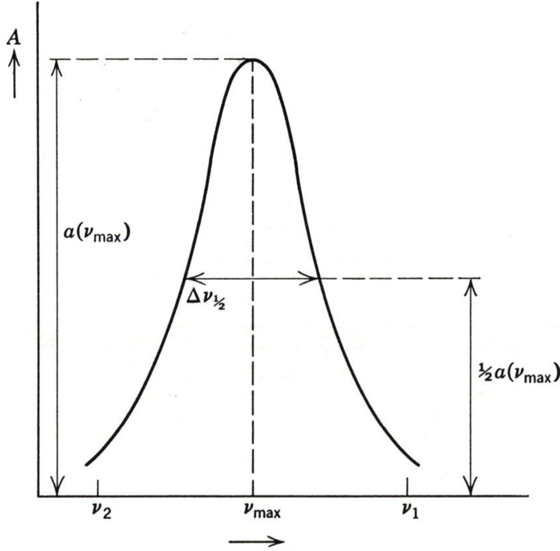

Fig. 10. A typical absorption band, where a = absorptivity, ν = wave number, ν_{max} = wave number at maximum absorption, $\Delta\nu_{1/2}$ = half-width of the absorption band.

Compensation for absorbed light is more difficult in other methods of sample preparation than in the solution method because of the significant scattering of light by polymers, which results in so-called "background." This effect is one reason for introduction of a factor K, representing all extraneous effects, into equation 3 to give equation 5; K is a factor depending on sample concentration, $c = \Sigma c_i$, but is inde-

$$\log I_0/I = \sum_{i=1}^{n} a_i c_i d + K \tag{5}$$

pendent of individual components, c_i. Elimination of K is accomplished by methods such as "difference of absorbances" or "base-line." These are described in detail in books on infrared spectroscopy dealing with quantitative analysis (16–18).

If more than one component is being analyzed, a set of equations corresponding to equation 3 or to 5 is solved for properly chosen wave numbers. In favorable cases the "analytical" bands are isolated and are not interfered with by other bands. For example, for butadiene–styrene rubber (Fig. 8) the analytical band is the phenyl band at 700 cm^{-1}; for nitrile rubber it is a band belonging to the C≡N group at 2250 cm^{-1}. The absorption bands of polymers are, however, usually wide and often overlap with other bands.

In addition to the choice of the most suitable absorption bands, another important factor is the precise determination of absorptivities of pure standards. In the case of mixtures, the individual components may serve as standards. For copolymers or for homopolymers containing more than one configuration, the situation is more complicated. If available, low-molecular-weight compounds or stereoregular polymers having a structure similar to the repeating units in the sample are used as standards for these cases. Another possibility is the measurement of several samples with qualitatively identical but quantitatively different compositions (19); however this method is applicable only when certain special requirements concerning the shapes of the absorption bands are fulfilled (20).

Complications may arise from the necessity of using unoriented and noncrystalline samples. It must also be kept in mind that intensities and wave numbers of some absorption bands can be influenced by alternation of monomer units. Thus, for example, the infrared spectrum of a block copolymer (**5**) need not be quite identical with that of an alternating copolymer (**6**) (21). Until all structural possibilities are known, it is advisable to use as standards only those polymers which were prepared with the same catalyst system.

~AAAABBBBAA~
(**5**)

~ABABABAB~
(**6**)

Unless the solution method is used, a further step in an analysis is the determination of sample thickness. If a potassium bromide pellet is used, the sample concentration in the pellet is calculated from the weight of the sample in the disc, using the known surface area F. The relationship $A = acd$ (from equation 3) is then transformed to the form $A = ag/F$, where g is the amount of polymer in the pellet. In the pellet or the film technique, the sample thickness can be determined from the absorption intensities of bands which occur in all structural forms considered and in the same amount per unit; this is, for example, the band of the CH_3 group at 1380 cm^{-1} in polyisoprene. By dividing the absorbance of the "analytical" band by the absorbance of such a band the values related to one concentration can be obtained. It is also possible to use an internal standard, ie, by adding a compound with a typical absorption band, by which the concentration or the thickness of the analyzed sample can be determined.

Another method of analysis follows from the assumption that the total concentration of all components, Σc_i, is known and only their relative content is to be determined. In this method, however, the possibility exists of overlooking a structural form not clearly indicated in the spectrum because of the absence of any specific absorption band, as, for example, is the case with cyclic forms of polyisoprene.

The methods discussed above are based on measurements of absorption at the wave number of maximum absorption. The higher the resolving power of the spectrometer, ie, the narrower the distribution of wavelengths in the beam of infrared radiation, the stronger is absorption at the wave number of maximum absorption. It follows that the intensities of the same absorption band recorded on different types of instruments are not necessarily identical. Better results in the determination of the concentration of groups (group analysis) on different instruments can be obtained from the determination of the area under the absorption bands. In many cases, the position of the absorption maximum of the same group present in different compounds is not identical, but the integrated absorptions are practically equal (22). It is reasonable to expect that current studies on integrated absorption, eg, Ref. 23, will widen the application of this method in polymer chemistry.

The accuracy of quantitative analysis by infrared-absorption spectroscopy varies from case to case. The accuracy of the analysis of the compounds with isolated absorption bands can reach several tenths of one percent. The sensitivity depends upon the absorptivity of the selected absorption band. Both accuracy and sensitivity are worsened by overlapping of the bands, by increasing the number of the components in the sample, and by use of the potassium bromide pellet or the film technique, in which the sample thickness is difficult to determine. In the case of potassium bromide pellets, the accuracy can reach about 1% in ten measurements or about 10% in a single measurement, although sometimes an absolute accuracy of 20% is the best that may be expected.

When the reflection method is used, the reliability of quantitative analysis is lessened by the unequal lengths of the paths and the manifold reflections of radiation on passage through the sample (24), as can be seen in Figure 6. The ATR method is, however, applicable for quantitative analysis (25).

Studies of the Chemical Nature of Polymers

The data obtained in structural, qualitative, and quantitative analyses can be used for studies of specific properties of polymers, both chemical and physical. Infrared spectroscopy is useful in the determination of molecular weight and of the degree of branching, as well as in studies of chemical reactions of the polymers, such as oxidation or decomposition. Changes in physical state of the polymers, such as crystallization or orientation, are also reflected in the infrared spectrum and can be studied by this method.

Number-Average Molecular Weight and Degree of Branching. The number-average molecular weight can be determined from infrared spectra provided that the polymer molecules have easily detectable end groups. Obviously, the lower the molecular weight, the more prominent are the absorption bands of the end groups. This measurement is of a quantitative character and the main problems are the determination of absorptivities, elimination of nonspecific absorptions, and interference of the absorption bands.

The method can be illustrated by the following example: The content of methyl groups in polyethylene was determined from the absorption intensity at 1380 cm^{-1}. The interference of methylene absorption at 1368 cm^{-1} was eliminated through use of the differential method; a film of high-density polyethylene inserted into the path of the reference beam compensated for this absorption band (ASTM D 2238-64T). The absorptivity was considered the same as in a low-molecular-weight standard which was, in this case, *n*-hexadecane. The calculated number of methyl end groups in the analyzed sample would correspond to the number of macromolecules only if branching did not occur in polymerization. If the number-average degree of polymerization is known from another method, eg, ebulliometric, cryoscopic, or osmometric, the difference between the two values would indicate the degree of branching (27). High-pressure polyethylene contains about 20–30 CH_3 groups per 1000 carbon atoms, low-pressure polyethylene, 3–10 CH_3 groups, and Phillips polyethylene, 1–2 CH_3 groups per 1000 carbon atoms. These values do not supply information about the length of the side groups; precise infrared study indicates that the branches are not formed by methyl groups (28). See also Polyethylene under ETHYLENE POLYMERS.

Infrared determination of the molecular weight of polymers by end-group analysis is especially instructive when the polymer contains more than one type of end group. Polyisobutylene, for example, exhibits two absorptions: the CH out-of-plane bending vibration in a trisubstituted double bond at 830 cm^{-1} and the vinylidene double bond at 892 cm^{-1}. From these absorption bands and a band of the monomeric unit at 855 cm^{-1}, not only the degree of polymerization but also the relative occurrence of both types of end group can be evaluated (29). Studies have also been made in which number-average molecular weight of polybutadiene was determined using the overtone CH vibration in the range of about 2–2.5 μ (5000–4000 cm^{-1}). It was found that polybutadienes obtained with various initiators such as sodium, butyllithium, and free-radical sources have different slopes of the plot of molar absorptivity vs molecular weight. The molecular weights in combination with the infrared data thus gave information about the microstructure (30). See also BUTADIENE POLYMERS.

Reactions of Macromolecules. The infrared spectra of monomers differ markedly from the spectra of their polymers. As a consequence, it is possible, by means of infrared spectroscopy, to follow the course of polymerization or copolymerization reactions, and, simultaneously, to analyze the structure of the polymer being formed.

Any chemical reaction, whether involving the main chain or side groups, results in a change of composition of one or more groups, and thus also in the infrared spectrum. This makes it possible to study oxidation, thermal degradation, cyclization, grafting, and other reactions of the polymer. The course of these reactions can be followed directly on a film of polymer deposited on an optical material. Evaluation of both qualitative and quantitative changes, as well as the determination of kinetic constants of the reaction are possible.

Before using infrared spectroscopy for the study of a chemical reaction, the question of whether the number of groups undergoing change is sufficient to bring about perceptible changes in the infrared spectrum must be considered. For example, the number of crosslinks formed in vulcanization is not sufficient to be detectable by infrared spectroscopy, yet causes profound changes in the solution properties of the polymer. Small degrees of oxidation of diene polymers, which cause significant changes in the mechanical properties of the rubber, are also invisible in the infrared spectrum.

In spite of these circumstances the *oxidation reactions* of polymers are among those most frequently studied by the infrared method. Quantitative examination of a series of oxidation reactions has been in accord with the proposed radical-chain mechanism (see also DEGRADATION). The occurrence of bands near 1700 cm^{-1}, typical of the C=O vibration and appearing in the spectrum of a polymer whose monomer did not contain this group, is an indication that oxidation has taken place. Quantitative evaluation of the extent of oxidation is possible by examination of the intensity of this band.

Thermal degradation results in an increase of the number of end groups, and, usually, in end groups containing double bonds, which are easily detectable in infrared spectra. For example, in the thermal degradation under vacuum of chlorotrifluoroethylene homopolymer or of its copolymer with vinylidene fluoride, a band from the $—CF=CF_2$ group appeared at 1782 cm^{-1}, whereas on degradation in air another band, typical of the —OCF group, appeared at 1880 cm^{-1} (26).

Cyclization is another type of reaction that is easy to follow. In the case of diene polymers, it results in a decrease of the number of double bonds. Whereas chemical methods do not differentiate between double bonds reacting in a cyclization and those newly formed, infrared analysis accomplishes this relatively easily. The formation of cyclic structures on interaction of pendant unsaturated groups is also connected with changes of the infrared spectrum that can be followed quantitatively.

Isomerization reactions can be followed either through the consumption of the original functional groups or through the increase of concentration of new structures. In some cases when the reaction proceeds through an intermediate, the latter can be detected in the infrared spectrum. In *grafting* and in polymer reactions such as *chlorination*, etc, substantial changes occur in the composition of the macromolecules and, therefore, also in the infrared spectra.

Some quantitative treatment of any reaction of a polymer requires careful differentiation of specific absorption bands from the background, and possibly also the separation of interfering absorption bands; the methods of the evaluation of the spectra will not be applicable when these conditions cannot be met.

Studies of the Physical Nature of Polymers

In the section devoted to the elucidation of structure, it was stated that the degree of regularity of the spatial arrangement of the isolated macromolecule appears in the infrared spectrum. Another type of regularity of the polymer is an ordered arrangement of a system of polymer chains which arises on orientation, eg, by stretching, or on crystallization.

Linear polymers, whether in the form of films or fibers, can be oriented by stretching. The main chains become oriented in a direction parallel to the direction of stretch. The type of space arrangement of the individual groups is analogous to crystallization (Fig. 11). Both effects, orientation and crystallization, have common features.

Orientation (qv). If the sample is put into the instrument in such a way that the direction of stretch is parallel to the slit and if the absorption spectrum is recorded using polarized radiation with the electric vector first parallel and then perpendicular to the stretching direction, two values of absorbances are determined, $A_{\parallel}$ for the par-

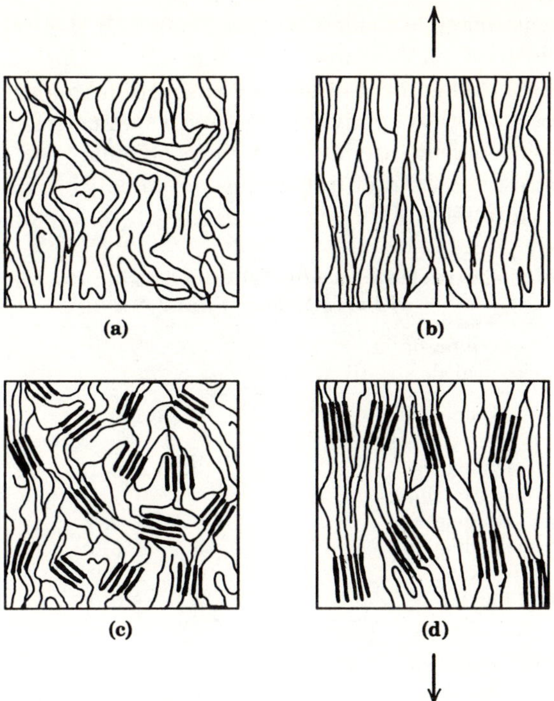

Fig. 11. (**a**) Unoriented amorphous, (**b**) oriented amorphous, (**c**) crystalline, and (**d**) oriented crystalline polymer.

allel and $A_\perp$ for the perpendicular radiation. Their ratio, $R = A_{||}/A_\perp$, is called the *dichroic ratio* and can theoretically range from zero to infinity; in practice it ranges from 0.1 to 10. If $R < 1$ the absorption band is called a *parallel band* and if $R > 1$ the band is *perpendicular*. When $R = 1$ either the sample is not oriented, or the absorption is unchanged for the following reason: The dichroic ratio is influenced by the degree of orientation as well as by the angle θ between the direction of the moment of transition and that of the chain axis, ie, the angle of the chemical group with respect to the chain axis. Theoretical considerations (see Refs. 3 and 14) give, depending upon the type of orientation, different relationships for R in which the angle θ participates in its trigonometric functions, so that in favorable cases R can reach unity. This means that even a sample with perfect orientation can have, at a certain angle, no dichroism. See also OPTICAL ROTATORY DISPERSION.

In partially oriented films the relationship for R involves the so-called orientation parameter, S, which is related to the distribution of chain orientations. For the case of partial axial orientation R is expressed in the form shown in equation 6. Using the

$$R = \frac{\sin^2\theta + S}{2\cos^2\theta + S} \tag{6}$$

values of R which can be found experimentally for a series of absorption bands, the angle of some groups with respect to the macromolecular chain can be determined if the distribution of chain orientation is known from another independent measurement, eg, from x-ray diffraction or electron-diffraction analysis.

In practice, various possible models are assumed, and the hypothetical dichroic ratios of these models are then compared with measured values in order to derive a "most probable" model. Assignment of various absorption bands to specific vibration modes is utilized, especially those of symmetric and asymmetric stretching vibration modes whose transition moments are mutually perpendicular.

Table 2 presents results of structural studies obtained from the dichroic ratio of polyethylene film drawn to tenfold its original dimension (3). It was assumed that the

Table 2. Results of Structural Studies Obtained Using the Dichroic Ratio of Polyethylene Film

Frequency, cm^{-1}	Vibration mode	R	θ, degrees
2919	asymmetric stretching	2.54 ± 0.15	68–71
2851	symmetric stretching	2.98 ± 0.15	68–71
1473	bending	5.9 ± 0.3	79
1463	bending	5.9 ± 0.3	79
731	rocking	8.9 ± 0.9	87–90
720	rocking	10.0 ± 0.05	87-90

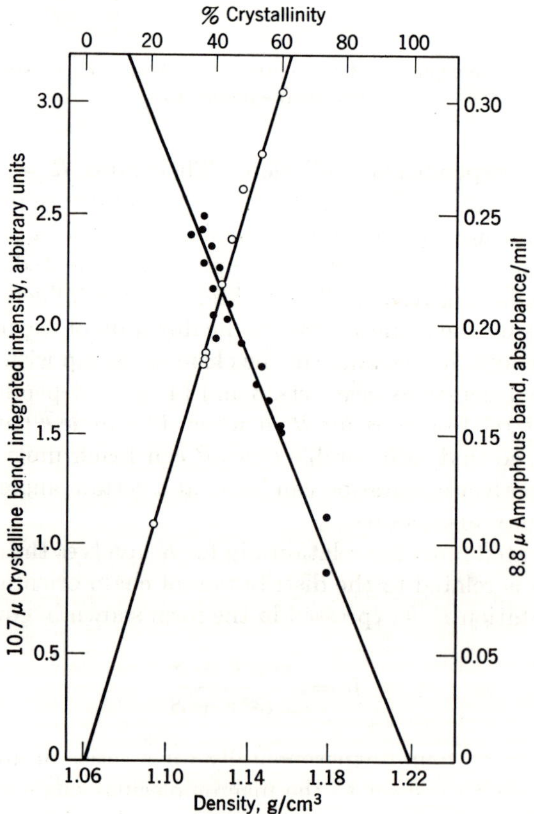

Fig. 12. Intensities of "crystalline" (O) and "amorphous" (●) bands of nylon-6,6 at various percents of crystallinity (31).

orientation is symmetrical around the stretching direction. The transition moments are well defined. The x-ray diffraction pattern showed that the sample is highly oriented. The Kratky-type distribution function was assumed (30a). The observed orientation parameter, S, was 0.115. Using equation 6, values of θ for measured R were obtained (Table 2). An explanation for values lower than 90° has also been presented (3).

Crystallinity (qv). In crystallization of polymers, a spatial arrangement similar to that in orientation occurs. The polymer forms "crystalline" and "amorphous" regions. The formation of crystalline regions is accompanied by an increase in new vibration modes caused by the crystal lattice interactions, so that "regularity" bands occur in the infrared spectrum. Splitting and frequency shifts are also observed. On crystallization, the wave numbers of some vibrations may change, or vibrations that are forbidden in the crystalline lattice may disappear.

The spectrum of a given polymer differs by various absorption bands depending on whether it is in the amorphous or the crystalline state; these bands are designated as "amorphous" or "crystalline," depending upon the state in which they occur. Other absorption bands are not affected by crystallization and remain the same in both cases. According to the definition proposed by Zerbi and co-workers (12) an absorption band may be called a *true crystallinity band* if (a) x-ray diffraction data prove that the material is crystalline, (b) the band disappears on melting, (c) other infrared studies (such as those of solid solutions in isomorphous matrixes or those by the isotopic-dilution method) show that the band depends on the existence of a crystal lattice, and (d) group theory predicts the possible existence of such a band. The crystalline and amorphous bands can be used in the determination of the degree of crystallinity; the independent bands are useful for the determination of the sample thickness.

Nylon-6,6 and nylon-6,10 have "crystalline" absorption bands at 936 cm^{-1} and 853 cm^{-1}, respectively (31). The "amorphous" band is at 1136 cm^{-1}. The base-line method is adequate for this band, but the 936-cm^{-1} band must be integrated since its shape depends on the method of crystallization. The calibration curves of the absorption intensity vs density or percent crystallinity are linear, and the values for entirely amorphous and for highly crystalline polymer can be obtained by extrapolation (Fig. 12).

The orientation of a partly crystalline polymer leads to more regular arrangement of the crystallites (see Fig. 11). In such a case polarized light gives a strong dichroic effect. For example, if the degree of disorientation of the crystallites corresponds to an average angle with respect to the fiber axis of 10°, the perpendicular bands have a dichroic ratio, R, of 15.

All types of regularity will necessarily appear in the vibrational–rotational states of macromolecules. Besides absorption bands of group frequencies and stereoregularity bands, both of which give information on the chemical nature of the polymers, there are the regularity bands and the crystallinity bands which supply information about the degree of regularity. Their identification and theoretical calculations are on a solid foundation. Theoretical considerations are also in progress for the identification and assignment of bands associated with noncrystalline components of polymers (32). See also Crystallinity.

Literature

Most problems solved by infrared spectroscopy, especially those in the field of polymer chemistry, are based on comparative studies. It is therefore very important to possess a complete documentation of infrared spectra, both those prepared by the investigator himself, and those published by other workers.

Qualitative analysis of commercial polymers is carried out by comparison of the spectrum of the compound being analyzed with a spectrum published in any collection of spectra of common polymers (33–36). In the investigation of the structure of new polymers it is advisable to use correlation tables (8–11) or the spectra of model compounds which appear in some collection of spectra (37–40). Some of these collections also contain the spectra of polymers.

The theory of the infrared-absorption spectra is the subject of a book by Herzberg (41) and of several monographs (10,14). General aspects of infrared spectroscopy are also discussed in many other books (18,50–53).

Infrared spectra of polymers are discussed in books by Zbinden (3), Kline (54), and Kendall (55), in a series of lectures presented at the Conference of Vibration Spectra of Polymers (42), and in a number of reviews (14,26,34,43,44). A list of publications concerning the infrared spectra of common polymers is presented in Zbinden's book (3) and in the Hershenson collection (46), as well as in some reviews (14,33,34,43–45). Every two years, a survey of new works in the field of infrared analysis of polymers is published in the review issue of Analytical Chemistry (47,48). A collection of spectra of polymers including the range of 200–700 cm^{-1} is presented in a book by Hummel (35). This region is also the subject of a previous publication by the same author (49).

Because of extremely small differences in the spectra of polymers with similar structures and discrepancies and inaccuracies in reproductions of published spectra, it is highly desirable for every laboratory to maintain its own collection of infrared spectra, recorded, if possible, on the same instrument and under identical conditions.

This article can be concluded by the statement that infrared spectroscopy is a very useful tool for the solution of many problems in polymer science and technology, but that not all of its potentialities have yet been utilized.

Bibliography

1. H. W. Thompson and P. Torkington, *Trans. Faraday Soc.* **41**, 246 (1945).
2. J. R. Nielsen, *J. Polymer Sci.* [C] **7**, 19 (1964).
3. R. Zbinden, *Infrared Spectroscopy of High Polymers*, Academic Press, Inc., New York, 1964.

3a. N. B. Colthup, *J. Opt. Soc. Am.* **40**, 397 (1950).

4. G. R. Wilkinson, "Low Frequency Infrared Spectroscopy," in M. Davies, ed., *Infrared Spectroscopy and Molecular Structure*, Elsevier Publishing Co., Amsterdam, 1963, Chap. III.
5. *Synthetic Optical Crystals*, The Harshaw Chemical Co., Cleveland.
6. A. Elliot, *J. Polymer Sci.* [C] **7**, 37 (1964).
7. E. R. Lippincott, F. E. Welsh, and C. E. Weir, *Anal. Chem.* **33**, 137 (1961).
8. L. J. Bellamy, *The Infra-Red Spectra of Complex Molecules*, Methuen & Co., Ltd., London, 1958.
9. K. Nakanishi, *Infrared Absorption Spectroscopy Practical*, Holden-Day, Inc., San Francisco, and Nankodo Comp. Ltd., Tokyo, 1964.
10. H. A. Szymanski, *Theory and Practice of Infrared Spectroscopy*, Plenum Press, New York, 1963.
11. M. St. C. Flett, *Characteristic Frequencies of Chemical Groups in the Infrared*, Elsevier Publishing Co., Amsterdam, 1963.
12. G. Zerbi, F. Ciampelli, and V. Zamboni, *J. Polymer Sci.* [C] **7**, 141 (1964).

13. P. W. Higgs, *Proc. Roy. Soc.* (*London*) *Ser. A* **220,** 472 (1953).
14. S. Krimm, "Infrared Spectra of High Polymers," *Fortschr. Hochpolymer. Forsch.* **2,** 51 (1960).
15. J. L. Binder, *J. Polymer Sci.* [B] **4,** 19 (1966).
16. G. H. Beaven, E. A. Johnson, H. A. Willis, and R. C. J. Miller, *Molecular Spectroscopy, Methods and Applications in Chemistry,* Heywood, London, 1961.
17. I. Kössler, *Methoden der Infrarotspektroskopie in der chemischen Analyse. Quantitative Analyse,* Akademische Verlagsgesellschaft Geest u. Portig K.-G., Leipzig, 1966.
18. R. P. Bauman, *Absorption Spectroscopy,* John Wiley & Sons, Inc., New York, 1962.
19. W. Kimmer and E. O. Schmalz, *Kautschuk Gummi, Kunststoffe,* **16,** 606 (1963).
20. I. Kössler and J. Cizek, *Z. Anal. Chem.* **220** (4), 272 (1966).
21. I. Kössler and J. Vodehnal, *J. Polymer Sci.* [B] **1,** 415 (1963).
22. J. N. Lomonte, *Anal. Chem.* **34,** 129 (1962).
23. A. S. Wexler, *Spectrochim. Acta* **21,** 1725 (1965).
24. H. J. Dannenberg, W. Forbes, and A. C. Jones, *Anal. Chem.* **32,** 365 (1960).
25. R. J. Harris and G. R. Svoboda, *Anal. Chem.* **34,** 1955 (1962).
26. J. P. Sibilia and A. R. Paterson, *J. Polymer Sci* [C] **8,** 41 (1965).
27. J. J. Fox and A. E. Martin, *Proc. Roy. Soc.* (*London*) *Ser. A* **175,** 208 (1940).
28. W. M. D. Bryant and R. C. Voter, *J. Am. Chem. Soc.* **75,** 6113 (1953).
29. K. Iimura, R. Endo, and M. Takeda, *Bull. Chem. Soc. Japan* **37,** 874 (1964).
30. A. J. Durbetaki and C. M. Miles, *Anal. Chem.* **37,** 1231 (1965).

30a. O. Kratky, *Kolloid-Z.* **64,** 213 (1933).

31. H. W. Starkweather, R. E. Moynihan, *J. Polymer Sci.* **22,** 363 (1956).
32. S. Krimm, *J. Polymer Sci.* [C] **7,** 3 (1964).
33. R. A. Nyquist, *Infrared Spectra of Polymers and Resins,* 2nd ed., The Dow Chemical Co., 1961.
34. D. Hummel, *Kunststoff-, Lack- und Gummi-Analyse,* Carl Hanser Verlag, München, 1958.
35. D. O. Hummel, *Infrared Spectra of Polymers in the Medium and Long Wavelength Regions,* Interscience Publishers, a division of John Wiley & Sons, Inc., New York, 1966.
36. S. Sadtler Collection, S. Sadtler, Philadelphia.
37. API Collection (American Petroleum Institute, Research Project 44), Carnegie Institute of Technology.
38. NRC Collection, U.S. National Research Council, National Bureau of Standards, Washington.
39. DMS Collection, Infrared Absorption Data Joint Committee, London, Institut für Spektrochemie und angewandte Spektroskopie, Dortmund.
40. IRDC Collection, Infrared Data Committee of Japan, Nankodo Co., Huraki-cho, Tokyo.
41. G. Herzberg, *Molecular Spectra and Molecular Structure II. Infrared and Raman Spectra of Polyatomic Molecules,* Van Nostrand, New York, 1949.
42. G. Natta and G. Zerbi, *Conference on the Vibrational Spectra of High Polymers, Milano, 1963; J. Polymer Sci.* [C] **7,** 1964.
43. A. Elliot, *Advan. Spectry.* **1,** 214 (1959).
44. S. Krimm, "Infrared Spectra of Solids: Dichroism and Polymers," in M. Davies, ed., *Infrared Spectrscopy and Molecular Structure,* Elsevier Publishing Co., Amsterdam, 1963, Chap. VIII, p. 270.
45. G. Schnell, *Ergeb. Exakt. Naturw.* **31,** 270 (1959).
46. H. M. Hersherson, *Infrared Absorption Spectra. Index for 1945–1957 and Index for 1958–1962,* Academic Press, Inc., New York, 1959 and 1964.
47. R. C. Gore, *Anal. Chem.* **23,** 7 (1951); **24,** 8 (1952); **26,** 11 (1954); **28,** 577 (1956); **30,** 570 (1958); **32,** 238 R (1960).
48. J. C. Evans, *Anal. Chem.* **34,** 225 R (1962); **36,** 240 R (1964).
49. D. O. Hummel, *The Infrared Spectra of Polymers, Resins and Related Materials in the 700-250 cm^{-1} Region and their Analytical Application,* ASD-TDR 63-712.
50. C. N. R. Rao, *Chemical Application of Infrared Spectroscopy,* Academic Press, Inc., New York, 1964.
51. N. B. Colthup, L. H. Daly, and S. E. Wiberly, *Introduction to Infrared and Raman Spectroscopy,* Academic Press, Inc., New York, 1964.
52. W. J. Potts, Jr., *Chemical Infrared Spectroscopy,* Vol. I, *Techniques,* John Wiley & Sons, Inc. New York, 1963.

53. M. Davies, ed., *Infrared Spectroscopy and Molecular Structure,* Elsevier Publishing Co., Amsterdam, 1963.
54. N. Tryton and E. Horowitz, "Infrared Spectrophotometry," in G. M. Kline, ed., *Analytical Chemistry of Polymers,* Part II, *Analysis of Molecular Structure and Chemical Groups,* Interscience Publishers, a division of John Wiley & Sons, Inc., New York, 1962.
55. S. L. Koenig, "Application of Infrared Spectroscopy to Polymers," in D. N. Kendall, ed., *Applied Infrared Spectoscopy,* Reinhold Publishing Corp. New York, 1966.

Ivo Kössler
Ústav Fysikalní Chemie
Československá Akademie Věd

NUCLEAR MAGNETIC RESONANCE

Nuclear magnetic resonance (NMR) spectroscopy has proved to be a method of considerable interest and importance for the study of polymers. Early studies dealt with solid polymers and the spectra obtained were of the so-called "broad-band" or "wide-line" type. In such polymer spectra, as in the corresponding "wide-line" spectra of nonpolymeric solid substances, the width and detailed shape of the resonance line, particularly if known as a function of temperature, can be interpreted to yield valuable information concerning the nature and frequency of molecular motions.

However, when dissolved in suitable solvents, polymers give so-called "high-resolution" NMR spectra, in which the resonance lines are narrow, being only a small fraction (10^{-3} to 10^{-4}) of the width produced by the solid state. More recently, a series of investigations has proved that these high-resolution NMR spectra of polymers are of great value in solving certain difficult problems of chain structure.

The NMR Phenomenon

When the spin quantum number of a nucleus is ½ or greater, the nucleus possesses a magnetic moment. The proton is an example of such a nucleus. It has a spin of ½ and when placed in a magnetic field H_0 it can occupy either of two energy levels, corresponding to alignment of its magnetic moment μ with or against the field. The two orientations differ in energy by

$$\Delta E = h\nu_0 = 2\mu H_0 \tag{1}$$

where ν_0 is the frequency of the resonant radio-frequency field which causes transitions between these energy levels, and is also the rate at which the nuclear magnetic moments precess about the field direction, called the Larmor frequency. The quantity of energy ΔE must be absorbed to raise the nuclei in the lower state up to the higher level, and is emitted in the reverse process. As equation 1 indicates, ν_0 is proportional to H_0 and to μ. For a field of 14,100 G, $\nu_0 = 60$ Mc/sec for protons. In a 23,400-G field, $\nu_0 = 100$ Mc/sec. These are the fields most commonly employed at present. Emerging from the development stage are instruments employing superconducting magnets and operating at about 52 kG, corresponding to $\nu_0 = 220$ Mc/sec.

Consider a collection of protons, for example, water or benzene, in a tube (A in Fig. 1) to which a field of 60 Mc/sec is applied by means of a coil (B) wrapped about the sample and excited by an oscillator. A magnetic field is applied in a direction perpendicular to the axis of the coil and, by means of the sweep coils (C), is slowly increased ("swept") until resonance occurs. At this point, energy is absorbed from the radio-frequency field, and this absorption may be detected and recorded. Alternatively, the nuclear moments as they are turned over at resonance induce a voltage in a second coil (D) placed so that its axis is at right angles to both the field and the exciting coil; this voltage is amplified and recorded.

In the 14–23 kG field now commonly used, separation of energy levels is only 0.02 small cal. This is very small compared to kT, and therefore causes no perturba-

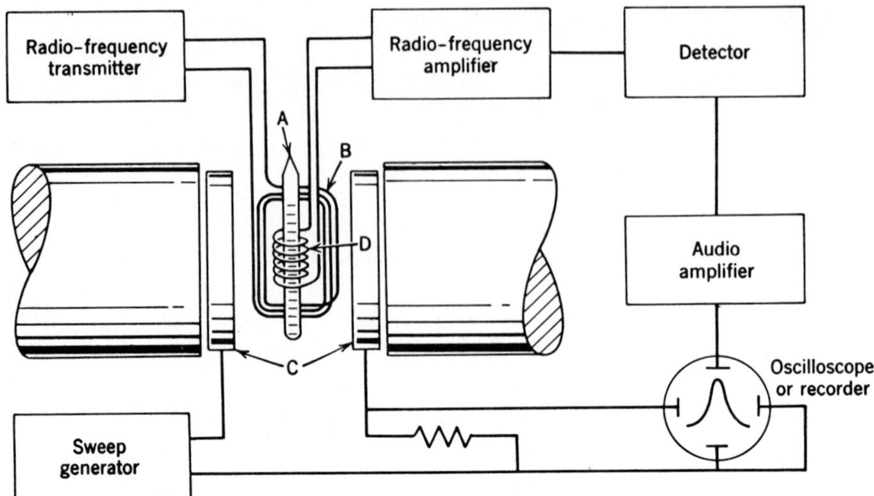

Fig. 1. Schematic diagram of a cross-coil NMR spectrometer; A, sample tube; B, radio-frequency input coil; C, magnet sweep coils; D, pickup coil (1a). Courtesy *Chemical and Engineering News*.

tion of the thermal energies of the molecules at ordinary temperatures. Since the population of energy levels usually follows a Boltzmann distribution, the lower level will exceed the upper in population by only a few parts per million. In other words, the net degree of polarization of the nuclear spins in a strong magnetic field is very small at ordinary temperatures, even for protons. For other nuclei, it is still smaller.

In time, the radio-frequency field, designated H_1, would cause the populations of the energy levels to become equalized. This is expressed by saying that the spin system would become saturated, corresponding to a very high Boltzmann spin temperature, unless there were some means by which the upper-level spins could relax to the lower level. Such relaxation may be thought of as a cooling down of the spin system. It is brought about by interaction of the spins with the fluctuating magnetic fields produced by the random motions of neighboring nuclei. This process is called spin–lattice relaxation, the neighboring spins being termed the "lattice" for both solid and liquid systems. The characteristic time for this process is called the *spin–lattice relaxation time*, T_1. Its dependence on molecular structure and motion will be considered in more detail later.

From what has been said so far, it might be supposed that at any particular radio frequency ν_0, all nuclei of a given species, for example, all protons, would resonate at the same value of H_0. That this is not true is what makes NMR important to the chemist. Actually, resonance occurs at different values of H_0 for each type of proton, depending upon its chemical binding and position in the molecule. The total range of variation of H_0 for protons in organic compounds is on the order of 10 ppm, corresponding to 600 cps in a 14,100-G field. The cause of this variation in resonant field strength is the cloud of electrons about each of the nuclei. When a molecule is placed in a magnetic field, a circulatory motion is induced in these electrons. This motion gives rise to a small local field which opposes H_0, even though all the electrons are paired so that they have no net magnetic moment per se. Such behavior is common to all molecules and gives rise to the universally observed diamagnetic properties of matter. Each nucleus is thus partially shielded from H_0 by the electrons, and a slightly higher value of H_0 is actually needed to achieve resonance than would be needed for a bare proton. Protons attached to or near electronegative groups such as OR, OH, OCOR, CO_2R, and halogens experience a lower density of shielding electrons and resonate at lower values of H_0. Protons removed from such groups resonate at higher H_0 values.

It is expected from fundamental considerations and observed experimentally that the relative shielding of two different nuclei, and thus their displacement along the H_0 axis of the spectrum, will be proportional to H_0. There is thus no natural zero of reference in NMR spectroscopy, and no fundamental scale unit. It has therefore become customary to use: (a) parts per million relative change in H_0 as the scale unit; (b) an arbitrary reference substance dissolved in the sample (an "internal" reference) and refer all displacements in resonance, called *chemical shifts,* to this standard.

The use of a dimensionless scale unit has the great advantage that the chemical shifts so expressed are independent of the value of H_0 of any particular spectrometer, and therefore a statement of chemical shift does not have to be accompanied by a statement of the frequency employed, as is the case when gauss or cycles per second (cps) are used as the scale unit. As a standard reference substance, tetramethylsilane (a volatile liquid, suggested for this purpose by Tiers in 1958) is widely accepted, and is the basis of the *τ scale.* On this scale, tetramethylsilane is assigned a value of 10.000τ, rather than zero, since its protons are more shielded than those of nearly all other organic compounds.

Another important parameter in determining the appearance of the NMR spectrum is the phenomenon of *nuclear coupling.* The direct coupling of nuclear magnetic dipoles through space is discussed in some detail in the section on Line Broadening by Local Fields. In addition to this interaction, which is highly dependent on molecular motion, magnetic nuclei may also transmit information to each other concerning their spin states through the intervening chemical bonds. This interaction occurs by slight polarizations of the spins and orbital motions of the valence electrons, and is unaffected by the tumbling of the molecules. If a nucleus has n sufficiently close, equivalently coupled neighbors, its resonance will be split into $n + 1$ peaks, corresponding to the $n + 1$ possible spin states of the neighboring group of spins. Intensities of the peaks are given by simple statistical considerations and are therefore proportional to the coefficients of the binomial expansion. Thus, one neighboring spin splits the observed resonance to a doublet, two produce a 1:2:1 triplet, three a 1:3:3:1 quartet, and so on. The strength of the coupling, denoted by J, is given by the spacing of the multiplets and is expressed in cycles per second, cps (sometimes called "Hertz," abbreviated Hz; this usage will not be employed in this article). The coupling of protons on adjacent saturated carbon atoms in open-chain compounds is commonly on the order of 5–8 cps. Couplings through more than three bonds are usually quite weak, and cannot be observed in polymer spectra.

It is a particularly noteworthy fact that the magnitude of the spin–spin coupling, unlike the chemical shift, is a fundamental property of the molecule and does *not* depend upon the strength of the magnetic field H_0. This has important consequences. Let us consider the simplest case of a two-spin system. If the chemical shift difference between these nuclei, $\Delta\nu$, expressed in cps, is large compared to the coupling J between them, the spectrum will be a pair of well separated doublets, the members of which will be equal (or nearly so) in intensity. The centers of these doublets will correspond to the true chemical shifts of the nuclei. Such a spectrum is designated an AX, or "first-order," spectrum, spacing in the alphabet being customarily employed to indicate roughly the magnitude of $\Delta\nu$ with respect to J. When the ratio $J/\Delta\nu$ exceeds about 0.1–0.2, the peak intensities in the two-spin system become seriously perturbed, the outermost two peaks being weaker than the inner peaks. The doublet centers do not correspond exactly to the chemical shifts, although the spacing in each still gives J. Such a pair of spins is called an AB system, and gives the simplest possible "strong-coupled" or "second-order" spectrum. Geminal protons commonly give AB spectra (in the absence of couplings to other nuclei), because such protons are relatively strongly coupled to each other (about -12 to -15 cps; J may be positive or negative, but this has no effect on the appearance of two-spin spectra).

Systems of several spins will give spectra which can be solved by "first-order" spectral spacing measurements, and examples of these can be seen among polymer spectra. Examples of "strong-coupled" polymer spectra will also be encountered; such spectra are best solved by computer.

There are several ways by which complex, strong-coupled spectra may be simplified. The most straightforward in principle is to decrease $J/\Delta\nu$ by making H_0 very large. As has been seen (p. 356), the largest presently available fields are about 52 kG, corresponding to 220 Mc/sec. As the magnetic field is increased, an increasing fraction of spectra become "first order" and can be solved by inspection. Complex spectra are strikingly simplified and in addition small chemical-shift differences become evident which cannot be discriminated in weaker fields.

Other methods of spectral simplification are *deuterium substitution* and *double resonance;* these will be described in the course of the later discussion.

Line Broadening by Local Fields: Wide-Line Spectra

For protons sufficiently removed from each other that they do not feel the effects of each others' magnetic fields, the local magnetic field at the nuclei will be essentially equal to H_0 (actually slightly smaller because of the screening discussed in the previous section). Therefore, if H_0 can be made very homogeneous over the sample, the width of the absorption peak may be as small as 10^{-4} G, ie, on the order of 1 part in 10^8 or 0.3–0.5 cps. In most substances, the protons are sufficiently close to each other for each to be appreciably influenced by the magnetic fields of its neighbors. Let us first imagine that we are dealing with a system composed of isolated pairs of protons. The field felt by each proton will be made up of H_0 plus the small additional field of the immediate neighboring proton H_{loc}, the other protons being too distant to be felt. The sign and magnitude of H_{loc} will depend upon the distance between the nuclei, r, and upon θ, the angle between the line joining the nuclei and the direction of H_0. This dependence is expressed by equation 2. The $\pm$ sign results from the

$$\Delta E = 2\mu(H_0 \pm H_{loc}) = 2\mu[H_0 \pm \tfrac{3}{2}\,\mu r^{-3}(3\cos^2\theta - 1)] \qquad (2)$$

fact that the local field may add to or subtract from H_0, depending on whether the neighboring magnetic dipole is aligned with or against the direction of H_0. The degree of polarization of the spins being very small, as has been seen, it follows that the chances of a given proton being aligned with or against the field are almost exactly equal. The spectrum of such protons thus consists of two lines (Fig. 2**a**) whose separation at a fixed value of θ will vary inversely as r^3. Only when $\cos^2 \theta = 1/3$ ($\theta = 54.7°$) will these lines coincide to produce a single line.

In many simple solids, for example, certain hydrated salts, there actually are pairs of protons in definite orientations and one may expect to find a twofold NMR resonance. This is indeed observed; the lines are not narrow and isolated, however, but appear instead as a broadened and partially blended pair of lines (Fig. 2**b**), the separation of which depends upon the orientation of the crystal in the magnetic field. Here the nearest-neighbor interactions dominate but the magnetic interactions between pairs are not negligible. These many interactions, varying in both r and θ, give rise to a multiplicity of lines which cannot be resolved but whose envelope is

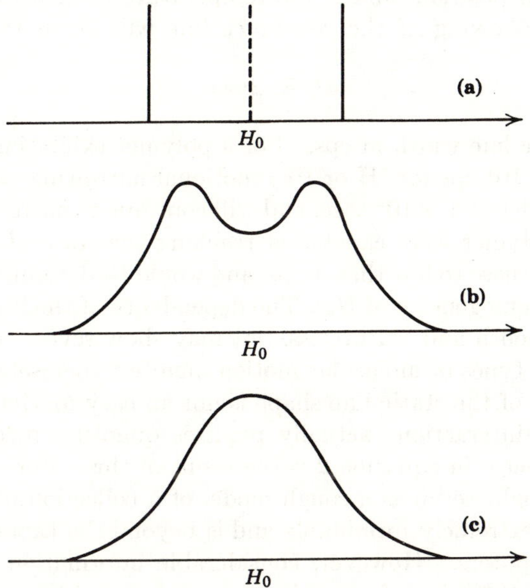

Fig. 2. Schematic NMR spectra of arrays of protons: (**a**) isolated pairs of protons; (**b**) semi-isolated pairs of protons; (**c**) a random or nearly random array of protons.

seen as a continuous curve. In most solids, including solid polymers, pairing of protons does not occur to this degree and instead a single broad peak, as in Figure 2**c**, is found. In most rigid-lattice solids, the distribution of local field strengths experienced by each proton is such that the half-height width of the resonance is on the order of 10 G.

It has so far been assumed that the nuclei are at rest. When *molecular motion*, which in polymers will be chiefly group rotation, takes place, the variables in equation 2 become functions of time. If we assume for the present discussion that r is constant and only θ varies with time (as for protons of methylene or methyl groups or benzene rings) then the time-averaged local field will be given by equation 3, where the averag-

$$H_{\text{loc}} = \mu r^{-3} T_2^{-1} \int_0^{T_2} (3 \cos^2 \theta - 1) dT \tag{3}$$

ing is carried out over the time, T_2, that the nuclei reside in a given spin state. If θ varies rapidly over all values, ie, if the local magnetic field variations are very short-lived, the time average can be replaced by a space average (eq. 4).

$$H_{\text{loc}} = \mu r^{-3} \int_0^{\pi} (3 \cos^2 \theta - 1) \sin \theta \, d\theta = 0 \tag{4}$$

The time t_c is called the *correlation time;* it will be taken as the average time required for the line joining two neighboring nuclei to rotate through an angle of one radian. Equation 4 shows that if t_c is so short that a space averaging is valid, then the net magnetic effect of the neighboring nuclei upon a given nucleus is effectively erased. If the original sample is a solid at a temperature so low that the protons are virtually fixed in position, and if the temperature is then raised until molecular motion begins, a narrowing of the resonance line will begin to be observed when

$$1/t_c \geqq 2\pi\delta\nu \tag{5}$$

where $\delta\nu$ is the static line width in cps. For a polymer exhibiting a static line width of 10 G (about 4×10^4 cps for ^1H or ^{19}F) motional narrowing can be expected when t_c becomes on the order of 4×10^{-6} sec, and will continue to narrow as the temperature is raised. If the polymer were capable of reaching the state of a mobile liquid, the line width would decrease to less than 1 cps, and would be determined not by molecular motion but by the homogeneity of H_0. The dependence of the line width on temperature need not be smooth and featureless, but may show several plateaus and abrupt decreases as various types of molecular motion manifest themselves.

The calculation of the static line shape is not an easy matter. The proper treatment of the dipole interactions actually requires quantum mechanical calculation, and in fact the $3/2$ factor in equation 2 is the result of this. For even a simple array of nuclei, such as might serve as a rough model of a collection of polymer chains, an exact calculation is extremely formidable and is beyond the capacity of the largest of modern digital computers. However, considerable information can be gained from the *second moment*, ΔH_2^2, which for a polymer containing only one species of magnetic nucleus (eg, ^1H or ^{19}F) is given by Van Vleck's (1) quantum mechanical treatment as

$$\Delta H_2^2 = 9/4 \, \mu^2 \sum_{i>j} (3 \cos^2 \theta_{ij} - 1)^2 r_{ij}^{-6} \tag{6}$$

The meaning of the terms is the same as in equation 2, except that now a typical nucleus in the postulated structure of the polymer is considered and the distances r_{ij} and angles θ_{ij} to all the other nuclei are determined. Because of the r^{-6} dependence, only comparatively close neighbors make significant contributions, and so the summation needs to be carried out over only a limited element of the lattice.

It is implied in equation 6 that the nuclei are all structurally equivalent, as in polyethylene or polytetrafluoroethylene. When this is not the case, as in polyisoprene or polystyrene (and in fact for nearly all polymers), a separate summation must be carried out for each type and these sums combined with appropriate weighting factors. The quantities r_{ij} must be evaluated not only within a given chain or structure but also between molecules. For unoriented polymers, even though crystalline, it is often a good assumption that θ_{ij} takes on all values with equal probability; in this case, $(3 \cos^2 \theta_{ij} - 1)^2$ averages to $4/5$.

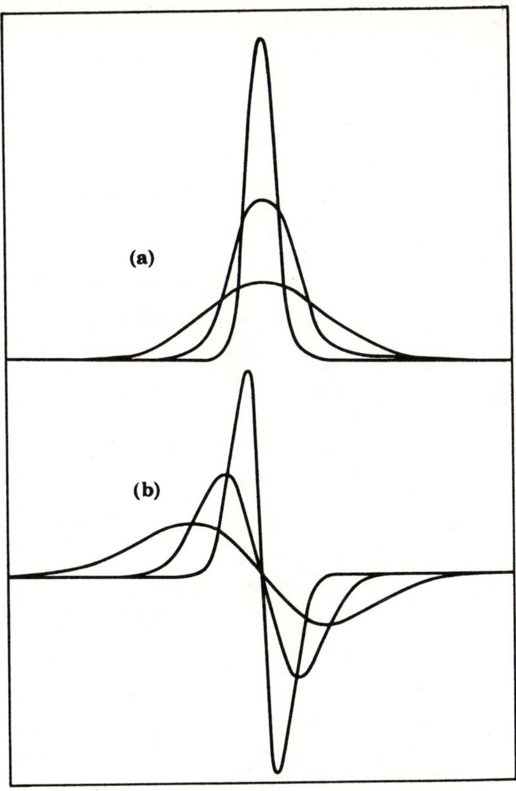

Fig. 3. NMR signals in the (**a**) absorption mode and (**b**) dispersion mode. Signals of decreasing width are shown, such as might be observed for the same polymeric system at increasing temperature.

The experimental second moment, with which the theoretically calculated value is to be compared, is obtained from equation 7. Here, $f(H)$ is the line shape, ie, the

$$\Delta H_2{}^2 = \frac{\int_{-\infty}^{\infty} (H-H_0)^2 f(H)dH}{\int_{-\infty}^{\infty} f(H)dH} \tag{7}$$

amplitude of the resonance signal as a function of the magnetic field H, and $(H - H_0)$ measures in gauss the departure of each point in the spectrum from the central resonance position H_0. The second moment can of course be measured for any line, whether it corresponds to a rigid-lattice structure or not, and will reflect both the shape and the width of the resonance. Comparison to the theoretically calculated value will be meaningful only at a temperature that is sufficiently low for a virtually motion-free structure to be assured.

For instrumental reasons, it is more desirable to display wide-line spectra not as an absorption signal, but as the derivative of the absorption signal. These line shapes are compared in Figure 3, which shows signals of decreasing width, such as might be expected for a polymer at increasing temperatures. The *width* of an NMR signal like those in Figure 3**b** is customarily reported as the distance between the

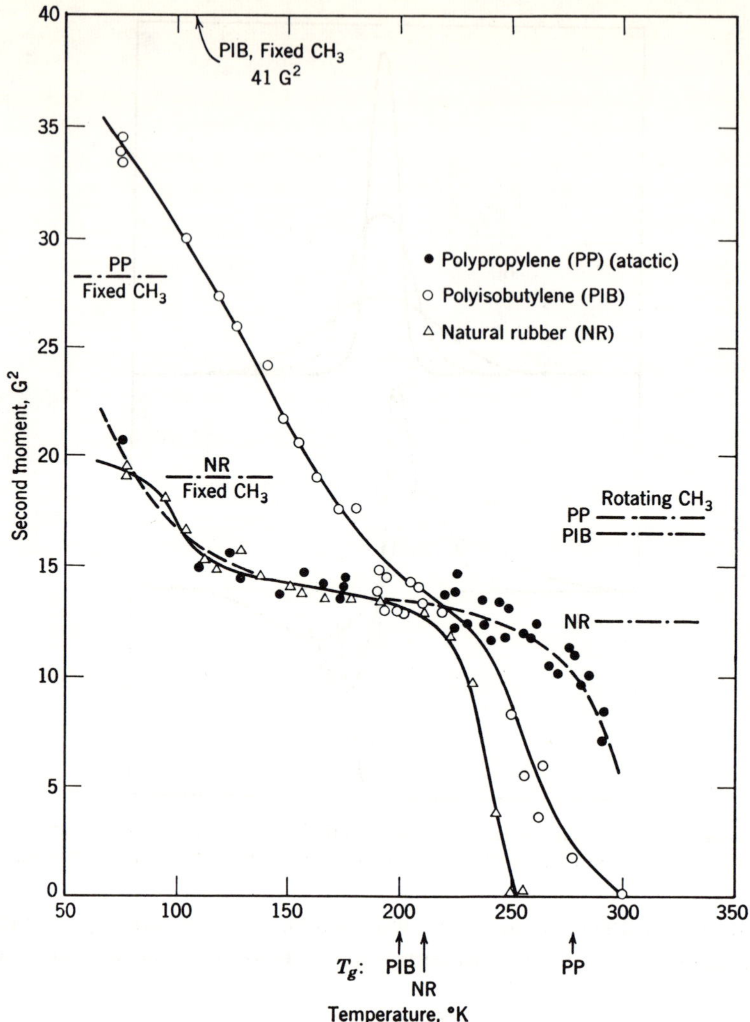

Fig. 4. The second moment of polypropylene (PP), polyisobutylene (PIB), and natural rubber (NR) as a function of temperature, °K.

two extrema. The narrow signals of high-resolution spectra usually approximate closely the form known as *Lorentzian*, which is, in general, characteristic of the resonance of damped oscillators. For such signals, the half-height width of Figure 3**a** and the distance between extrema in Figure 3**b** are equal and can be shown to be equal to $1/\pi T_2$ (in cps), where T_2 is the spin lifetime as previously defined. T_2 is also commonly called the *spin–spin relaxation time;* it depends upon the interactions of the nuclear spins, as discussed above, in such a way that T_2 is at a minimum when molecular motion is at a minimum. Wide-line signals, however, do not have a Lorentzian form and usually cannot be described by any simple analytic function, and therefore the separation of extrema of the derivative signal is only an approximate measure of T_2.

The results observed for a number of typical elastomeric and rigid polymers will be discussed in the following paragraphs.

Elastomers. Figure 4 (2) shows the second moment for three elastomers over a 225° temperature range: (a) unvulcanized natural rubber; (b) polyisobutylene, molecular weight (viscosity average) 1×10^6; (c) atactic polypropylene. The calculated values of the second moment for the motionless polymer are indicated at the left. Natural rubber attains this value at the lowest temperature; then $\Delta H_2{}^2$ falls rather abruptly near 100°K owing to the onset of methyl group rotation, which should require a very small activation energy. Following a long plateau, there is another abrupt decrease near 220°K corresponding to the onset of segmental motion. This decrease coincides rather closely with the glass temperature, indicated on the abscissa.

Polypropylene does not attain the calculated static second moment at 77°K; the methyl groups are evidently still able to rotate. The shape of the curve suggests, however, that this motion would be quenched at 50°K. There is a plateau similar to that shown by natural rubber, followed by a decrease above 250°, corresponding to the substantially higher glass temperature of this polymer. The calculated second-moment values corresponding to methyl group rotation for both polymers are also shown in Figure 4 (on the right-hand side). The value for natural rubber agrees quite well with observation, but that for polypropylene is somewhat larger.

In contrast, the second moment of polyisobutylene does not fall to the value calculated for methyl group rotation until the temperature is quite high, about 200°K. Although there is indication of an inflection around 225°K, there is no clear separation of methyl group rotation from segmental motion, as in the other polymers. It appears that methyl group rotation is dependent on main-chain motion, possibly because the methyl groups are interlocked and do not become sufficiently free to rotate until segmental motion begins (3).

Rigid Polymers. Polymers which are rigid at ordinary temperatures show behavior generally similar to that of elastomers, but the curves are displaced to higher temperatures. At temperatures below about 150°K, the spectra appear to consist of a narrow component superimposed on a broad component. These probably correspond to amorphous and crystalline portions, respectively, at least for the highly stereoregular materials (p. 372). Polymers of methyl methacrylate (4) show a decrease in the width of the broad component at about 100°K, regardless of their predominant stereochemical configuration (Fig. 5). This probably corresponds to the onset of α-methyl group rotation, the ester methyl groups being capable of rotation well below 100°K. (In the narrow-line portion of the spectra, presumably both types of methyl group are capable of rotation.) In contrast, the onset of segmental rotation and chain translation, signaled by the line-width decrease at high temperature, is highly sensitive to stereochemical configuration. The most highly isotactic polymer (A in Fig. 5c) begins to show line narrowing at about 360°; polymer B, which is less highly isotactic, shows a displacement to slightly higher temperatures. The predominantly syndiotactic polymer, C, which corresponds closely to commercial material in tacticity, exhibits a displacement of about 40° to higher temperature, and the highly syndiotactic material, D, a still greater displacement. This behavior has not been fully explained in structural terms, but correlates well with the measured "glass-transition temperatures" of these materials.

The behavior of a highly crystalline, rigid polymer, polytetrafluoroethylene (Du Pont Teflon), is shown in Figure 6 (5). The calculated static value of the second

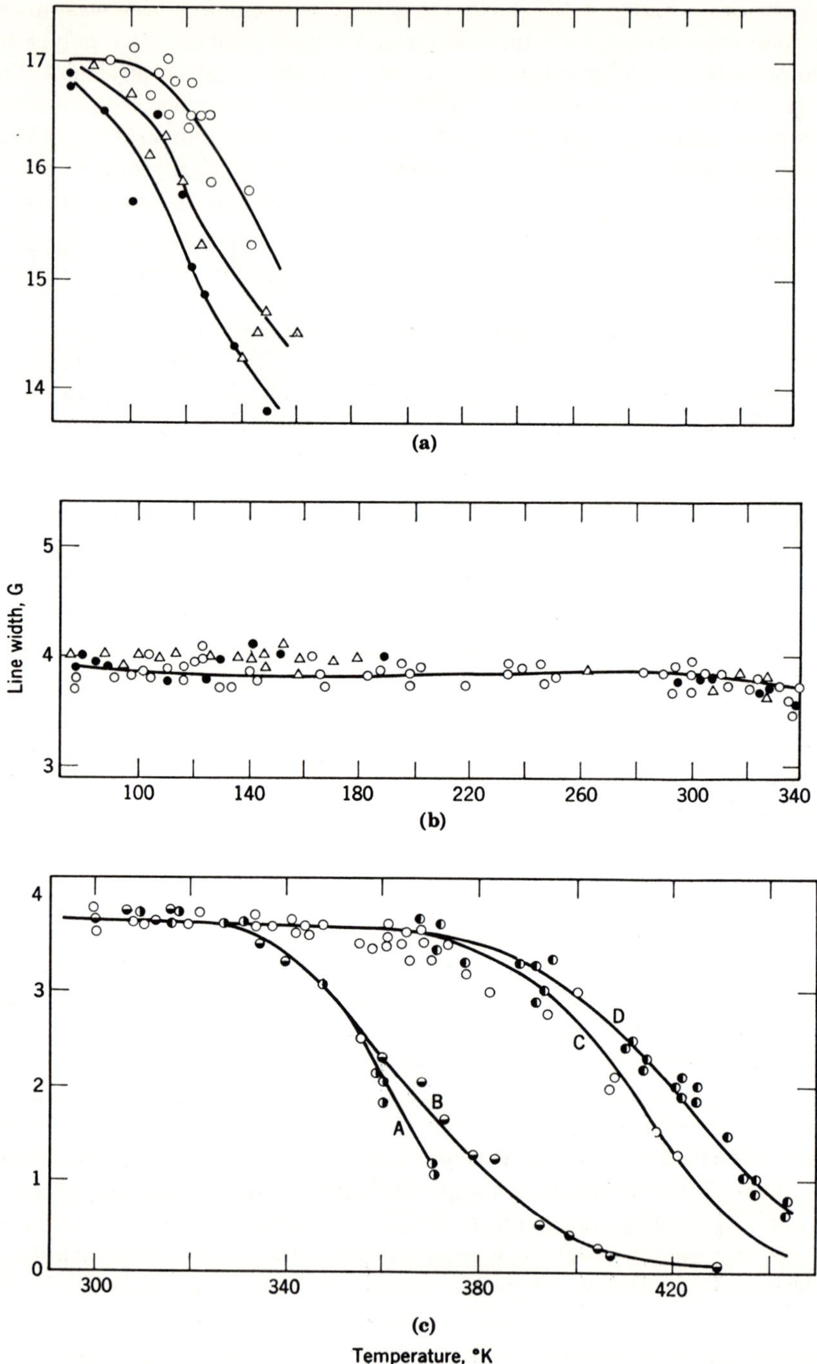

Fig. 5. Line width of poly(methyl methacrylate) NMR spectra. Spectra (**a**): broad component of low-temperature spectrum, observable to 150°K: ○, 60% syndiotactic; ●, >90% isotactic; △, 78% syndiotactic. Spectrum (**b**): narrow component, 77–340°K; identification of samples as in (**a**). Spectra (**c**): line width from 300–450°K: A, >90% isotactic (◑); B, 63% isotactic (◒); C, commercial sample (about 60% syndiotactic) (○); D, 78% syndiotactic (◐).

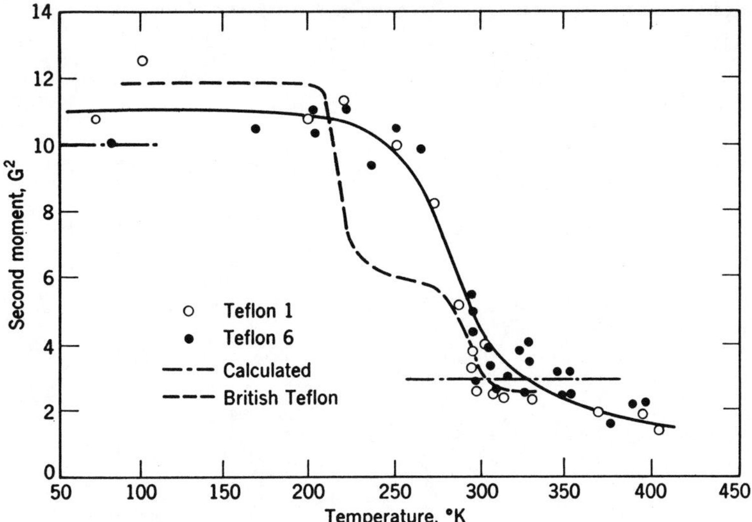

Fig. 6. The second moment of polytetrafluoroethylene (Teflon) as a function of temperature, °K.

moment, 10.0 G², is composed of an intramolecular contribution, calculated from the helical structure of the molecules (established by x-ray diffraction), of 7.5 G² and, in addition, an estimate of dipole–dipole interaction of 2.5 G between chains. It was assumed that in the powdered sample there was no preferred orientation of the crystallites. The observed low-temperature value of 10.9 G² (Fig. 6) is in adequate agreement. This figure shows the temperature dependence of $\Delta H_2{}^2$ for two varieties of Du Pont Teflon which differ in molecular weight; their behavior appears to be the same within experimental error, both exhibiting a marked decrease in $\Delta H_2{}^2$ in the interval of 240–310°K. In this interval, marked changes in crystal structure, specific volume, and specific heat are well established. There is little doubt that it corresponds to the onset of chain rotation. Also shown in Figure 6 is the behavior of a sample of British Teflon studied by Smith (6). The markedly earlier onset of rotation in this polymer is probably caused by a plasticizing impurity or low-molecular-weight fraction.

A more revealing view of the molecular motion in this polymer may be obtained by observing the second moment of a stretched fiber as a function of its angle with respect to the direction of the magnetic field H_0. The fiber is composed of crystallites oriented so that the long axes of the helical polymer molecules are parallel to the fiber axis. If chain motion consists of rotation about the axis of orientation, the second moment will depend upon θ', the angle between the axis of orientation of the fiber and the direction of H_0, according to equation 8 (7). The constant k need not be specified

$$\Delta H_2{}^2 = k(3\cos^2\theta' - 1)^2 \tag{8}$$

in detail here. This expression, derived by an extension of the considerations leading to equations 2 and 6, predicts a zero value of the second moment when $\theta' = 54.7°$. In Figure 7, the experimental dependence is shown (5), together with the above function arbitrarily adjusted to fit at $\theta' = 0°$ and 90°. The results show that the principal mode of chain motion is indeed rotation about the molecular axis. The fact that $\Delta H_2{}^2$ does not narrow to zero when the fiber is oriented at 54.7° is a reflection of the existence of an amorphous fraction, and of imperfect orientation of the crystallites.

Relaxation Measurements. It was seen in the section on the NMR phenomenon that the process of *spin–lattice relaxation*, characterized by T_1, tends to maintain a Boltzmann distribution of spins in the two magnetic energy levels characteristic of a system of nuclei such as protons or ^{19}F nuclei. The spins and the "lattice" of

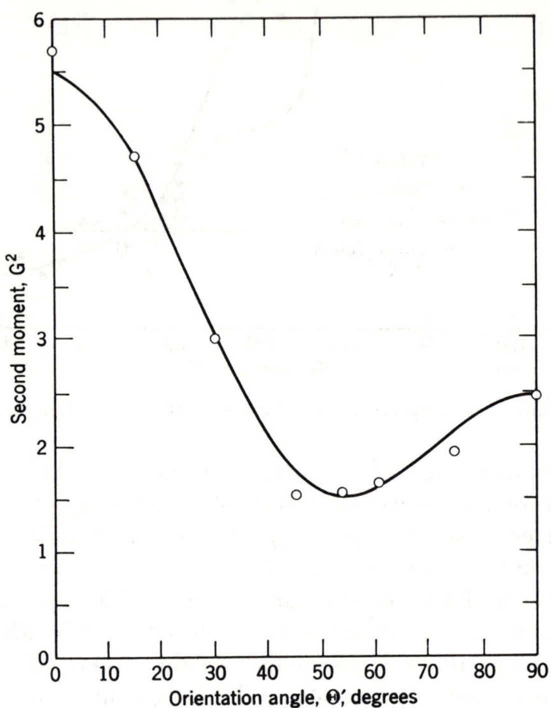

Fig. 7. The second moment of a stretched polytetrafluoroethylene fiber as a function of θ', the angle between the direction of H_0 and the axis of the fiber.

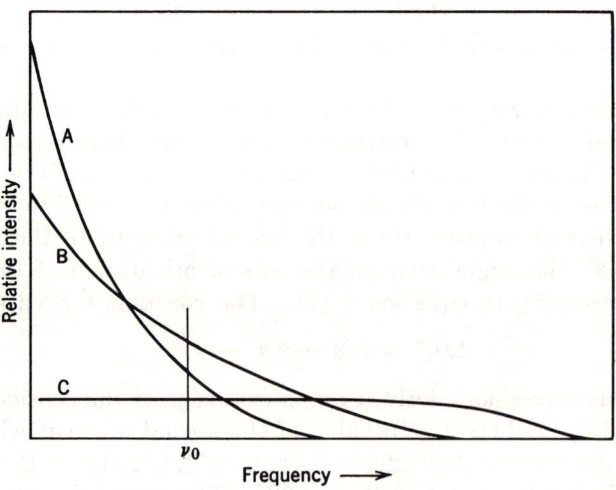

Fig. 8. Motional frequency spectrum at A, high viscosity; B, moderate viscosity; C, low viscosity; ν_0 is precession frequency.

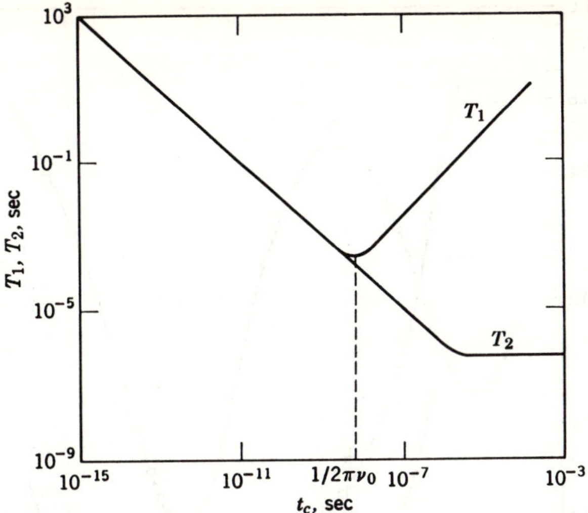

Fig. 9. Theoretical dependence of spin–lattice relaxation time, T_1, and spin–spin relaxation time T_2, on the correlation time, t_c, assuming all interactions to be characterized by the same correlation time.

neighboring spins may be considered to be essentially separate coexisting systems with a very inefficient, but nevertheless very important, link by which thermal energy may be exchanged. This link is provided by molecular motion. Each nucleus sees a number of other nearby magnetic nuclei, both in the same molecule and in other molecules. These neighboring nuclei are in motion with respect to the observed nucleus, and this motion gives rise to fluctuating magnetic fields. The observed nuclear magnetic moment will be precessing about the direction of the applied field H_0, and will also be experiencing the fluctuating fields of its neighbors. Since the motions of each molecule and of its neighbors are random or nearly random, there will be a broad range of frequencies describing them. To the degree that the fluctuating local fields have components in the direction of H_1 and at the precession frequency ν_0, for example 60 Mc/sec, they will induce transitions between energy levels because they provide fields equivalent to H_1. In solids or very viscous liquids, the molecular motions are relatively slow, and therefore the component at ν_0 will be weak. The frequency spectrum will resemble curve A in Figure 8. At the other extreme, in liquids of very low viscosity, the motional frequency spectrum may be very flat, ie, the magnetic noise may be nearly "white," and so no one component, in particular that at ν_0, can be very intense (curve C in Fig. 8). It can then be expected that at some intermediate condition, probably that of a moderately viscous liquid (curve B), the component at ν_0 will be at a maximum, and thermal relaxation of the spin system can occur with optimum efficiency. Since T_1 is usually on the order of 1–10 sec in mobile liquids, spin–lattice relaxation ordinarily does not contribute observably to line broadening. As the temperature is lowered and viscosity increases, the component of the local magnetic noise spectrum at the Larmor frequency will increase, pass through a maximum, and decrease again. Correspondingly, T_1 will decrease, pass through a minimum, and increase again. For most nonassociated molecules, T_1 is governed mainly by interactions of nuclei within the same molecule, rather than by motions of neighboring

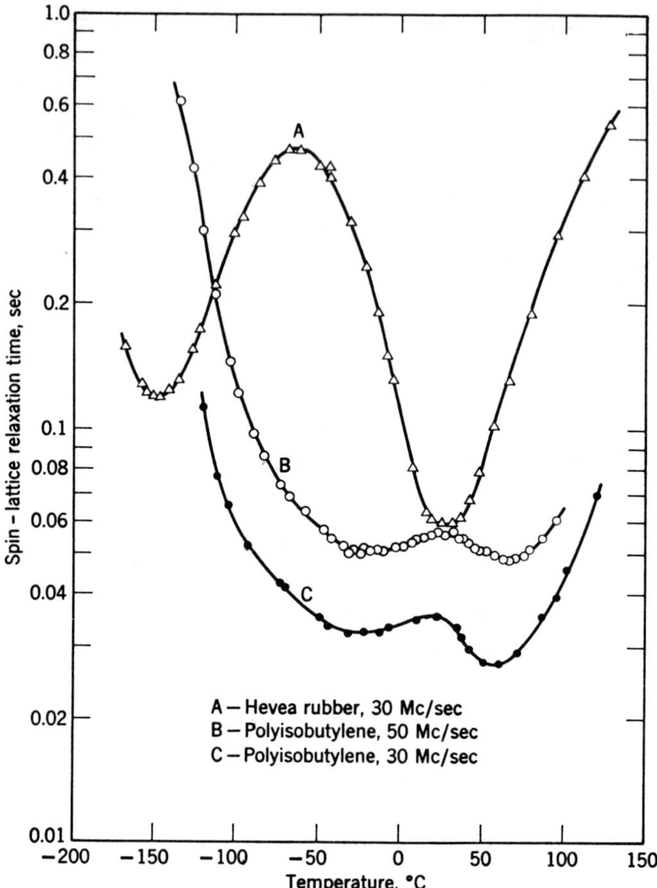

Fig. 10. The spin–lattice relaxation time, T_1, for polyisobutylene and natural rubber.

molecules. Under these conditions, it is found from the detailed theory of Bloembergen, Purcell, and Pound (8) that the rate of spin–lattice relaxation for a given nucleus in a molecule composed of j identical spin-½ nuclei will be given by equation 9,

$$\frac{1}{T_1} = \frac{6}{5}\left(\frac{\pi}{h}\right)^2 \mu^4 \sum_j r_j^{-6} \left[\frac{t_c}{1 + 4\pi\nu_0^2 t_c^2} + \frac{4t_c}{1 + 16\pi^2\nu_0^2 t_c^2}\right] \tag{9}$$

where μ is the magnetic moment, r_j are the distances from the observed nucleus to each of its magnetic neighbors, and t_c is the correlation time, as previously defined. Figure 9 is a log–log plot of T_1 vs t_c, according to equation 9. As a measure of t_c, the quantity η/T is often useful for liquids composed of at least approximately spherical molecules (see eq. 10). The predicted dependence of T_1 on η/T has been demonstrated (8).

Figure 9 also shows the theoretical dependence of T_2 on t_c. Note that for short correlation times, ie, when $t_c \ll 1/\nu_0$, $T_2 \simeq T_1$, but that after T_1 passes through its minimum, T_2 continues to decrease as molecular motion becomes slower and levels out as the system approaches a rigid lattice. This is in accord with the previous qualitative discussion of the dependence of line width on t_c.

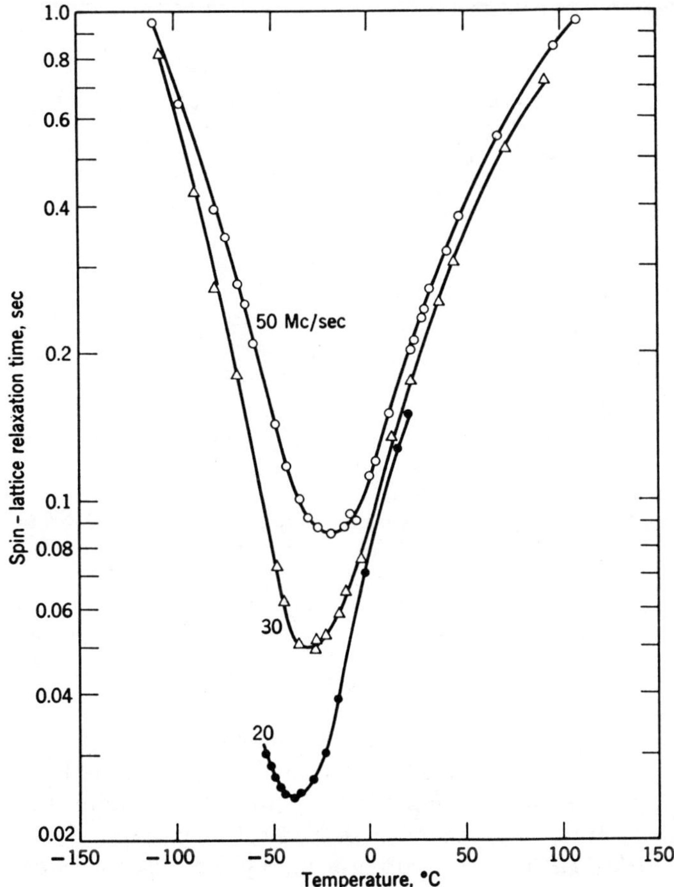

Fig. 11. The spin–lattice relaxation time, T_1, of *cis*-1,4-polybutadiene.

T_2 is readily obtained from the wide-line spectrum, but the measurement of T_1 requires a separate experiment. A direct method involves heating the nuclear spin system with a strong radio-frequency field at the resonant frequency ν_0 until saturation occurs, as indicated by the disappearance of the signal. Upon abruptly reducing the field strength to a low value, the signal will recover exponentially with a time constant, T_1. This method cannot be used for the short ($\ll$1 sec) relaxation times of solid polymers. For such measurements it is preferable to employ special apparatus by which the radio-frequency field can be applied in short (about 1 μsec) pulses. As a matter of fact, it is advantageous to measure T_2 also by a pulse method. For T_2, the method of choice is that of "spin echoes" (Hahn, Ref. 9), whereas for T_1 the so-called "null method" (Carr et al., Ref. 9) is preferred. Detailed descriptions of theory and methods can be found in the literature cited. Examples of the results obtained on polymeric systems will be given below.

Figure 10 shows the dependence of T_1 upon temperature for polyisobutylene (observed at 30 Mc/sec and at 50 Mc/sec) and for natural rubber (10). It is a consequence of equation 9 that at a given temperature, the T_1 minimum must move toward higher temperatures as the observing frequency ν_0 is increased; careful examination

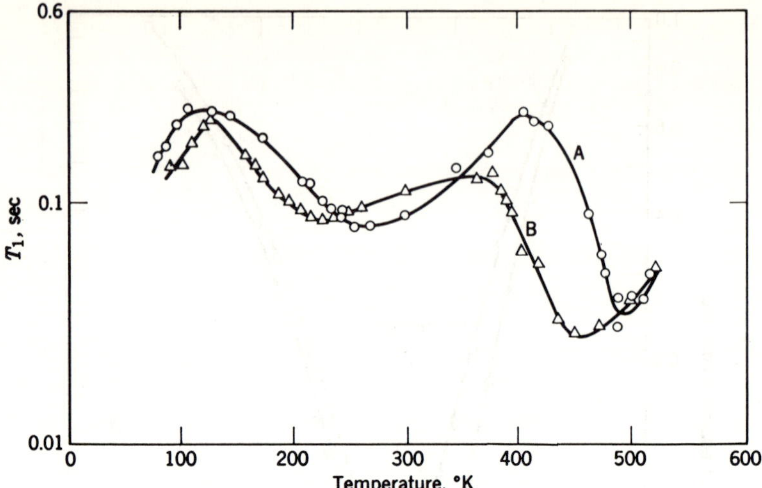

Fig. 12. The spin–lattice relaxation time, T_1, of poly(methyl methacrylate) at 21.5 Mc/sec. A, syndiotactic; B, isotactic.

of Figure 10 shows that this is the case. This plot differs in several obvious ways from the theoretical plot of Figure 9. The abscissa is °C, not η/T, for η is unknown for these systems over most of the observed temperature range. More important is the occurrence in each curve of *two* distinct minima in T_1. It is hardly to be expected that the motion of polymer molecules such as these can be characterized by a single frequency, and it is not surprising that their T_1–temperature dependence is more complex than that of a system composed of small, nearly spherical molecules. The low-temperature T_1 minimum, occurring at $-150°$ (at 30 Mc/sec) for natural rubber, is believed to correspond to the rotation of methyl groups, the motional frequency spectrum of which must have a maximum component at 30 Mc/sec at this temperature. Again at 30° the natural rubber chain attains a maximum component at 30 Mc/sec; this corresponds to extensive motion of whole chain segments, involving both rotational and translational displacements. For polyisobutylene, both minima are displaced to higher temperatures than for natural rubber. The displacement of the first minimum, by nearly 125°, is particularly striking; it correlates with the similar displacement of the drop in ΔH_2^2, discussed above (p. 364), attributed to interlocking of the methyl groups. The greater breadth of the first minimum in polyisobutylene suggests that the methyl groups experience a broad range of restraints. The T_1–temperature dependence of *cis*-polybutadiene confirms this interpretation. Lacking methyl groups, it shows no low-temperature minimum (Fig. 11). The CH_2 and CH backbone protons evidently relax at similar rates, giving only a single minimum. This minimum occurs at a much lower temperature than the corresponding minima in natural rubber and polyisobutylene. The reason for this is not clear. It correlates with the substantially lower glass temperature of *cis*-polybutadiene (about $-110°$) compared to that of rubber and polyisobutylene (both about $-70°$).

Similar measurements have been made on a number of rigid polymers. Figure 12 shows the results obtained by Powles and co-workers (11) for methyl methacrylate polymers of widely different stereochemical configuration. Both polymers exhibit two minima in T_1. The explanation is entirely parallel to that for the elastomers, and

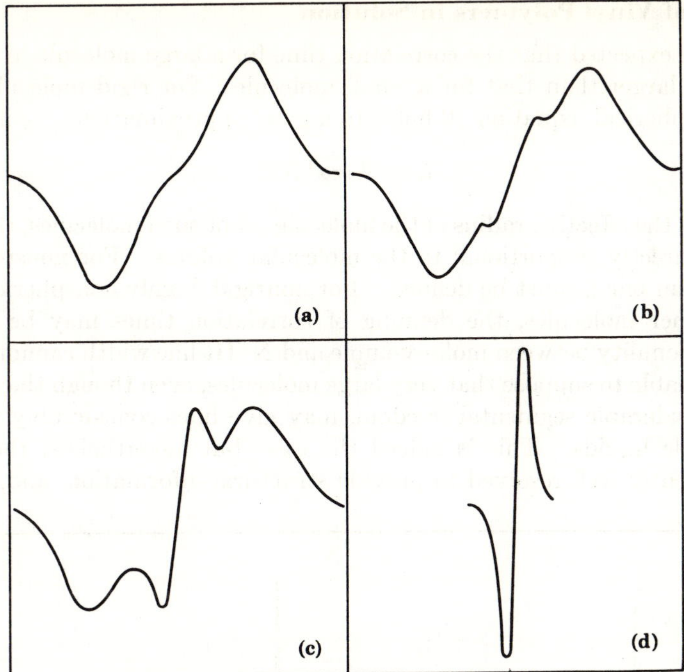

Fig. 13. Wide-line NMR polymer spectra exhibiting two components: (**a**) both components broad, corresponding to low temperature; (**b**) and (**c**) at intermediate temperature, wide and narrow components distinguishable; narrow component greater in (**c**); (**d**) both components narrow, as at elevated temperature.

correlates with the behavior of ΔH_2^2 discussed earlier. Curve A is that of a predominantly syndiotactic polymer, similar in configuration to commercial material; curve B is that of a predominantly isotactic polymer. The lower-temperature minimum, corresponding to methyl group rotation, and the high-temperature minimum, corresponding to segmental motion, are both about 50° lower for the isotactic polymer. Thus, stereochemical configuration affects both types of motion.

Measurement of Crystallinity. It has been observed (12) that certain partially crystalline polymers, such as polyethylene and polytetrafluoroethylene, give composite wide-line spectra, consisting of a broad and a narrow component, ascribed to the crystalline and amorphous portions, respectively (Figs. 13**b** and 13**c**). For this behavior to be observed, motions in the amorphous portions must occur at frequencies large compared with the frequency corresponding to the onset of motional narrowing for the structure concerned (usually about 10^4 cps) and, at the same temperature, motions in the crystalline portions must occur at frequencies smaller than this (13). At temperatures above the appropriate range, motional narrowing affects both the amorphous and crystalline phases (Fig. 13**d**); at low temperatures, the signals from both phases are so broadened that they cannot be distinguished (Fig. 13**a**). Under appropriate circumstances, this method can furnish an additional useful measure of crystallinity. Thus, by this means it is found (13) that in linear polyethylene the degree of crystallinity changes little over a broad temperature range, whereas in branched polyethylene it changes rapidly over a relatively short temperature range.

Spectra of Vinyl Polymers in Solution

It is to be expected that the correlation time for a large molecule in solution will, in general, be larger than that for a small molecule. For rigid molecules which are not far from spherical, equation 10 holds to a good approximation; η is the solution

$$t_c = 4\pi\eta a^3/kT \tag{10}$$

viscosity and a the effective radius of the molecule. For such molecules, the correlation time is thus directly proportional to the molecular volume. For nonspherical molecules, more than one t_c must be defined. For nonrigid, highly nonspherical molecules, such as polymer molecules, the defining of correlation times may be complex. A direct proportionality between molar volume and NMR line width cannot be expected, but it is reasonable to suppose that very large molecules, even though they are nonrigid and have considerable segmental freedom, may give lines considerably broader than those of mobile liquids. This is indeed the case, but nevertheless, the spectra are usually sufficiently well resolved to provide structural information, and, as expected,

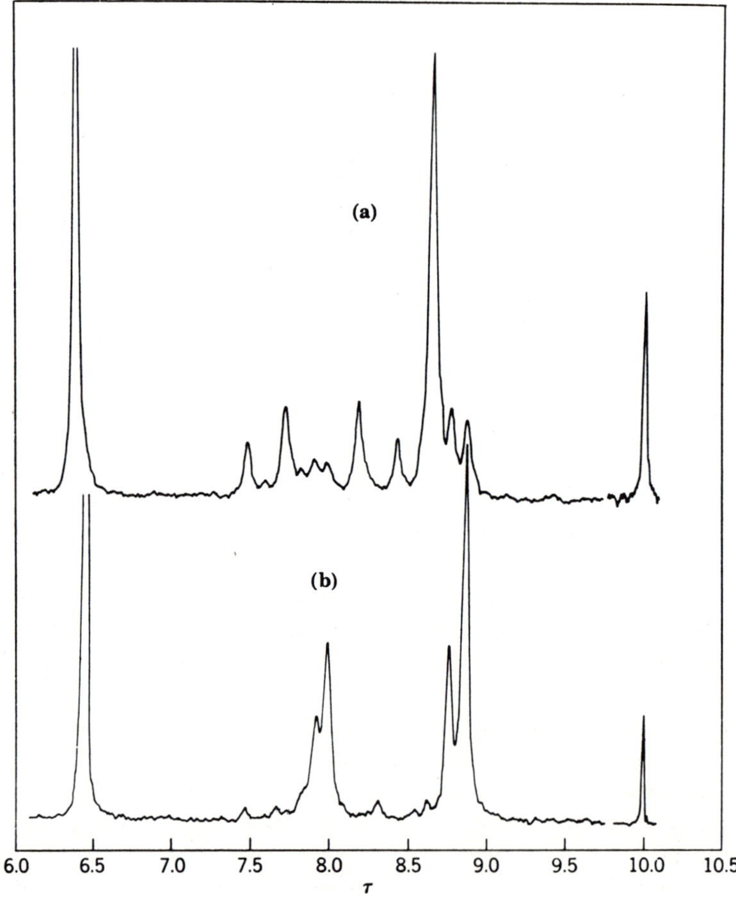

Fig. 14. High-resolution NMR spectra (60 Mc/sec, 15% in chlorobenzene, observed at 150°C) of polymers of methyl methacrylate prepared with (**a**) phenylmagnesium bromide initiator and (**b**) a free-radical initiator (benzoyl peroxide) (13a). Courtesy *Chemical and Engineering News.*

the resolution is improved at higher temperatures. It is further found that the line width is only weakly dependent on the molecular weight of the polymer and on the solution viscosity, indicating that "local" viscosity, ie, segmental motion, is the determining parameter.

Poly(methyl Methacrylate). Figure **14a** shows the spectrum of a 15% solution of poly(methyl methacrylate) in chlorobenzene at 150°. The peaks can be readily identified by reference to related small molecules. The peak at 6.42τ is the ester methyl group. The α-methyl resonance appears at 8.7–8.8τ and the backbone methylene resonance in the 7.3–8.4τ region. (The tetramethylsilane reference peak appears at 10.0τ.) This polymer was prepared using phenylmagnesium bromide, a free-radical initiator. The spectrum of a polymer prepared with benzoyl peroxide, an anionic initiator, is shown in Figure **14b**. There is some similarity in peak positions, but the two spectra are very different. In both, there are three α-methyl peaks, at 8.67τ, 8.79τ, and 8.90τ, whose relative heights vary greatly with the method of polymer preparation. The polymer prepared with the Grignard reagent shows a very prominent peak at 8.67τ, the others being much smaller. The polymer prepared with benzoyl peroxide initiator shows the same three peaks, but now the peak at 8.90τ is the most prominent. There are also marked differences in the methylene region.

The interpretation of these spectral features furnishes a powerful means for the determination of the stereochemical configuration of these polymers (see also ACRYLIC ESTER POLYMERS). Poly(methyl methacrylate) prepared with Grignard initiators in hydrocarbon solvents is believed to be predominantly isotactic (14). The peak at 8.67τ must therefore be due to the α-methyl groups of monomer units which are flanked on both sides by units of the same configuration. This is termed an isotactic triad, the central unit being termed an i unit (see also MICROTACTICITY). The most prominent peak (at 8.90τ) in the polymer prepared with free-radical catalyst is attributed to α-methyl groups of central monomer units in syndiotactic configurations (s units), ie, the central monomer unit sees monomer units of opposite configuration on each side; in polymers prepared with free-radical catalysts, at least in those prepared at low temperatures, this structure tends to predominate. The peak at 8.79τ is due to α-methyl groups of central monomer units in heterotactic configuration

Table 1. Structure of Polymers of Methyl Methacrylate Prepared with Free-Radical Initiators

Polymer no.	Polymerization conditions	$P \times 100$			
		i	h	s	σ
1	gamma irradiation in bulk, 0°C	7.5	30.0	62.5	0.21
2	benzoyl peroxide in bulk, 100°C	8.9	37.5	53.8	0.26
3	lauroyl peroxide in 10% hexane soln, 50°C	10.5	35.8	53.8	0.26
4	benzoyl peroxide in 10% nitromethane soln, 100°C	4.9	30.8	64.2	0.20
5	azoisobutyronitrile in 10% toluene soln, 50°C	6.3	37.6	56.0	0.25
6	ultraviolet irradiation in bulk with $ZnCl_2$, 25°C	7.0	29.0	64.0	0.20

(h units), ie, the central monomer unit is flanked by a monomer unit of the same configuration on one side and by one of opposite configuration on the other side. The peaks at 8.67τ, 8.79τ, and 8.90τ will be proportional to the numbers of i, h, and s units, respectively.

Let us designate by σ the probability that a polymer chain will add a monomer unit to give the same configuration as that of the last unit at its growing end. Let us assume that σ is independent of the configuration of the growing polymer chain and that therefore the propagation can be described by a single value of σ, ie, that the process follows Bernoulli-trial statistics. Then

$$P_i = \sigma^2 \tag{11}$$

$$P_s = (1 - \sigma)^2 \tag{12}$$

and

$$P_h = 1 - P_i - P_s = 2(\sigma - \sigma^2) \tag{13}$$

where P_i, P_s, and P_h represent the probabilities of forming i, s, and h units, respectively. In Figure 15, these relationships are plotted. It will be noted that the proportion of h units rises to a maximum at $\sigma = 0.5$, which corresponds to random propagation. For a random polymer, the proportion $i{:}h{:}s$ will be 1:2:1.

Table 1 shows the analyses of the spectra of a number of poly(methyl methacrylates) prepared with free-radical catalysts under conditions varying as indicated;

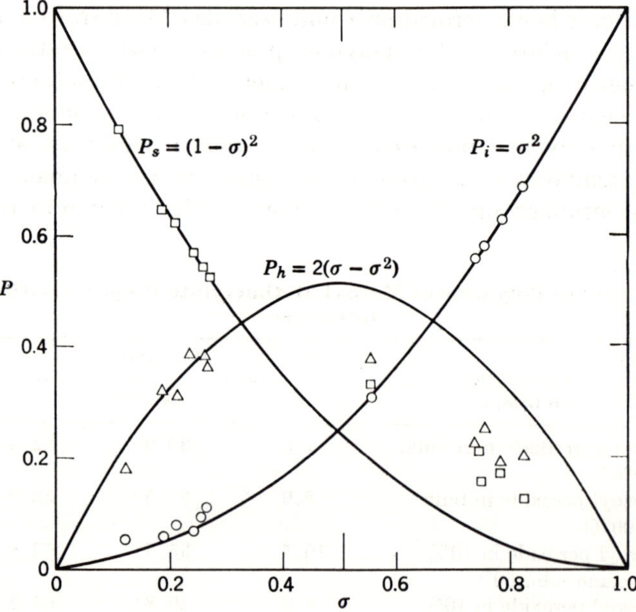

Fig. 15. The probabilities P_i, P_s, and P_h of formation of isotactic, syndiotactic, and heterotactic triads, respectively, as a function of σ, the probability of isotactic monomer placement during propagation. Experimental points at the left are for polymers prepared with free-radical initiators; those at the right for polymers prepared with anionic initiators; ○, isotactic peaks; △, heterotactic peaks; □, syndiotactic peaks.

the spectra are interpreted in terms of the proportions of *i*, *h*, and *s* units. In Figure 15 these results have been fitted to the probability curves by placing the *s* points on the curve and letting the other points fall where they may. It will be observed that for the polymers prepared with free-radical catalysts, the proportions of *h* and *i* units fall satisfactorily close to the calculated curves. This indicates that the free-radical propagation can be described by a single value of σ. There is a noticeable effect of temperature on σ (compare polymers 1 and 2), but the syndiotactic tendency is always predominant. The effect of temperature is treated in more detail below.

Table 2 shows results obtained for polymers prepared with anionic initiators. These polymers appear to be predominantly isotactic, provided effective complexing solvents are not employed. (An interpretation in terms of σ cannot be offered for reasons to be discussed below.) It has been found that in glyme alone, with 9-fluorenyllithium as initiator, syndiotactic polymer is obtained (14) and that in the presence of dioxane, block polymers consisting of alternating syndiotactic and isotactic regions are obtained.

Table 2. Structure of Polymers of Methyl Methacrylate Prepared With Anionic Initiators

Polymer	Polymerization conditions	$P \times 100$		
		i	*h*	*s*
7	*n*-BuLi in toluene, −62°C	63	19	18
8	*n*-BuLi in toluene, 25°C	68	20	12
9	*n*-BuLi in 60:40 toluene: dioxane, 62°C	56	23	21
10	*n*-BuLi in 50:50 toluene: glyme,[a] −62°C	31	37	32
11a	C_6H_5MgBr in toluene, 0°C	58	26	16
11b	C_6H_5MgBr in toluene, 0°C	100	0	0
12	sodium naphthalenide in glyme, −70°C	$i + h \cong 10$		90

[a] Ethylene glycol dimethyl ether.

Phenylmagnesium bromide (polymer 11a) and *n*-butyllithium appear to function in essentially the same manner. Polymer 11b was obtained from 11a by fractionation, and indicates the presence of a low-molecular-weight fraction of low stereoregularity in the latter. Sodium naphthalenide, which is stable only in an ether solvent and decomposes in toluene, gives a highly syndiotactic polymer.

The "backbone" methylene resonance would be expected to be a single peak in a syndiotactic polymer because from simple geometrical considerations both protons must, on a time average, experience the same magnetic environment. In an isotactic polymer, however, this is not true; the two protons will be differently shielded and may therefore be expected to show electron-coupled spin–spin splitting to give a fourfold resonance. This will consist of two doublets having nearly equal peaks if the difference in shielding considerably exceeds the coupling constant, J, whereas if the shielding difference is approximately equal to J, the peaks will be more closely spaced and the outer peaks will be weaker, ie, it will be an AB quartet. In Figure 14**a**, it can be seen that the methylene resonance of the predominantly isotactic polymer, centered at about 7.90τ, is indeed an AB quartet. The methylene spectrum of the syndiotactic polymer is clearly not a quartet, but predominantly a single peak, which, however, shows a splitting. This splitting is not expected on the basis of the simple triad analysis so far presented, and shows that this picture is somewhat oversimplified. This will be discussed later.

It is important to note that the methylene resonance provides an ***absolute and independent measure of the predominant configuration*** of the polymer, and one that does not rest on empirical correlations with x-ray results or any other method.

It is sometimes convenient to refer to methylene groups in which the protons are nonequivalent as *heterosteric* and those in which they are equivalent as *homosteric*, a more general terminology which will be useful in the subsequent discussion. By analogy to small molecules of related configuration, heterosteric methylene groups are also referred to as *meso* (abbreviated m), and homosteric methylene groups as *racemic* (abbreviated r). The proportions of *meso* and *racemic* methylene groups are designated as P_m and P_r, respectively, and for a Bernoulli-trial chain are

$$P_m = \sigma = P_i^{1/2} \tag{14}$$

$$P_r = 1 - \sigma = P_s^{1/2} \tag{15}$$

and of course for any chain

$$P_m + P_r = 1 \tag{16}$$

For a non-Bernoulli chain, equations 14 and 15 will not in general hold. Whereas polymers prepared with free-radical catalysts appear to be Bernoullian, as has been seen, polymers prepared with anionic initiators, particularly in the presence of complexing solvents, are not. For example, in the presence of 40 volume % of dioxane (polymer 9 in Table 2), the s peak is comparable to the h peak in area, which is not in accord with expectation for a "single-σ" propagation. In the presence of 50 volume % of glyme (polymer 10), a large deviation from expectation occurs, as can be seen from the conspicuous failure of the peak areas for this polymer to fit the theoretical curves (see the points near about 0.55 on the abscissa in Fig. 15). It has been seen that in the presence of complexing solvents, isotactic–syndiotactic block copolymers may be formed. Such block chains would in general be formed if any given chain stereoregularity, once formed, tended to propagate itself preferentially but not exclusively. It can readily be shown that if lengths are assigned to the blocks in the manner illustrated by

$$\begin{array}{ccc} \ldots\text{iiiii} & \text{hsssssss} & \text{hiiii}\ldots \\ \bar{n}_i & \bar{n}_s & \end{array}$$

then

$$(\bar{n}_i - 1)/(\bar{n}_s - 1) = i/s \tag{17}$$

$$\bar{n}_i/\bar{n}_s = m/r \tag{18}$$

$$\bar{n}_i/(\bar{n}_i + \bar{n}_s) = m/(r + m) \tag{19}$$

$$(\bar{n}_i + \bar{n}_s)/2 = 1/h = \bar{n} \tag{20}$$

where $\bar{n}_i$ and $\bar{n}_s$ are number-average lengths of the isotactic and syndiotactic blocks, respectively, and $\bar{n}$ is the number-average length of all blocks. If the propagation *can* be described by a single value of σ, then

$$\bar{n}_i/\bar{n}_s = m/r = \sigma/(1 - \sigma)$$

$$\bar{n}_s = 1/\sigma$$

$$\bar{n}_i = 1/(1 - \sigma)$$

Thus, a prominent h peak means that block lengths must be relatively short.

A more detailed study has been made of the effect of temperature on the free-radical-catalyzed polymerization of methyl methacrylate (15). The structures of polymers prepared with free-radical initiators over a 178° temperature range are summarized in Table 3. A trend toward increasing syndiotactic character with decreasing temperature is evident.

Table 3. Effect of Polymerization Temperature on Stereochemical Configuration of Poly(methyl Methacrylate) Prepared with Free-Radical Catalysts

Temp, °C	$[\eta]$ in benzene at 25°, dl/g	$M_w \times 10^{-3}$	$P \times 100$			σ	$\sigma/(1 - \sigma)$
			i	h	s		
−78	0.10	19	4.8	17.2	78.0	0.12	0.13_6
0	0.58	200	7.5	30.0	62.5	0.21	0.26_6
50	0.27	69	8.5	31.5	60.0	0.23	0.30_0
100	0.46	140	8.9	37.5	53.9	0.27	0.37_0

If the propagation steps are expressed in terms of absolute reaction rate theory, the rate constant for isotactic propagation is given by equation 21

$$k_i = (kT/h)\exp[(\Delta S_i^{\ddagger}/R) - (\Delta H_i^{\ddagger}/RT)] \tag{21}$$

and for syndiotactic propagation by equation 22

$$k_s = (kT/h)\exp[(\Delta S_s^{\ddagger}/R) - (\Delta H_s^{\ddagger}/RT)] \tag{22}$$

from which

$$\sigma = k_i/(k_i + k_s) = \frac{\exp[-(\Delta F_i^{\ddagger} - \Delta F_s^{\ddagger})/RT]}{1 + \exp[-(\Delta F_i^{\ddagger} - \Delta F_s^{\ddagger})/RT]} \tag{23}$$

and

$$\sigma/(1 - \sigma) = k_i/k_s = \exp[(\Delta S_i^{\ddagger} - \Delta S_s^{\ddagger})/R - (\Delta H_i^{\ddagger} - \Delta H_s^{\ddagger})/RT] \tag{24}$$

Therefore the difference in activation enthalpies for propagation is given by equation 25

$$\Delta(\Delta H_p^{\ddagger}) = \Delta H_i^{\ddagger} - \Delta H_s^{\ddagger} = -R\,\partial\ln[\sigma/(1 - \sigma)]/\partial(1/T) \tag{25}$$

and the difference in activation entropies by equation 26

$$\Delta(\Delta S_p^{\ddagger}) = \Delta S_i^{\ddagger} - \Delta S_s^{\ddagger} = R\ln[\sigma/(1 - \sigma)] + \Delta(\Delta H_p^{\ddagger})/T \tag{26}$$

A plot of the data of Table 3 according to equation 25 gives a value of 775 ± 75 cal for $\Delta(\Delta H_p^{\ddagger})$ and 0.0 ± 0.1 eu for $\Delta(\Delta S_p^{\ddagger})$. Redetermination of these quantities (16,17) indicates that more accurate values are about 1 kcal and 1 eu, respectively. Thus, the preference for syndiotactic placement is due to a small additional energy required for isotactic placement; the latter is favored by entropy.

Poly(vinyl Chloride). The spectrum of poly(methyl methacrylate) is relatively simple because, although there is a strong geminal coupling of the protons of the *meso* methylene groups, there is no observable coupling of the methylene and α-methyl protons. In polymers of monosubstituted monomers, such as vinyl chloride, the *vicinal* coupling of the α- and β-methylene protons produces a splitting of the observed resonances. On the simplest basis one would expect the α protons, being coupled to four neighboring β protons, to appear as a pentuplet, and the β protons, coupled to two α protons, as a triplet. The observed spectrum (Fig. 16**a**) does indeed apparently show a pentuplet centered at 5.53τ (in chlorobenzene solution) and a group of five peaks centered at about 7.9τ for the methylene protons (18,19). The methylene resonance consists essentially of two overlapping triplets centered at 7.78τ and 7.96τ corresponding to *meso* and *racemic* methylene groups, respectively.

This interpretation is confirmed by *double resonance.* In this technique, one multiplet corresponding to a group of coupled spins is observed using the usual weak radio-frequency field while simultaneously a second multiplet, corresponding to another nucleus or group of nuclei coupled to the first group, is irradiated with a much stronger radio-frequency field. This has the effect of abolishing the coupling between the spins. For optimum effect, the difference in frequency between the two fields, $\Delta\nu$, must correspond to the chemical shift difference between the two groups of spins. As shown in Figure 16, when the β protons are decoupled from the α protons ($\Delta\nu$ = 137 cps), two peaks are observed (18,19), separated by 0.20 ppm (Fig. 16**c**). A very similar spectrum is shown by the polymer of α-deuterovinyl chloride (**1**). The substitution

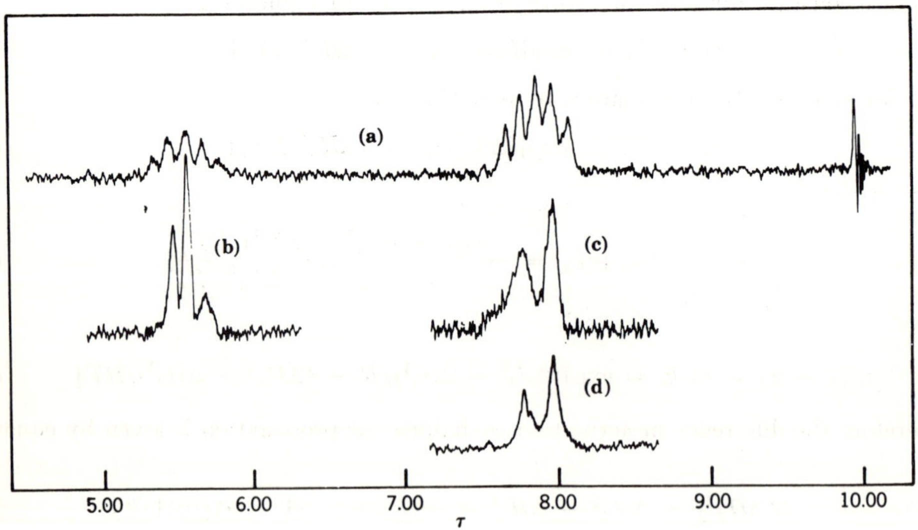

Fig. 16. Normal and decoupled spectra at 60 Mc/sec of poly(vinyl chloride) (in chlorobenzene): (**a**) normal spectrum; (**b**) α protons of decoupled spectrum; (**c**) β protons of decoupled spectrum; (**d**) spectrum of poly(α-d_1-vinyl chloride).

$$\underset{\mathrm{D}}{\overset{\mathrm{Cl}}{>}}\mathrm{C{=}CH_2}$$

(1)

of hydrogen by deuterium represents another means of spectral simplification, for the resonance of deuterium is very far removed from that of hydrogen; in addition, the H–D coupling has less than one-sixth the magnitude of the corresponding H–H coupling, and produces no observable multiplicity in polymer spectra (Fig. 16**d**).

When the CH_2 protons are irradiated ($\Delta\nu$ = 133 cps), the α-proton resonances show peaks at 5.48τ, 5.59τ, and 5.71τ (Fig. 16**b**); the dependence of the decoupled α-proton spectrum upon $\Delta\nu$ has been employed (19) to demonstrate that these peaks correspond to syndiotactic, heterotactic, and isotactic triads, respectively, assuming the correctness of the above assignment of the β-proton resonances. The normal α-proton spectrum thus actually consists of three overlapping pentuplets.

As discrimination increases, through improved techniques and advances in instrumental design, finer structure becomes observable. With respect to α substituents, one might expect to resolve *pentads* of monomer units, appearing as a fine structure on the *i, h,* and *s* peaks. One should also be able to resolve *tetrads* of β-methylene groups. The *meso* resonance should be split into three tetrad resonances, all heterosteric, giving six different chemical shifts; the *racemic* resonance should be split into two homosteric resonances and one heterosteric resonance, giving an

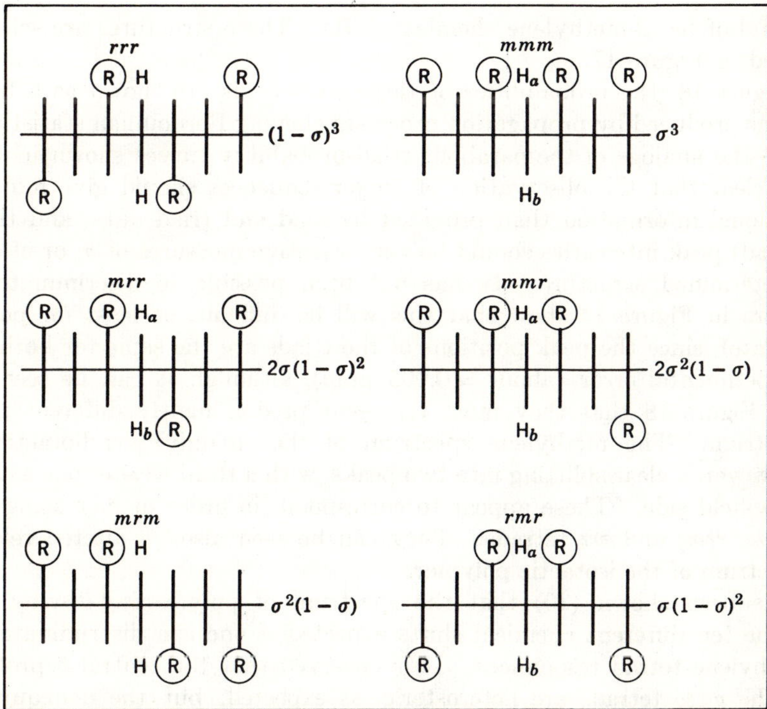

Fig. 17. Tetrad configurations in projection, viewed along the plane of the planar zigzag. Bernoulli-trial probabilities indicated at the right of each tetrad.

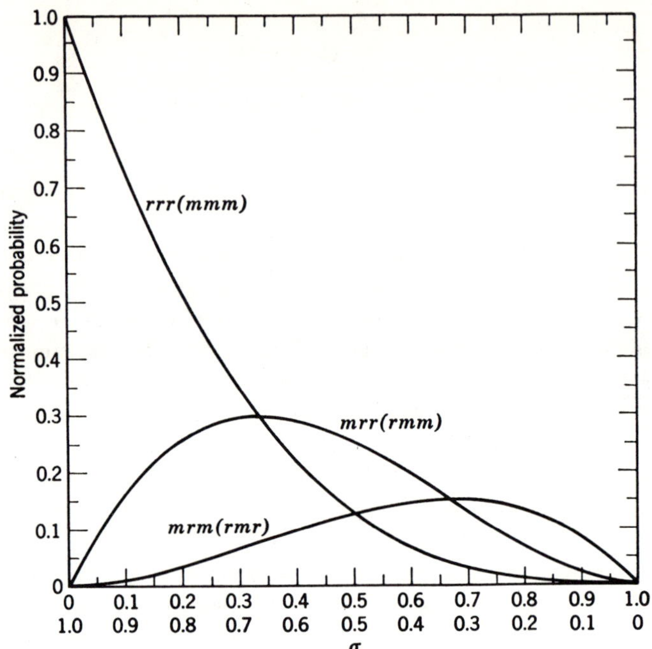

Fig. 18. Tetrad probabilities for Bernoulli-trial vinyl polymer chains. For *rrr, mrr,* and *mrm,* the upper σ scale is used; for *mmm, mmr,* and *rmr* the lower σ scale is used.

overall total of ten β-methylene chemical shifts. These structures are schematically represented in Figure 17.

In Figure 18, the probabilities of these six tetrads are shown as a function of σ for chains produced by propagation processes obeying Bernoullian statistics. These curves are the analogs of the parabolic triad-probability curves shown in Figure 15.

It is clear that the observation of longer sequences should give more detailed configurational information than provided by diad and triad data, and that tetrad (and pentad) peak intensities should be very sensitive measures of σ, or of deviations from σ-determined structure. It has not been possible to discriminate pentads. The spectra in Figure 14 show that this will be difficult, at least for poly(methyl methacrylate), since the peak positions of the triads are the same for both polymers within experimental error (about ± 0.005 ppm), although it can be seen from the curves of Figure 18 that they must represent predominantly different pentads in each spectrum. The methylene spectrum of the (mainly) syndiotactic polymer shows, however, a clear splitting into two peaks, with a third weaker one as a shoulder on the low-field side. These appear to correspond, in order of increasing shielding, to the *mrm, rrm,* and *rrr* tetrads. They can be seen also, in altered relative size, in the spectrum of the isotactic polymer.

It has been shown (20) that the spectrum of poly(α-*cis*-β-d_2-vinyl chloride) exhibits the ten different chemical shifts expected if one can discriminate all six of the β-methylene tetrad resonances. It is observed that the central β protons in all three of the *meso* tetrads are heterosteric, as expected, but the nonequivalence is substantial only for the *rmr* tetrad. The presence of geminal coupling, as in normal poly(vinyl chloride) or poly(α-d_1-vinyl chloride) (Fig. 16), tends to concentrate

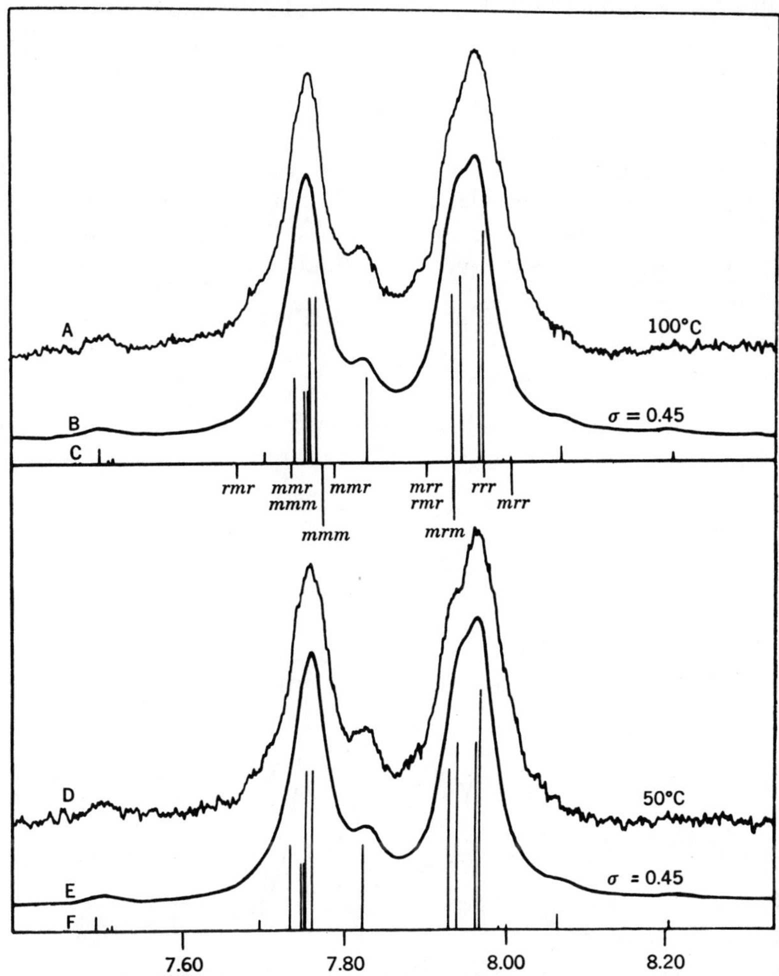

Fig. 19. Calculated and experimental NMR spectra at 60 Mc/sec of poly(α-d_1-vinyl chloride) prepared at varying temperatures of polymerization: A, observed spectrum of 100°C polymer; B, calculated spectrum of 100°C polymer, with line width of 2.5 cps; C, "stick" spectrum corresponding to B; D, observed spectrum of 50°C polymer; E, calculated spectrum of 50°C polymer with line width of 2.5 cps; F, "stick" spectrum corresponding to E. On the abscissa of spectra A, B, and C are shown the chemical shifts of the six tetrads (see text).

intensity at the center of the band, and this effect is accentuated by the fact that the *mean* chemical shift for each heterosteric pair of protons is nearly the same for all *meso* tetrads.

In these terms, the spectrum of poly(vinyl chloride) can be quite fully interpreted. In Figure 19, curve A, the spectrum shown in Figure 16**d** is presented in greater detail. This polymer was prepared at 100°C. Curve C shows calculated AB quartets (for heterosteric CH_2 groups) and singlets (for homosteric CH_2 groups) corresponding to the six methylene tetrads. The assignments of the chemical shifts are shown in this figure; these assignments are the same for the other three polymers, prepared at 50°, 0°, and −78°C (Figs. 19 and 20). The relative intensities of the calculated resonance lines are obtained from Figure 18, ie, from Bernoullian statistics, choosing values of σ which give the best match of experimental and calculated spectra at each temperature. The calculated spectrum in Figure 19 (curve B) and the corresponding calculated spectra for the other three polymers, are obtained from a computer program which allows the introduction of any chosen line width (2.5 cps in these spectra) and the summation of the resulting spectra in any desired proportion. By visual inspection and matching of relative peak heights in calculated and experimental spectra, the following values of σ are deduced: −78°C, 0.37; 0°C, 0.43; 50°C, 0.45; 100°, 0.46. From these data and equations 25 and 26 it is found that

$$\Delta(\Delta H_p)^{\ddagger} = 310 \pm 20 \text{ cal}$$

$$\Delta(\Delta S_p)^{\ddagger} = 0.6 \pm 0.1 \text{ eu}$$

There is thus a measurable tendency for vinyl chloride to give more syndiotactic polymers at lower temperatures, but the tendency is much smaller than that for methyl

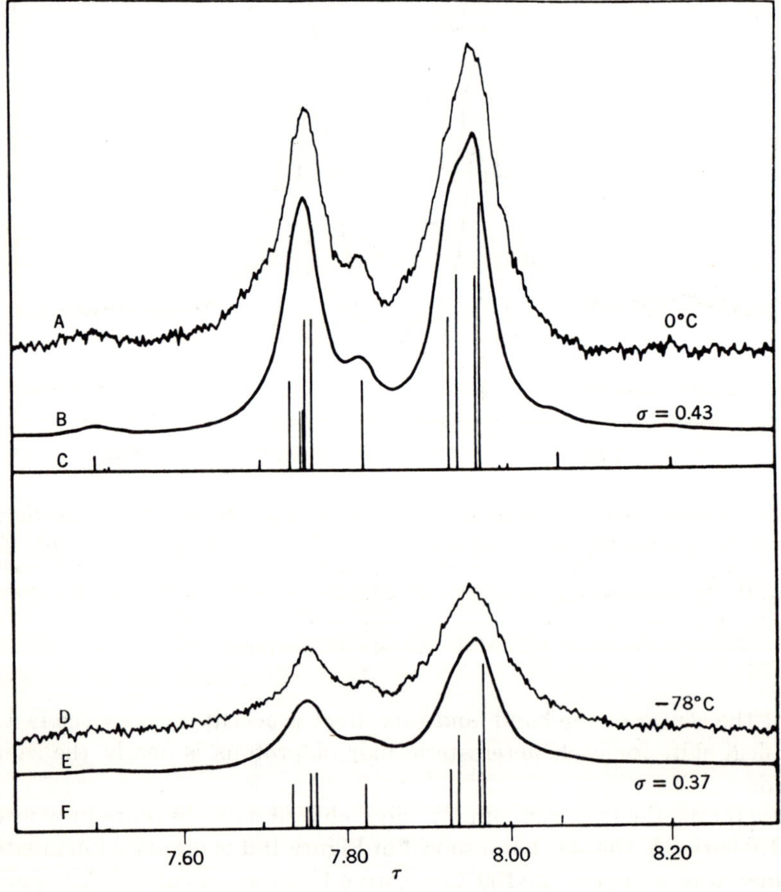

Fig. 20. NMR spectra (60 Mc/sec) of poly(α-d_1-vinyl chloride): A, observed spectrum of polymer prepared at 0°C; B, calculated spectrum of the 0°C polymer, with line width of 2.5 cps; C, "stick" spectrum corresponding to B; D, observed spectrum of −78°C polymer; E, calculated spectrum of −78°C, with line width of 3.2 cps; F, "stick" spectrum corresponding to E.

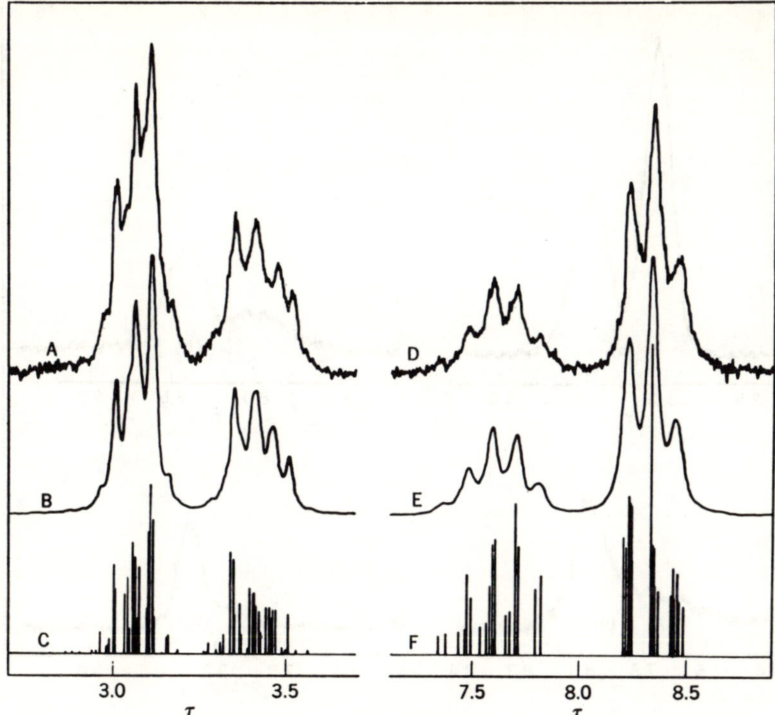

Fig. 21. Spectra of isotactic polystyrene (60 Mc/sec): A, aromatic proton spectrum, in tetrachloroethylene at 135°C; B, calculated spectrum: $\nu_{(\text{ortho})} - \nu_{(\text{meta})} = 22$ cps; $\nu_{(\text{para})} - \nu_{(\text{meta})} = 2$ cps; all $J_{\text{ortho}} = 6.5$ cps; both $J_{\text{meta}} = 2.0$ cps; $J_{\text{para}} = 1.0$ cps; line width = 1.0 cps; C, "stick spectrum" corresponding to B; D, backbone-proton spectrum, in *o*-dichlorobenzene at 200°; E, calculated spectrum: $\nu_\beta - \nu_\alpha = 43.8$ cps; both $J_{\text{vic}} = 7.0$ cps; $J_{\text{gem}} = 14.0$ cps; line width = 3.0 cps; F, "stick spectrum" corresponding to E.

methacrylate and appears hardly sufficient to account for the alteration in properties observed for the low-temperature polymers, particularly the enhanced crystallizability (21).

Polystyrene. The earliest reported NMR spectrum of a polymer was that of polystyrene (22), which in carbon tetrachloride solution gave an aromatic resonance showing a separate peak for the *ortho* protons, upfield from the *meta–para* proton resonance, and a single broad peak for the aliphatic protons. Curves A and D in Figure 21 show the spectrum of isotactic polystyrene at 60 Mc/sec obtained using 15% solutions in tetrachloroethylene at 128°C for the aromatic protons and in *o*-dichlorobenzene at 200°C for the main-chain protons (23). Since there is no significant coupling of ring and main-chain protons, they may be treated as separate systems. Such "strong-coupled" spectra, ie, those for which the chemical shift differences are comparable in magnitude to the couplings between the protons, cannot be solved for values of τ and J by inspection, since the spacings and line positions do not provide these parameters directly. Instead, quantum mechanical calculations are necessary. In the case of a polymer, some simplified spin model is also necessary, since one obviously cannot include *all* the protons present in a long chain. Following the suggestion of Tincher (24), the polystyrene backbone protons are treated as if the chain were only a cyclic dimer. Curve F in Figure 21 shows the line spectrum resulting from a six-spin

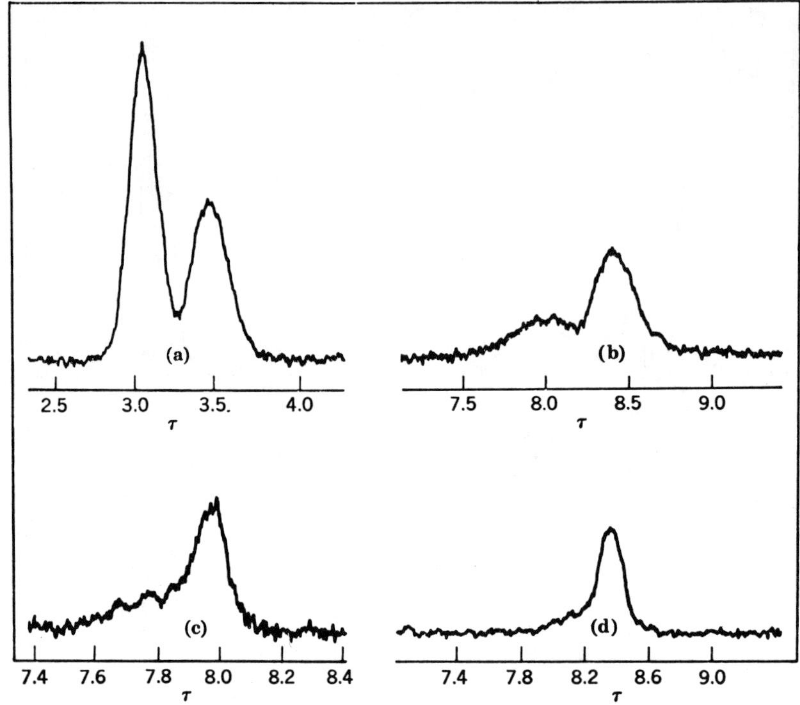

Fig. 22. Spectra of "atactic" polystyrene (60 Mc/sec): (**a**) aromatic-proton spectrum, in tetrachloroethylene at 135°C; (**b**) backbone-proton spectrum, in *o*-dichlorobenzene at 200°C; (**c**) poly-β,β-d_2-styrene in *o*-dichlorobenzene at 200°C with deuterium irradiation; (**d**) poly-α-d_1-styrene in *o*-dichlorobenzene at 200°C with deuterium irradiation.

"dimer" calculation, using the parameters given in the figure title. The results of a five-spin calculation for the aromatic protons are also shown. Curve E shows the calculated spectrum with the experimental line widths included. It can be seen that the matching is excellent. From this result the following conclusions can be drawn:

1. The phenyl spin system is symmetrical about a bisecting plane through the *para* carbon atom, and thus gives no evidence of a high (exceeding about 15 kcal) barrier to rotation.

2. The two *meso* β protons are fortuitously equivalent in chemical shift within experimental error (± 0.05 ppm).

3. The two β protons show couplings (7.00 ± 0.20 cps) to the α protons, which are equal within experimental error. This observation is consistent with the assumption that the polymer chain has a *local* conformation in solution which is the same as in the 3_1 helix which it is known to have in the crystal.

"Atactic" polystyrene, such as is normally obtained by free-radical and ionic polymerization, gives a much less satisfactory spectrum. Brownstein and co-workers (25) have observed that in the α-proton spectrum of "atactic" poly-$\beta\beta$-d_2-styrene, resonances corresponding to isotactic, heterotactic, and syndiotactic triads can be distinguished, and this observation has been confirmed (23). Figures 22**a** and 22**b** show the spectra for "atactic" polystyrene prepared by free-radical polymerization, in tetrachloroethylene at 135°C and in *o*-dichlorobenzene at 200°C. Figure 22**c** shows the α-proton spectrum of "atactic" poly-β-d_2-styrene and Figure 22**d** shows the β-proton spectrum of "atactic" poly-α-d_1-styrene, both prepared in the same way and observed at 200°C in *o*-dichlorobenzene. Three peaks can be discerned in the α-proton spectrum The peak at lowest field (7.68τ) is in approximately the same position (allowing for the increased shielding which is known to be produced by neighboring deuterium nuclei) as that for isotactic polystyrene (7.61τ), indicating an assignment of the predominant peak to syndiotactic triads. These polymers appear to be about 80%

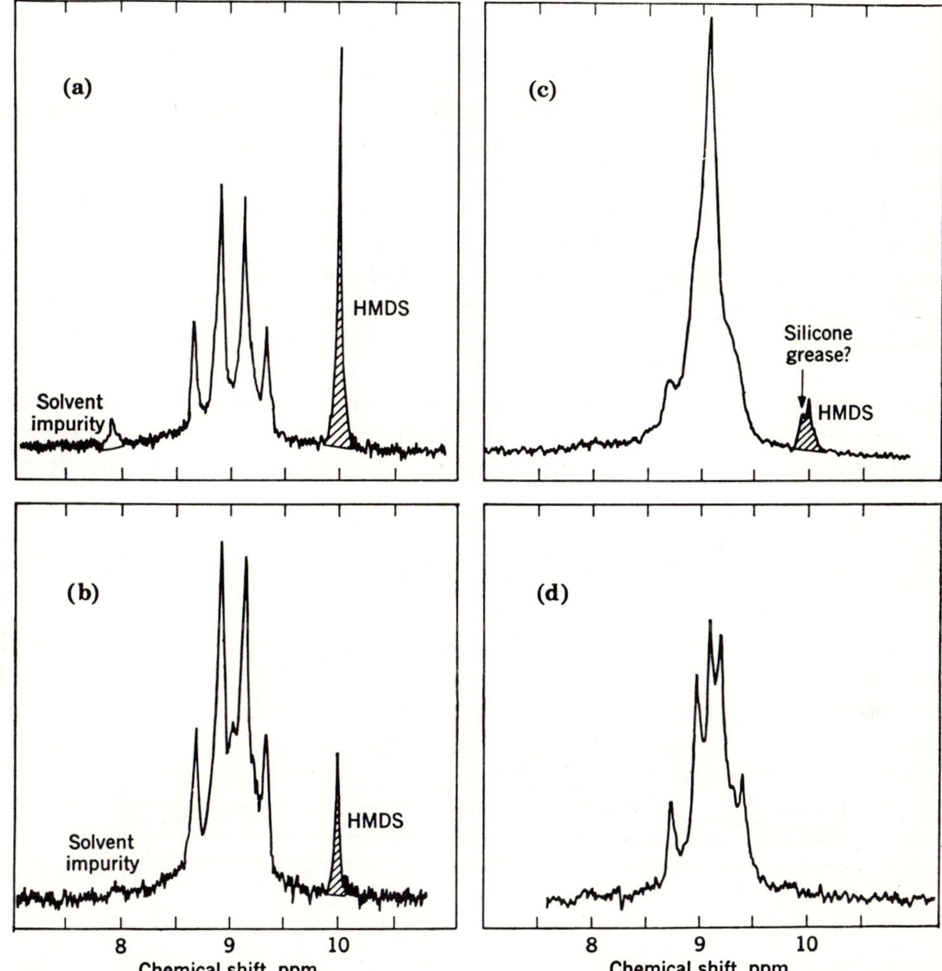

Fig. 23. Spectra of poly-2,3,3,3-d_4-propylene in 2-chlorothiophene at 110°C (60 Mc/sec): (**a**) isotactic fraction; (**b**) stereoblock fraction; (**c**) amorphous fraction; (**d**) unfractionated. Internal reference is hexamethyldisiloxane (HMDS) at 9.94τ (26). Courtesy *Journal of Polymer Science.*

syndiotactic. The β-proton spectrum shows only a broad singlet, the *meso* protons not being distinguishable from each other (as expected from the foregoing discussion) or from the much larger *racemic* singlet.

The atactic polymer spectra always appear to be much more poorly resolved than the isotactic. This can very probably be attributed to the overlapping of small but appreciable chemical shift differences of pentads and tetrads.

Polypropylene. A further step in complexity is presented by polypropylene, in which the coupling of side-chain protons to the main-chain α protons, not heretofore observable, strongly affects the spectrum. The first useful results (26) were obtained by employing the deuteromonomer (**2**). Solvent fractionation of a polymer prepared

$$\begin{array}{l} CD_3 \\ \quad \diagdown \\ \quad\; C{=}CH_2 \\ \quad \diagup \\ D \end{array}$$

(**2**)

with a Ziegler-Natta type catalyst from this monomer gave the results shown in Figure 23. The crystallizable fraction, which is insoluble in ethyl ether, gives an AB quartet for the methylene resonance, and appears thus to be almost purely isotactic. The mostly soluble fraction gives largely a singlet, and is predominantly syndiotactic. What appeared to be a "block" isotactic–syndiotactic fraction was also obtained, although this could not be demonstrated from the CH_2 spectra alone.

The spectrum of the undeuterated polymer has also yielded to analysis (27,28). Figure 24**a** shows (27) the spectra of isotactic and Figure 24**b** of syndiotactic polypropylene, observed in *o*-dichlorobenzene at 150°C. Eight broad peaks, not all clearly visible, centered at about 8.50τ, are given by the α proton, split by three methyl and four methylene protons. Centered at about 8.94τ in the spectrum of Figure 24**b** is the distorted triplet of the methylene protons. This region is of course much more

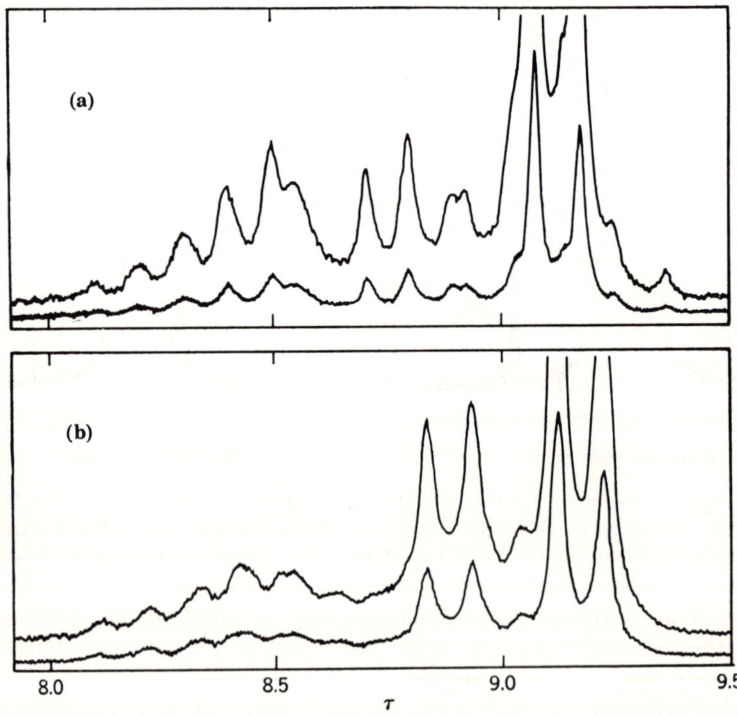

Fig. 24. Spectra of polypropylene (60 Mc/sec) in *o*-dichlorobenzene at 150°C; (**a**) isotactic; (**b**) syndiotactic (27). Courtesy *Journal of Polymer Science.*

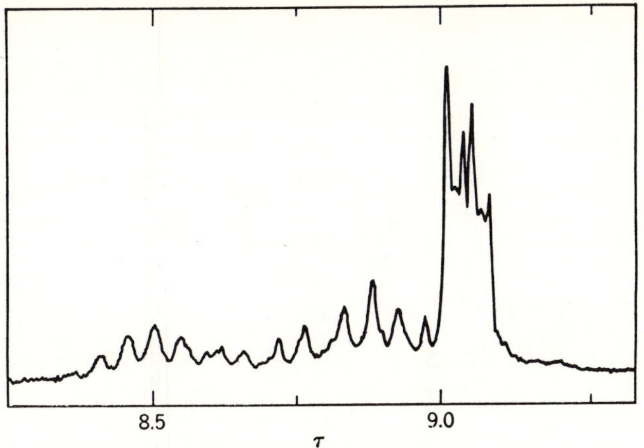

Fig. 25. Spectrum of stereoblock polypropylene at 100 Mc/sec in *o*-dichlorobenzene at 150°C (27). Courtesy *Journal of Polymer Science.*

complex in the spectrum of Figure 24**a** and overlaps the α-proton resonance. Most useful for analytical purposes are the strong resonances of the methyl groups, split into doublets by the α protons. In the isotactic spectrum this doublet appears at 9.13τ and in the syndiotactic spectrum at 9.18τ. Figure 25 shows the spectrum obtained at 100 Mc/sec for a polymer containing both isotactic and syndiotactic sequences, as indicated by two overlapping methyl doublets. At an intermediate position, as expected, appears the weak doublet of the heterotactic sequences, representing the junctions of the isotactic and syndiotactic "blocks." The polymer departs very far from Bernoulli-trial structure. From the intensity of the heterotactic doublet, $\bar{n}$ in equation 20 can be estimated at about 16. The better resolution in the spectrum at 100 Mc/sec is evident. In fact, the heterotactic doublet cannot be distinguished at 60 Mc/sec. Figure 26 shows the striking improvement and simplification that is achieved at 220 Mc/sec (29) using an instrument with a 52-kG superconducting solenoid magnet, such as mentioned at the beginning of this article.

Configuration Measurements on Other Vinyl Polymers. Many other polymers have been observed by NMR and the spectra at least partially interpreted in terms of the principles described above. These include poly-α-methylstyrene (30,31) poly(vinyl fluoride) (19), polytrifluorochloroethylene (32), poly(vinyl methyl ether) (19), poly(α-methylvinyl methyl ether) (33), poly(vinyl acetate) (19,34), poly(vinyl alcohol) (34,35), poly(isopropyl acrylate) (36,37), poly(methyl acrylate) (37), poly-2-vinylpyridine (38), polymethacrylonitrile (39), polyacrylonitrile (40,41), poly(vinyl formate) (42), polyacetaldehyde (43–45), and copolymers (46,47).

Head-to-Head and Head-to-Tail Measurements. High-resolution NMR spectroscopy is well adapted to the detection of head-to-head units in vinyl polymers, provided these occur in appreciable proportions. For most polymers, the proportion of such units is probably much less than 1% because resonance stabilization strongly favors a single form of the propagating species. Ethylene derivatives with substantial fluorine substitution are an exception (48), and offer the additional advantage that one can observe the ^{19}F resonance. ^{19}F has a spin of ½, like ^{1}H, and a magnetic moment nearly as large as that of ^{1}H. Like most nuclei other than hydrogen, its resonance is much more sensitive to environmental influence than is that of ^{1}H, and

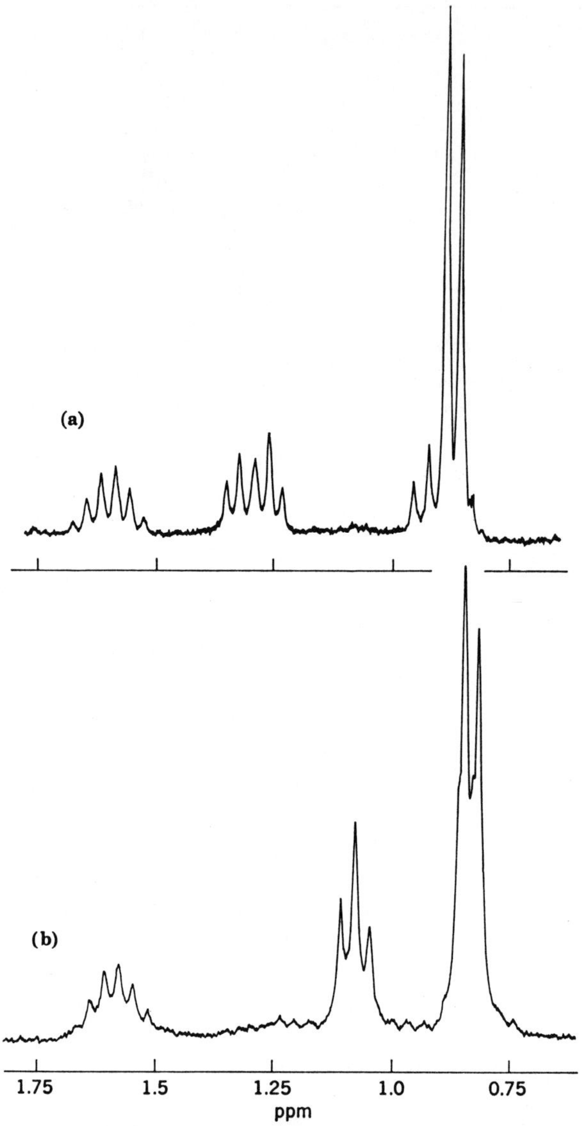

Fig. 26. Spectra of polypropylene at 220 Mc/sec in *o*-dichlorobenzene at 165°C; (**a**) isotactic; (**b**) syndiotactic. Courtesy R. C. Ferguson.

its observed range of chemical shifts extends over at least 300 ppm, a range about 30-fold greater than that of ^1H. Wilson (49) has observed the ^{19}F spectrum of poly(vinylidene fluoride) at 56.4 Mc/sec, employing *N*,*N*-dimethylacetamide as solvent and $CFCl_3$ as internal reference. Despite the absence of any possible stereochemical complications, the spectrum is complex, as can be seen from Figure 27, in which chemical shifts are expressed in ppm upfield from $CFCl_3$. The only logical explanation for this observation is that head-to-head and tail-to-tail monomer units must occur with considerable frequency. The chain units, as identified by Wilson, are indicated on the spectrum and in the representative chain segment (**3**). Here, the letters refer

```
 →    →    →    →    →    ←    →    →    →    →    →    →
H F  H F  H F  H F  H F  F H  H F  H F  H F  H F  H F  H F
—C—C—C—C—C—C—C—C—C—C—C—C—C—C—C—C—C—C—C—C—C—C—C—C—
H F  H F  H F  H F  H F  F H  H F  H F  H F  H F  H F  H F
       A    A    A    C    D    B    A    A    A    A
```

(3)

to the spectral lines in Figure 27. The sequence associated with line D can occur only if the reversed monomer unit (indicated by the arrow ←) is immediately followed by a tail-to-tail link. Since B, C, and D have about equal intensity, it is evident that this reversed unit has little or no tendency to propagate further units of its kind. Quantitative estimation indicates that 5–6% of the monomer units are reversed.

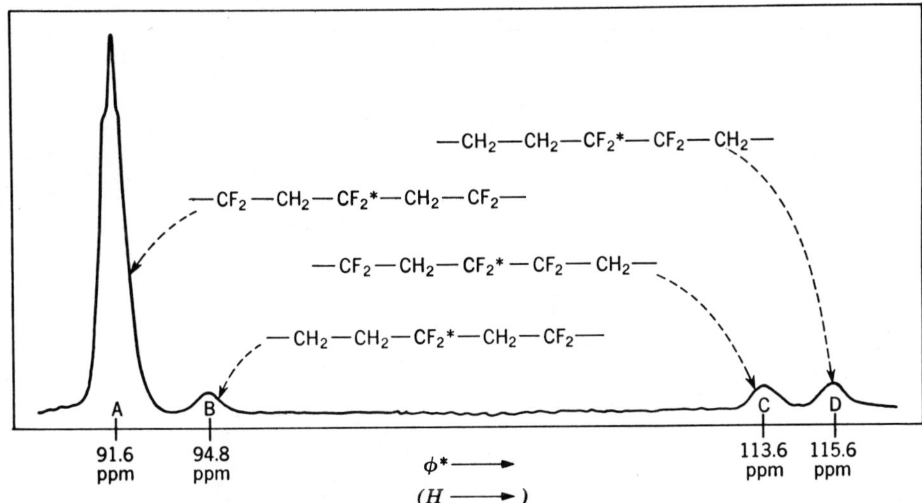

Fig. 27. [19]F high-resolution spectrum of poly(vinylidene fluoride) at 56.4 Mc/sec in *N,N*-dimethylacetamide at room temperature, with $CFCl_3$ as the internal reference. The ^{19}F chemical shifts are expressed on the ϕ* scale, which is in ppm upfield from this reference (not extrapolated to infinite dilution). A, B, C, and D are discussed in the text (49). Courtesy *Journal of Polymer Science.*

Isomerism in Diene Polymers. High-resolution NMR spectroscopy is useful in the determination of isomeric structures in polydiene chains. Thus, the methyl groups of *cis*-1,4-isoprene units (**4**), appear at 8.33τ (in carbon tetrachloride (50) or carbon disulfide (51)), and those in *trans*-1,4 units (**5**), appear at about 8.40τ. The proportion of 3,4-units (**6**), can be estimated from the characteristic vinylidene proton

```
  CH3       H            CH3       CH2—          —CHCH2—
     \     /                \     /                  |
      C=C                    C=C                     C
     /     \                /     \                //  \
 —CH2       CH2—        —CH2       H            CH2    CH3
      (4)                    (5)                   (6)
```

peak (5.33τ in CS_2, upfield from the vinyl proton peak of the 1,4 units, appearing at 4.92τ). The *cis*-1,4 units in polychloroprene appear at 4.49τ (CS_2), and the *trans*-1,4 units appear at 4.56τ (52). High-resolution NMR is now the method of choice for such measurements. It is particularly valuable because it counts protons directly, and does not require calibration.

Biopolymers

This section includes both polymers of biological origin, proteins and nucleic acids, and synthetic polymers of related structure. This appears to be one of the most promising applications of NMR to the study of polymers. Dipolar broadening, generally an obstacle, is here exploited as a means of studying stages in the unfolding of macromolecular structures.

Polypeptides. The transition between the random coil and α-helical conformation of synthetic polypeptides has yielded particularly interesting results (53–55). In helix-supporting solvents, such as trichloroethylene (53) or chloroform (55), poly-(γ-benzyl L-glutamate) gives a spectrum in which only the phenyl and benzylic resonances can be clearly discerned, and these appear as broad peaks. The peptide and

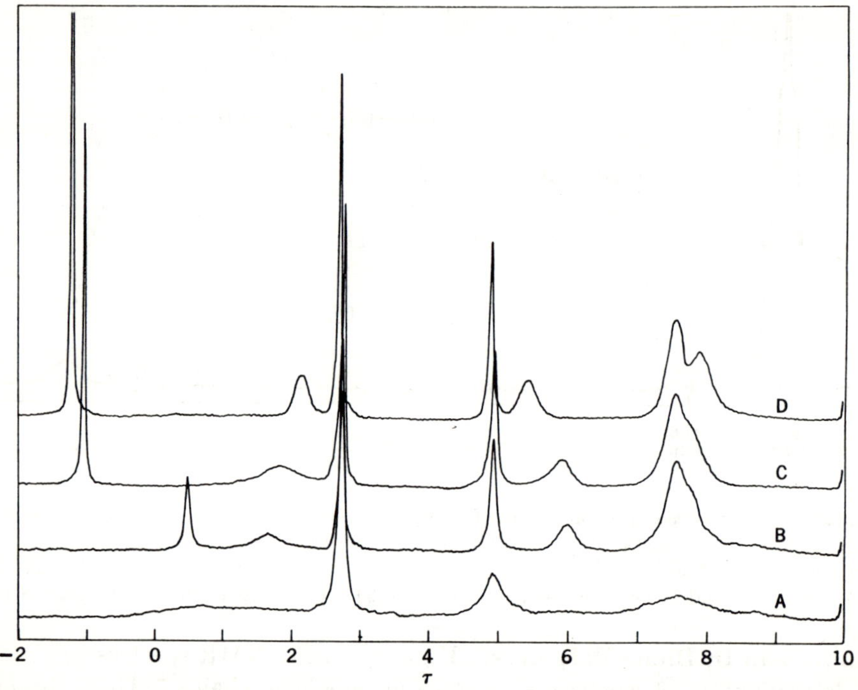

Fig. 28. Spectra of poly(γ-benzyl L-glutamate) at 60 Mc/sec: A, in $CDCl_3$; B, in $CDCl_3$ with 2.5% trifluoroacetic acid; C, in $CDCl_3$ with 15.0% trifluoroacetic acid; D, in $CDCl_3$ with 25.0% trifluoroacetic acid.

α-proton peaks cannot be seen, and the glutamyl methylene peaks near 7.5τ are much broadened (Fig. 28, spectrum A). Upon addition of a small amount of trifluoroacetic acid, there is a marked decrease in viscosity and most of the polymer peaks become narrow enough to be clearly seen. The NH peak appears at about 1.6τ, but is very broad and still hardly discernible. The α-proton peak can be seen at about 5.8τ. Measurements of optical rotation indicate that the polymer is still in the α-helical conformation, so this transition evidently corresponds to the breaking up of associations between helixes. When 15% of trifluoroacetic acid is added, the spectrum is about the same. At slightly higher acid concentration, the helix-to-random coil transition occurs, as indicated by optical rotation, and the peaks become somewhat

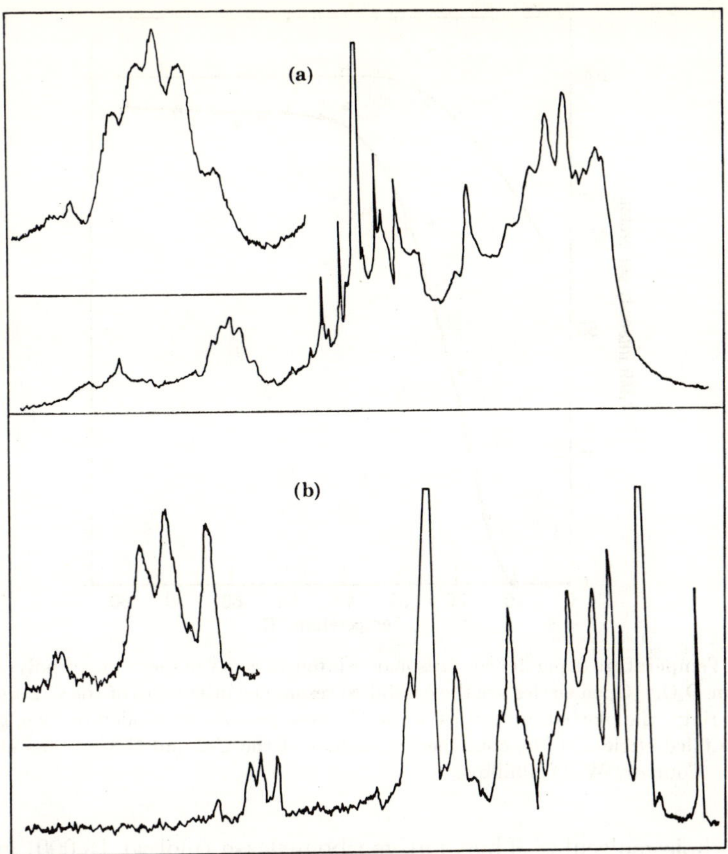

Fig. 29. Spectrum of ribonuclease (220 Mc/sec) in D_2O at pH 6.8; (**a**) native state, 22°C; (**b**) denatured state, 72.5°C (53a). Courtesy *Science;* copyright 1967 by the American Association for the Advancement of Science.

narrower. The α proton undergoes a marked downfield shift. Thus, three distinct regions of chain mobility can be recognized. The association of the helixes probably gives rise to a nematic liquid crystal phase at the polymer concentrations employed in these studies (10% or greater); the bundles of helixes in this phase rotate so slowly that the peaks, particularly those of the peptide backbone, are greatly broadened. Upon addition of a small amount of acid, the nematic phase is replaced by dissociated helixes. The further narrowing upon disruption of the helixes is much less marked (except for the NH peak), and may indicate that segmental freedom is not so very much greater in the random coil than in the helix.

Proteins. Globular proteins in the native state are composed of polypeptide chains, partly helical, which are rather tightly intertwined to form compact spheres or ellipsoids. The complete structure of several such proteins is now known from x-ray diffraction studies: myoglobin, hemoglobin, lysozyme, and ribonuclease (see also PROTEINS). There is little or no free segmental motion in such a structure and so the rotation of the molecule as a whole determines the line width. If the molecular weight is high, the reorientation rate will be insufficient to permit high resolution, except perhaps for certain side-chain protons whose correlation times are shorter than

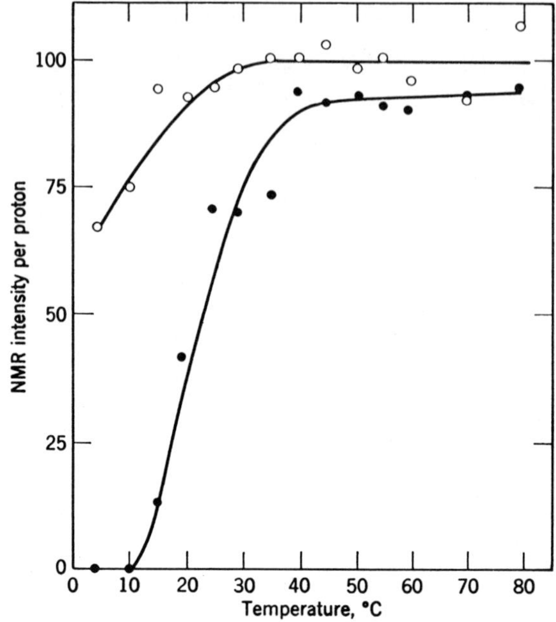

Fig. 30. Temperature dependence of resonance intensities in the spectrum of poly(adenylic acid) at 60 Mc/sec in D_2O. Open circles are the combined resonance intensities of the C-2 and C-8 protons of the adenine ring; the plotted intensities of the 1′-ribose proton, multiplied by two, give the curve represented by filled circles. (The combined intensities of the C-2 and C-8 protons above 35% are taken as 100.) Courtesy W. D. Phillips.

that of the whole molecule. Thus, native ribonuclease (mol wt 13,000) gives a moderately well resolved spectrum in aqueous solution (56,57), whereas native bovine serum albumin (mol wt 65,000) gives a much more poorly resolved spectrum which is hardly more than a single very broad asymmetrical band (53). Upon unfolding the latter in 8 *M* urea solution, the spectrum shows a number of narrow resolved peaks (53), since segmental motion is now possible. The behavior is entirely analogous to that of polypeptides, discussed in the previous section. In such denaturants as aqueous urea, some α-helical or other fixed structures may be retained. In trifluoroacetic acid, which is a good solvent for many proteins, all organized structure is probably lost, and resolution is still further improved (53). However, line widths are still on the order of 10–15 cps, and so finer details tend to be obscured by peak overlap when spectra are obtained at 60 Mc/sec. Substantial improvement in apparent resolution is observed at 100 Mc/sec; a marked further advance in discriminating spectral detail is achieved at 220 Mc/sec. Figure 29 shows spectra of ribonuclease at 220 Mc/sec in D_2O solution at pH 6.8, as observed by McDonald and Phillips (57). The use of D_2O as solvent reduces greatly the magnitude of the water peak, prominent near the center of the spectra; a peak for residual HDO can still be seen at this position. Because of deuterium exchange, the carboxyl, amino, hydroxyl, and most of the peptide NH protons do not appear as such in the spectra. The spectrum in Figure 29**a** represents the protein in the native, folded state at 22°C; the spectrum in Figure 29**b** was taken at 72.5°C, at which temperature the protein is believed to be fully unfolded. A marked narrowing of the peaks can be seen in the latter. There are also qualitative changes in the peak patterns and positions. In a general way, the resonances can be assigned by reference to the known spectra of amino acids and peptides. For example, the cluster of peaks at 2.2–3.1τ represent the aromatic ring protons of tyrosine, phenylalanine, tryptophan, and histidine, whereas those at the other extreme of the spectrum, near 9τ, represent methyl groups of valine, leucine, and isoleucine. An increase in resolution upon denaturation is particularly obvious for these last. Other, less highly organized proteins, such as trypsin and pepsin, show narrow peaks in this region even in the native state, an indication of greater internal freedom. It is clear that a full interpretation of these spectra will lead to great advances in the understanding of protein structure and behavior.

Polynucleotides. The spectra of polynucleotides furnish the clearest illustration of the way in which dipolar broadening in polymer solutions can be an important structural tool rather than a hindrance, if appropriately exploited. For example, it is believed that poly(adenylic acid), which is composed of the repeating unit (**7**) exists

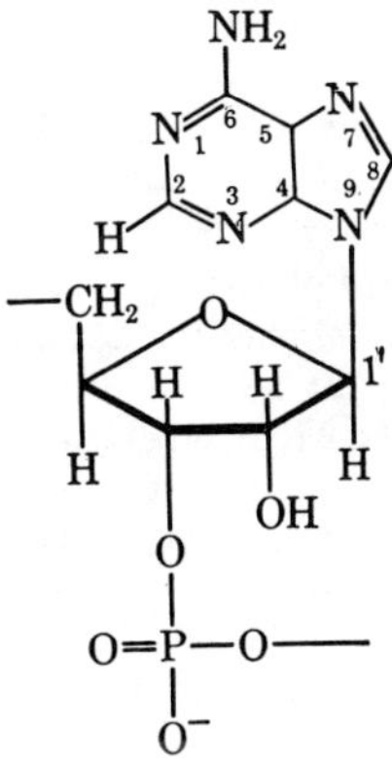

NH$_2$ N 6 5 N 1 7 8 —H 2 3 4 9 H N N —CH$_2$ O H H 1′ H H OH O O=P—O— O$^-$

(**7**)

in neutral aqueous solution in a single-stranded rodlike conformation, stabilized by stacking of the adenine bases. The "backbone" protons in such a structure give no observable intensity in a high-resolution spectrum. Upon raising the temperature, a "melting" of the structure occurs (57–59) and resonance peaks abruptly appear and grow in intensity as the temperature is raised. Figure 30 shows the integrated intensities of the C-2, C-8, and C-1′, proton peaks as a function of temperature. The H-1′ proton peak vanishes below 10°C, and the C-2 and C-8 intensities show a parallel but less complete reduction. There is also significant deshielding of these protons as structure is lost.

When poly(adenylic acid) and poly(uridylic acid) (the latter having a uracil unit (**8**) in place of the adenine unit above) are mixed in solution, the transition temperature

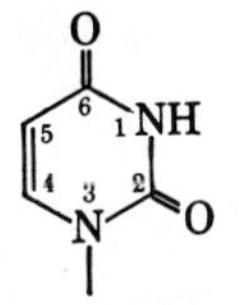

O 6 NH 5 1 4 3 2 N O

(**8**)

is raised by 50° because these two polynucleotides form a 1:1 helical complex which is more stable than the single-stranded structure. Similar "melting" curves are exhibited by RNA and DNA. The NMR method of following this process has the advantage that it is possible to identify which bases are attaining conformational freedom at each stage of the process.

Bibliography

1. J. H. Van Vleck, *Phys. Rev.* **74,** 1168 (1948).
1a. *Chem. Eng. News* **43** (35), 99 (Aug. 30, 1965).
2. W. P. Slichter, *Rubber Chem. Technol.* **34,** 1574 (1961).
3. J. G. Powles, *Proc. Phys. Soc. (London)* **B69,** 281 (1956).
4. A. Odajima, A. E. Woodward, and J. A. Sauer, *J. Polymer Sci.* **55,** 181 (1961).
5. W. P. Slichter, *J. Polymer Sci.* **24,** 173 (1957).
6. J. A. S. Smith, *Discussions Faraday Soc.* **19,** 207 (1955).
7. H. S. Gutowsky and G. E. Pake, *J. Chem. Phys.* **18,** 162 (1950).
8. N. Bloembergen, E. M. Purcell, and R. V. Pound, *Phys. Rev.* **73,** 679 (1948).
9. E. L. Hahn, *Phys. Rev.* **80,** 580 (1950). H. Y. Carr and E. M. Purcell, *Phys. Rev.* **94,** 630 (1954).
10. W. P. Slichter and D. D. Davis, *J. Appl. Phys.* **35,** 3103 (1964).
11. J. G. Powles, J. H. Strange, and D. J. H. Sandiford, *Polymer* **4,** 401 (1963). J. G. Powles, B. I. Hunt, and D. J. H. Sandiford, *Polymer* **5,** 585 (1964).
12. C. W. Wilson, III, and G. E. Pake, *J. Polymer Sci.* **10,** 503 (1953).
13. W. P. Slichter and D. W. McCall, *J. Polymer Sci.* **25,** 230 (1957).
13a. *Chem. Eng. News* **43** (35), 113 (Aug. 30, 1965).
14. T. G Fox, W. E. Goode, S. Gratch, C. M. Huggett, J. D. Kincaid, A. Spell, and J. D. Stroupe, *J. Polymer Sci.* **31,** 173 (1958). T. G Fox, B. S. Garrett, W. E. Goode, S. Gratch, J. F. Kincaid, A. Spell, and J. D. Stroupe, *J. Am. Chem. Soc.* **80,** 1768 (1958). J. D. Stroupe and R. E. Hughes, *J. Am. Chem. Soc.* **80,** 2341 (1958).
15. F. A. Bovey, *J. Polymer Sci.* **46,** 59 (1960).
16. T. G Fox and H. W. Schnecko, *Polymer* **3,** 575 (1962).
17. T. Otsu, B. Yamada, and M. Imoto, *J. Macromol. Chem.* **1,** 61 (1966).
18. U. Johnsen, *J. Polymer Sci.* **54,** S6 (1961).
19. F. A. Bovey, E. W. Anderson, D. C. Douglass, and J. A. Manson, *J. Chem. Phys.* **39,** 1199 (1963).
20. T. Yoshino and J. Komiyama, *J. Polymer Sci.* [B] **3,** 311 (1965).
21. J. W. L. Fordham, P. H. Burleigh, and C. L. Sturm, *J. Polymer Sci.* **20,** 251 (1956).
22. F. A. Bovey, G. V. D. Tiers, and G. Filipovich, *J. Polymer Sci.* **38,** 73 (1959).
23. F. A. Bovey, F. P. Hood, E. W. Anderson, and L. C. Snyder, *J. Chem. Phys.* **42,** 3900 (1965).
24. W. C. Tincher, *J. Polymer Sci.* **62,** S148 (1962); *Makromol. Chem.* **85,** 20 (1965).
25. S. Brownstein, S. Bywater, and D. J. Worsfold, *J. Phys. Chem.* **66,** 2067 (1962).
26. F. C. Stehling, *J. Polymer Sci.* [A] **2,** 1815 (1964).
27. J. C. Woodbrey, *J. Polymer Sci.* [B] **2,** 315 (1964).
28. G. Natta, E. Lombardi, A. L. Segrè, A. Zambelli, and A. Marinangeli, *Chim. Ind.* **47,** 378 (1965).
29. R. C. Ferguson, private communication.
30. S. Brownstein, S. Bywater, and D. J. Worsfold, *Makromol. Chem.* **48,** 127 (1961).
31. K. C. Ramey and G. L. Statton, *Makromol. Chem.* **85,** 287 (1965).
32. G. V. D. Tiers and F. A. Bovey, *J. Polymer Sci.* [A] **1,** 833 (1963).
33. M. Goodman and Y. Fan, *J. Am. Chem. Soc.* **86,** 4922 (1964).
34. K. C. Ramey and N. D. Field, *J. Polymer Sci.* [B] **2,** 62, 69 (1965).
35. W. C. Tincher, *Makromol. Chem.* **85,** 20 (1965).
36. C. Schuerch, W. Fowells, A. Yamada, F. A. Bovey, F. P. Hood, and E. W. Anderson, *J. Am. Chem. Soc.* **86,** 4481 (1964).
37. T. Yoshino, J. Komiyama, and M. Shinomiya, *J. Am. Chem. Soc.* **86,** 4482 (1964); *J. Phys. Chem.* **70,** 1059 (1966).
38. G. Geuskens, J. C. Lubikulu, and C. David, *Polymer* **7,** 63 (1966).

39. H. Sobue, T. Uryu, K. Matsuzaki, and Y. Tabata, *J. Polymer Sci.* [B] **1,** 409 (1963).
40. R. Yamadera and M. Murano, *J. Polymer Sci.* [B] **3,** 831 (1965).
41. K. Matsuzaki, T. Uryu, and M. Takeuchi, *J. Polymer Sci.* [B] **3,** 835 (1965); K. Matsuzaki T. Uryu, K. Ishigure, and M. Takeuchi, *ibid.*, [B] **4,** 93 (1966); K. Matsuzaki, T. Uryu, M. Okada, K. Ishigure, T. Ohki, and M. Takeuchi, *ibid.*, [B] **4,** 487 (1966).
42. K. Fujii, Y. Fujiwara, and S. Fujiwara, *Makromol. Chem.* **89,** 278 (1965).
43. J. Brandrup and M. Goodman, *J. Polymer Sci.* [B] **2,** 123 (1964).
44. M. Goodman and J. Brandrup, *J. Polymer Sci.* [B] **3,** 327 (1965).
45. E. G. Brame, Jr., R. S. Sudol, and O. Vogl, *J. Polymer Sci.* [A] **2,** 5337 (1964).
46. F. A. Bovey, *J. Polymer Sci.* **62,** 197 (1962).
47. H. J. Harwood and W. M. Ritchey, *J. Polymer Sci.* [B] **3,** 419 (1965).
48. R. E. Naylor and S. W. Lasoski, *J. Polymer Sci.* **44,** 1 (1960).
49. C. W. Wilson, III, *J. Polymer Sci.* [A] **1,** 1305 (1963).
50. H. Y. Chen, *Anal. Chem.* **34,** 1793 (1962).
51. M. A. Golub, S. A. Fuqua, and N. S. Bhacca, *J. Am. Chem. Soc.* **84,** 4981 (1962).
52. R. C. Ferguson, *J. Polymer Sci.* [A] **2,** 4735 (1964).
53. F. A. Bovey, G. V. D. Tiers, and G. Filipovich, *J. Polymer Sci.* **38,** 73 (1959).
53a. R. C. Ferguson and W. D. Phillips, *Science* **157,** 257–267 (July 21, 1967).
54. M. Goodman and Y. Masuda, *Biopolymers* **2,** 107 (1964).
55. D. I. Marlborough, K. G. Orrell, and H. N. Rydon, *Chem. Comm.* **1965,** 518.
56. M. Saunders, A. Wishnia, and J. Kirkwood, *J. Am. Chem. Soc.* **79,** 3289 (1957).
57. C. C. McDonald and W. D. Phillips, *Second International Conference on Magnetic Resonance in Biological Systems*, Pergamon Press, Inc., New York, 1967.
58. C. C. McDonald, W. D. Phillips, and S. Penman, *Science* **144,** 1234 (1964).
59. J. P. McTague, V. Ross, and J. H. Gibbs, *Biopolymers* **2,** 163 (1964).

General References

N. S. Bhacca and D. H. Williams, *Application of NMR Spectroscopy in Organic Chemistry*, Holden-Day, Inc., San Francisco, 1964.

J. W. Emsley, J. Feeney, and L. H. Sutcliffe, *High Resolution Nuclear Magnetic Resonance Spectroscopy*, Vols. I and II, Pergamon Press, Ltd., London, 1965.

L. M. Jackman, *Applications of Nuclear Magnetic Resonance Spectroscopy in Organic Chemistry*, Pergamon Press, Ltd., London, 1959.

J. A. Pople, W. G. Schneider, and H. J. Bernstein, *High Resolution Nuclear Magnetic Resonance*, McGraw-Hill Book Co., Inc., New York, 1959.

F. A. Bovey
Bell Telephone Laboratories

DIFFERENTIAL THERMAL ANALYSIS

Differential thermal analysis (DTA), or thermal spectrometry, measures the heat-energy change occurring in a substance as a function of temperature. Experimentally, the sample is heated side by side with an inert reference material at a uniform rate, and the temperature difference between them is measured as a function of temperature, as shown by Figure 1. The resulting curve is called a thermogram or a thermal spectrum, from which information on the temperature, heat, and rate of transformation can be derived. With the more recently developed DTA instruments, even quantitative measurements of heats of transformation and specific heat can also be carried out.

As DTA directly measures the heat-energy change occurring in a substance, it is theoretically possible to detect and measure any physical transition and chemical reaction that is accompanied by a heat-energy change. This implies an extremely broad scope for the DTA technique, as most changes in state and chemical reactions are accompanied by heat-energy changes.

DTA was first used by Le Châtelier in 1887 for the study of thermal transformation in clays. Since then it has been widely used in such fields as mineralogy, soil science, metallurgy, fuel technology, and the chemistry of inorganic and organic compounds (1,2). More recently, DTA has been applied to the study of a wide variety of problems involving high polymers (3–5). For many transitions and reactions, DTA is often simpler and more rapid than many other techniques for yielding a comparable amount of useful information. Since the basic principle and experimental technique of DTA can be found in standard works, the present review attempts to deal only with types of problems involving high polymers to which DTA may be applied.

Transformations may be classified into three basic types: endothermic, exothermic, and the so-called second-order or glass transition, in which not the enthalpy but the specific heat undergoes a sudden change. Schematic presentations for the enthalpy, specific heat, and temperature differential vs temperature are shown in Figure 2 for the three types of transformations. Transformation in high polymers may be divided into those involving physical transitions and those involving chemical reactions. The following treatment will be distributed among these two categories, each illustrated with pertinent examples.

Physical Transitions

Physical transitions that can be studied or have been studied by DTA are: glass transition, "cold" crystallization, crystal–crystal transition, crystallization from the melt, crystalline disorientation, and melting. See also CRYSTALLINITY.

Glass transition involves motion of short-chain segments in the amorphous region, and is related to the brittleness of a polymer. The so-called "cold" crystallization, which occurs far below the melting temperature but above the glass temperature, is

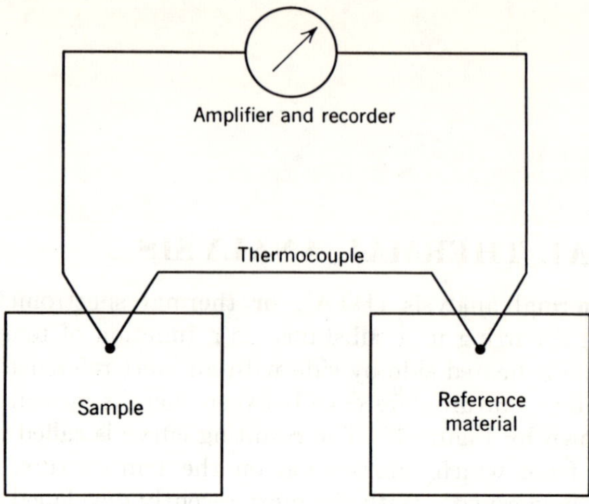

Fig. 1. Basic DTA circuit.

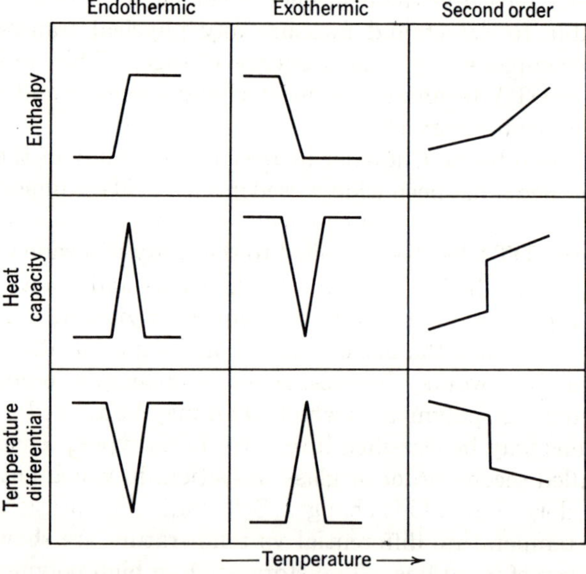

Fig. 2. Changes in enthalpy, heat capacity, and temperature for first- and second-order transitions.

well known in linear polyesters and polyurethans. Crystal–crystal transition has been found only for a few high polymers. Crystallization from the melt is of great practical importance, because it is a major process in practically all fabrications of crystalline thermoplastics. Disorientation takes place in the oriented regions, presumably immediately before melting. Melting, the transition from the crystalline to the amorphous state, has been studied by DTA more extensively than any other transitions.

Melting

Crystallizable polymers are only partially crystalline. Current concepts of the two-phase structure visualize a partially crystalline polymer as containing geometrically perfect regions or crystallites surrounded by amorphous regions. Because of a distribution of crystallite sizes, melting in semicrystalline high polymers invariably takes place over a range of temperatures. A sharply defined temperature at which the total disappearance of crystallinity takes place is usually taken as the "melting point."

A number of phenomena concerning polymer structure and properties may be studied through the melting process. *Properties:* melting temperature and melting

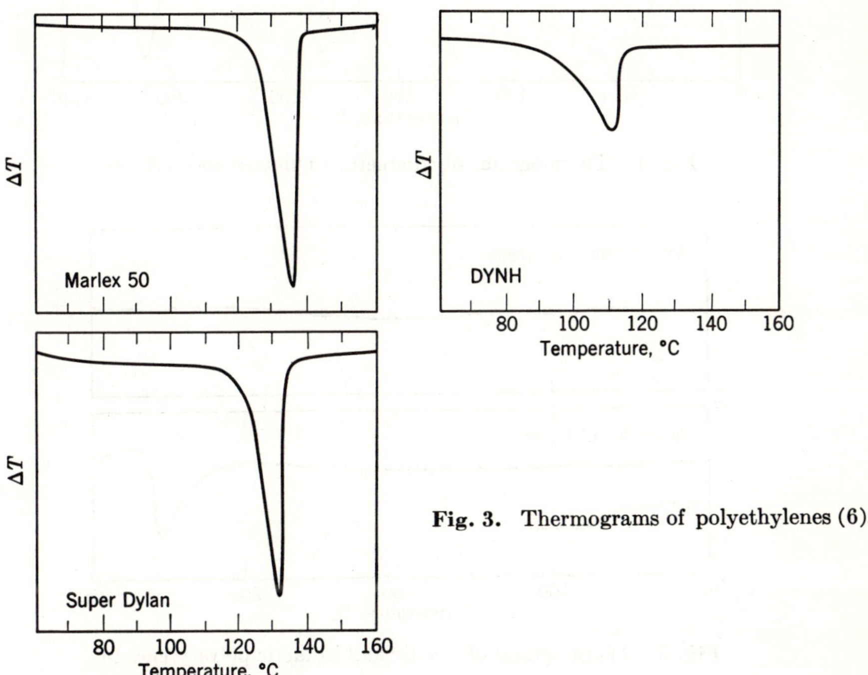

Fig. 3. Thermograms of polyethylenes (6).

range; heat of fusion; melting-point depression. *Structure:* degree of crystallinity; random-copolymer; structure block-copolymer structure; and stereoregularity. *Identification of mixtures.*

Melting Thermograms. Because melting occurring in the semicrystalline polymers is accompanied by a heat effect, it is amenable to study by DTA. Thermograms obtained over the melting region can furnish information about the melting point and heat of fusion as well as the structure and properties of a polymer.

Most polyolefins are highly crystalline thermoplastics. Among polyolefins polyethylene is the simplest in chemical composition and is especially amenable to study by DTA because of its high degree of crystallinity. However, sometimes quite different physical and chemical properties can arise among polyethylenes because of variations in molecular weight, molecular-weight distribution, and chain branching.

Thermograms of two high-density polyethylenes—Marlex (Phillips Chemical Co.) and Super Dylan (Koppers Co.)—and one low-density—DYNH (Union Carbide Co.)—polyethylene are shown in Figure 3 (6). Each polyethylene thermogram yields an endothermic peak corresponding to melting of the crystalline portion of the polymers. The melting point may be read from the position of the peak in the thermogram.

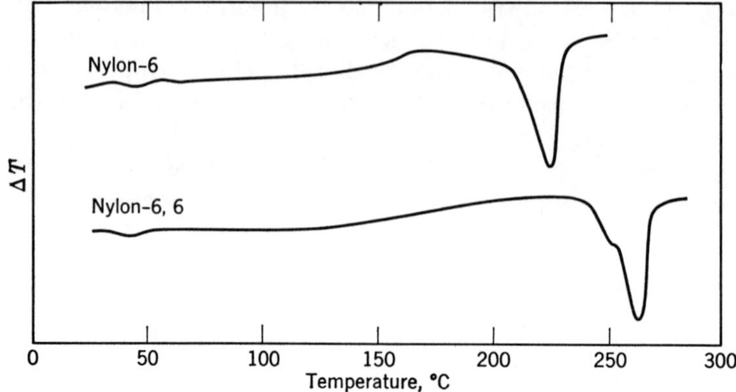

Fig. 4. Thermograms of commercial nylons-6 and -6,6 (8).

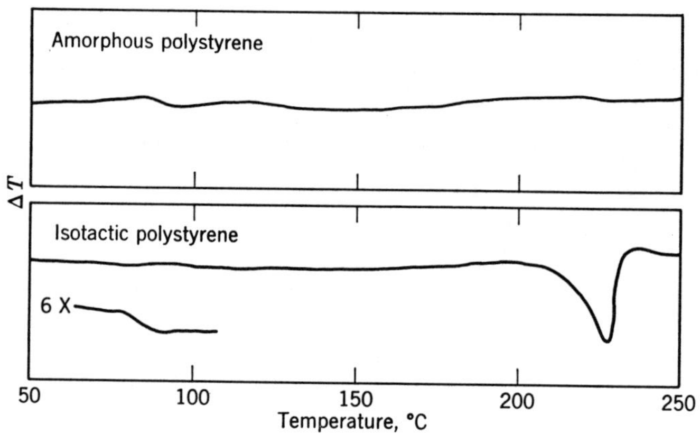

Fig. 5. Thermograms of atactic and isotactic polystyrenes (4).

This point, at which the heat effects are balanced, most closely approximates the temperature at which all crystallinity is lost. The melting points indicated by the peaks agree within 1° with the literature values of 136°, 129°, and 111°C, respectively. The two high-density polyethylenes have nearly the same 15° melting range, as indicated by the distance between the initial departure from the baseline and the peak. The lower melting temperature and the broader melting range of the low-density polyethylene result directly from a wide distribution of crystallite sizes.

Isotactic polypropylene gives a melting peak at 169°C (see Fig. 15) (6). Poly-(4-methyl-1-pentene) contains bulky pendant groups along the chain; however, it has a high melting point. Highly crystalline, isotactic poly(4-methyl-1-pentene) stabilized with approximately 1% Santonox antioxidant gives a well-defined melting peak at 238°C (see Fig. 27) (7).

Figure 4 shows melting thermograms for two commercial polyamides, nylon-6 and -6,6 (8). Melting points of the common polyamides 6 and 6,6 agree well with literature values. The inflection near 50°C is caused by glass transition.

In oriented nylon-6,6 monofilament, a doublet endothermic peak results. The first endothermic peak has been attributed to disorientation, which is followed by normal melting (8,10). When the same monofilament was heated to the disorientation temperature and then cooled slowly to room temperature, it yielded a thermogram with only a single melting peak upon reheating (8).

Crystalline poly(ethylene terephthalate), representing Dacron fiber and Mylar film (Du Pont), shows a melting peak at 260°C (see Figs. 9 and 17). The new polyester, represented by Kodel fiber (Eastman Kodak), poly(1,4-cyclohexylenedimethylene terephthalate), has a thermogram with a melting peak at 299°C (9).

Figure 5 shows melting thermograms of polystyrenes prepared by two different catalysts (4). Ordinary polystyrene prepared by the free-radical process is amorphous

Table 1. Melting Points and Actual Heats of Fusion of Commercial Polyethylenes

	Mp, °C	ΔH_f^*, cal/g	Crystallinity, %
Marlex	135	58.6	91
Super Dylan	130	52.2	81
DYNH	112	33.6	52

even in the stretched state. Isotactic polystyrene prepared by a stereospecific catalyst, because of its regular chain conformation, can crystallize and thus yields a melting thermogram. As shown in Figure 5, atactic or amorphous polystyrene yields an inflection near 90–100°C corresponding to glass transition, but no other peak occurs up to 300°C. The melting peak at 230°C for isotactic polystyrene indicates the crystalline nature of the polymer. The inflection due to glass transition occurs a few degrees lower in the isotactic material.

Heat of Fusion and Degree of Crystallinity or Isotacticity. The heat of fusion is the amount of energy necessary in transforming a polymer from a crystalline or a partially crystalline state to a completely disordered amorphous state. Valid heat effects of a reaction can be estimated by measuring the area under a well-defined thermogram peak. To obtain actual values of heats of fusion, the instrument may be calibrated with a substance such as benzoic acid, for which the heat of fusion is well known (6).

Melting points and actual heats of fusion (ΔH_f^*) measured from thermogram for the three polyethylenes in Figure 3 are listed in Table 1.

If the heat of fusion, ΔH_f, of a perfectly crystalline polyethylene is known, the percentage crystallinity x may be derived from equation 1. The heat of fusion of

$$x = (\Delta H_f^*/\Delta H_f)\ 100 \qquad (1)$$

crystalline dotriacontane ($C_{32}H_{66}$) was used as the true heat of fusion of a completely crystalline polyethylene (6). Resulting crystallinity values, listed in the last column of Table 1, agree well with those obtained by other methods reported in the literature. It is also obvious that, if the degree of crystallinity of a polymer is known through some independent measurements such as x ray or infrared, the heat of fusion obtained by DTA would allow one to calculate the true heat of fusion of the hypothetically perfect crystalline material.

Once the degree of crystallinity of a given polymer is determined, the relationship of crystallinity to temperature can also be determined directly from the thermogram. From measured areas between successive temperature intervals, $\int_{T_1}^{T_2} \Delta T\, dT$, curves showing the decrease of crystallinity with increasing temperature can be constructed. Such curves for the three polyethylenes are shown in Figure 6. For high-density polyethylenes, these curves indicate that 90% of the crystallinity disappears in a temperature range of about 15°. Because of its crystallite-size distribution, DYNH melts over a somewhat wider range.

Since only the isotactic species can crystallize and contribute heat of fusion to the melting peak, the peak area has been used to measure the isotactic content in polypropylene (11). Whole polypropylene was separated by cold toluene into isotactic and amorphous fractions, which were then used to reconstruct artificial mixtures containing 10, 20, 30, and 40% by weight of amorphous material. Thermogram peak areas for the mixtures are plotted against the amorphous content as curve I in Figure

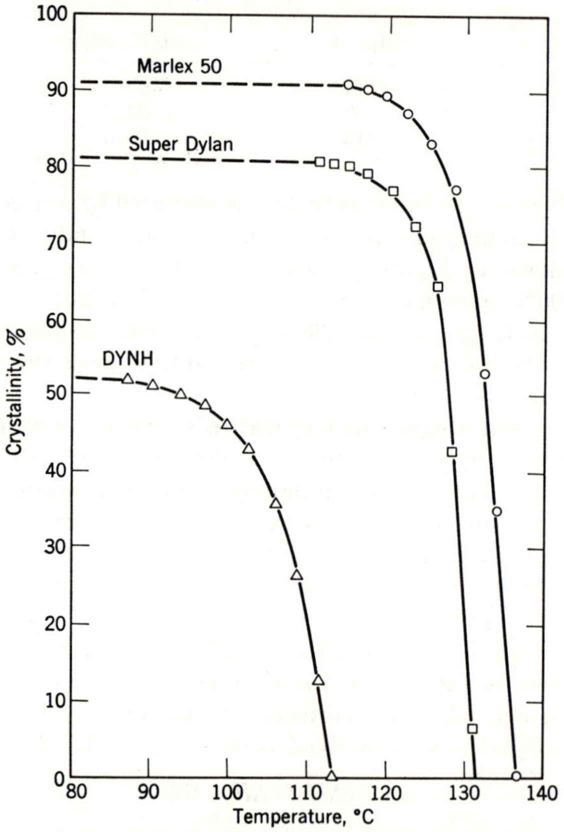

Fig. 6. Effect of temperature on crystallinity of polyethylenes (6).

7. At low amorphous content, the relative change in peak area is large. At 10% amorphous content or greater, the slope becomes linear and extrapolates to zero area at 100% amorphous content. Curve II was plotted from thermogram peak areas of a series of polypropylene samples with different amorphous contents. The numerical difference between the two curves was due to slightly different experimental conditions. For a given set of conditions, the empirical correlation obtained should be valid and consistent for estimating the isotactic content of an unknown sample.

Effect of Thermal History and Molecular Weight. Although crystallizability of a polymer is inherently determined by its molecular structure, the degree of crystallinity and the crystallite-size distribution can be considerably different, depending on the previous thermal history of the sample. High-density polyethylene sample rapidly quenched from the melt in a cold-air stream gives a thermogram noticeably different from that of an annealed sample (7). The effect of thermal history on the crystallinity and hence the thermogram is especially noticeable in polyesters (9). (See under Cold Crystallization for a detailed description of the effect of thermal history on poly(ethylene terephthalate).)

The effect of molecular weight on crystallinity is usually smaller than that of thermal treatment. However, such effects have been observed in low-density polyethylenes, as well as in high-density polyethylene fractions. Unfractionated polymethylenes of inherent viscosities 1 and 12, or weight-average molecular weights of approximately 40,000 and 2,000,000, were found to give noticeably different thermogram peaks (7).

Random Copolymer Structure. Incorporating a comonomer into a polymer profoundly affects its physical properties; it has been used to modify the properties of the parent polymer for particular end uses (12). In general, the effect of copolymerization depends on the nature, amount, and distribution of the comonomer. Comonomer distribution may be either random or ordered. Although random copolymers are most common, many ordered copolymers have also been prepared recently.

Random incorporation of a comonomer into polyethylene results in the formation of pendant groups along the otherwise linear chain. The degree to which the crystallinity and melting point of copolymers are reduced depends on the nature and number of these pendant groups. An early DTA study of a series of polyethylenes containing increasing amounts of propylene showed qualitatively that pendant methyl groups lower the melting point and decrease the peak area (6).

In a more recent DTA investigation (13), two series of olefinic copolymer models—polymethylenes containing up to 20 per 1000 carbon atoms of pendant methyl or

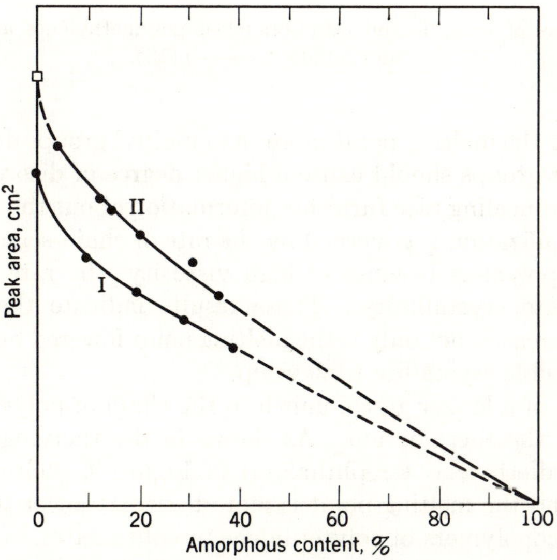

Fig. 7. Relationship between thermogram peak area and amorphous content of isotactic polypropylene (11).

ethyl groups—have been studied. Thermograms of the two series of branched polymethylenes are shown in Figure 8. The solid lines represent copolymers annealed slowly, whereas the dashed lines are for the same copolymers annealed more rapidly. As the number of pendant groups increases, the melting point of the copolymer decreases. As slower annealing promotes the development of fuller crystallinity and larger crystallites, the thermograms of the slowly annealed copolymers always have a higher melting peak position and a larger peak area than those of the polymers annealed fast.

Within the narrow range of comonomer concentration studied, the relationship between melting-point depression and branch concentration is practically linear.

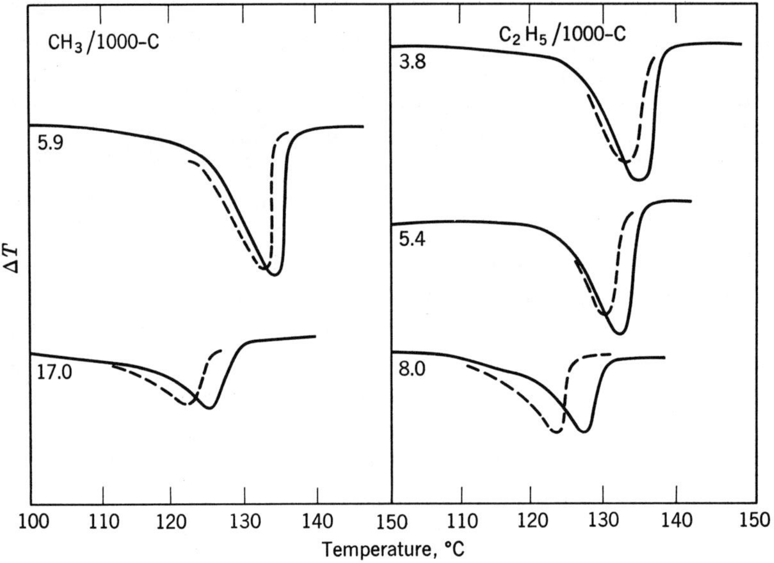

Fig. 8. Thermograms of methyl- and ethyl-branched polymethylenes annealed slowly (———) and rapidly (---------) (13).

Ethyl groups lower the melting point more than methyl groups do. This is reasonable, as bulkier pendant groups should cause a higher degree of disorder.

The effect of annealing rate furnishes information about the time effect on crystallization. As crystallization is governed by the rate of chain-segment relaxation, which is rather slow in polymers because of high viscosity, the rate of cooling should influence the ultimate crystallinity. These results indicate that, as the number of pendant groups increases, not only is the melting point lowered but more time is needed for the largest possible crystallite to develop.

Incorporation of a longer glycol unit into the chain of polyethylene terephthalate markedly changes the thermogram. As shown in the thermogram for the annealed poly(ethylene–oxydiethylene terephthalate) in Figure 9, melting begins nearly 20° lower, at 215°C, and the melting point is more than 20° lower than in the homopolymer. Four other copolymers of poly(ethylene terephthalate), in which either a second acid or a second alcohol was the comonomer, have also been studied by DTA (9).

Block Copolymer Structure. Ordered copolymers, such as block or alternating

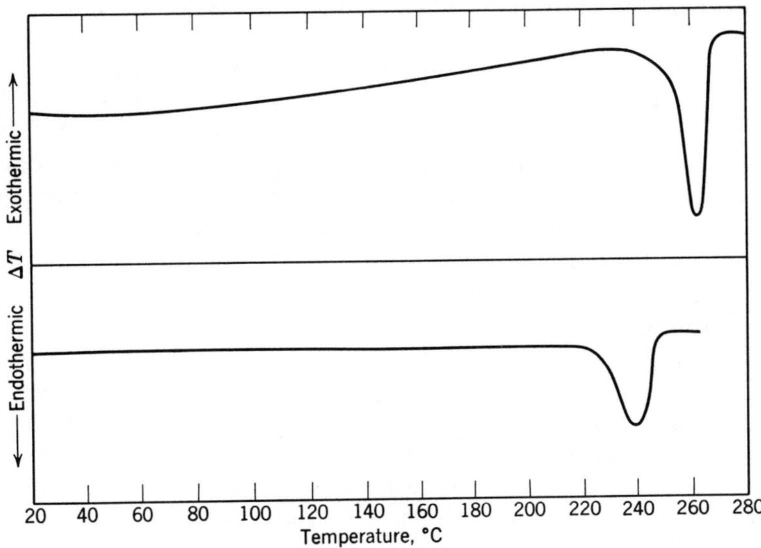

Fig. 9. Thermograms of annealed poly(ethylene terephthalate) (top) and poly(ethylene–oxydiethylene terephthalate) (bottom) (9).

copolymers, have long been predicted, and many have been reported. However, ordered copolymers of olefins have not been prepared until recently (14). Block-type crystalline copolymers of ethylene and propylene are reported to show interesting differences in their chemical and physical properties from random copolymers or physical mixtures (15).

The thermogram of a crystalline ethylene–propylene block copolymer, as shown in Figure 10, consists of two separate melting peaks, indicating the presence of two distinct types of crystallites, one attributable to ethylene and the other to propylene sequences. Conceivably polyethylene and polypropylene blocks either from the same molecule or different molecules can aggregate into separate crystalline regions.

For the sample annealed and cooled at 1°C/min, the crystalline melting points are 130 and 165°C, respectively, which are very close to the melting points of corresponding homopolymers. Thus, one may conclude that the blocks consist of units

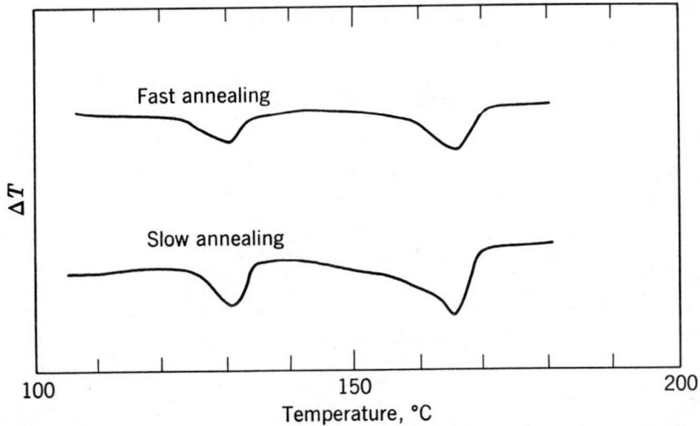

Fig. 10. Thermograms of a crystalline ethylene–propylene block copolymer (13).

that are conventionally considered as homopolymeric rather than random-copolymer units.

Slower annealing raised the melting points of the polyethylene and polypropylene peaks to 132 and 166°C, respectively. Also substantially higher crystallinity was developed by slower annealing. Total crystallinity estimated from the thermogram peak was 57%. With the estimated crystallinity and an independently determined density value (0.917) for the same slowly annealed sample, the percentage composition of the block copolymer was calculated to be 53% by weight of propylene (13).

If we assume the block copolymer contains 47% by weight of ethylene, the crystallinity of 12% for the polyethylene portion is less than half of what would be expected from the pattern developed from the copolymer models. It is conceivable that the block units are somewhat more restricted than in a homopolymer or a random copolymer. Evidence for this suggestion may be found in the crystallization behavior of the block copolymer, discussed in the section on Crystallization.

Similar multiple melting peaks observed in copolyamides prepared by interfacial polycondensation were used as evidence to suggest block copolymer formation (8). Thermograms of the adipamide–sebacamide copolymers prepared by interfacial polycondensation are shown in Figure 11.

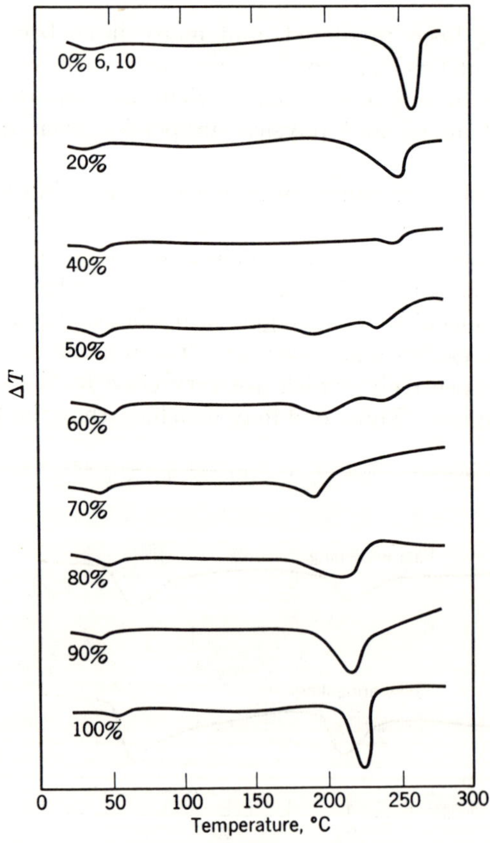

Fig. 11. Thermograms of copolyamides from hexamethylenediamine and adipic and sebacic acids (8).

In adipamide–sebacamide copolymers, the peak positions are shifted to lower temperatures than that of the adipamide parent polymer. The peaks are broadened, and the peak areas decrease as the concentration of the second component is increased. Furthermore, thermograms of copolyamides containing 50 and 60 mole % sebacamide show two peaks. A plot of peak temperatures against copolymer composition results in a eutectic-like phase diagram with the "eutectic" point corresponding to copolyamide containing 70% sebacamide.

The gradual melting-point depression of copolymers containing a small percentage of the second component may be interpreted as due to interference with the crystalline structure by this component. In copolyamides prepared by melt polymerization, substantially lower melting temperatures have been reported and also only one melting point was found for each (16). According to the Flory melting-point equation (17) (eq. 2, p. 49), the smaller melting-point depression found with the copolymers prepared by interfacial polycondensation indicates that the monomers may occur in long sequences, which result in block-type copolymers.

True random copolymers can be prepared by melt polymerization because the reaction conditions allow free interchange of species. Ordered copolymers have been prepared by using such special techniques as melt blending (18) or by starting with macrosegments containing noninterchangeable groups (19). In interfacial polycondensation, several factors may lead to order in the resulting copolymer (20). Among them are differences in the reaction rates of the intermediates and in the transfer rates of the reactants to the polymerization sites. The interplay of these factors can result in copolymers ranging in order from alternating to block type.

Polymer Mixtures. DTA can detect a physical mixture of polymers melting sufficiently wide apart. For instance, it is quite easy to obtain thermograms for a mixture of high-density and low-density polyethylenes. Such blends containing varying proportions of low-density and high-density polyethylenes are commercially available and are said to furnish intermediate physical properties more desirable for particular end uses.

Thermograms of two commercial polyethylene blends are shown in Figure 12 (4). Both blends, Dylan 3014 and 4014 (Koppers), show two separate peaks corresponding to the two components in the blend. Obviously Dylan 3014 contains a higher weight % of low-density component than Dylan 4014. If the exact nature of the components were known, a routine method could be developed for quantitative analysis of such blends using the component materials as calibration standards.

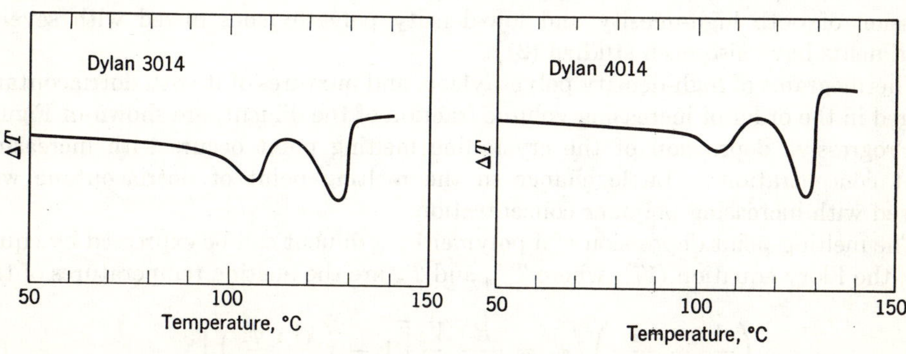

Fig. 12. Thermograms of two commercial polyethylene blends (4).

The thermogram of a physical mixture of equal amounts, by weight, of high-density polyethylene and isotactic polypropylene is shown in Figure 13 (6). Each component produces a peak that retains its characteristic shape and area. The peak areas are proportional to the amount of material present. The peaks for both polyethylene and polypropylene indicate slightly lower melting points, possibly due to an interplasticizing effect.

Melting-Point Depression by Diluents. Because DTA can detect the transition temperatures of individual components in a mixture, the effects of a diluent on the melting behavior of a crystalline polymer can be conveniently studied (21). Knowledge of the nature and magnitude of the interaction between a polymer and a

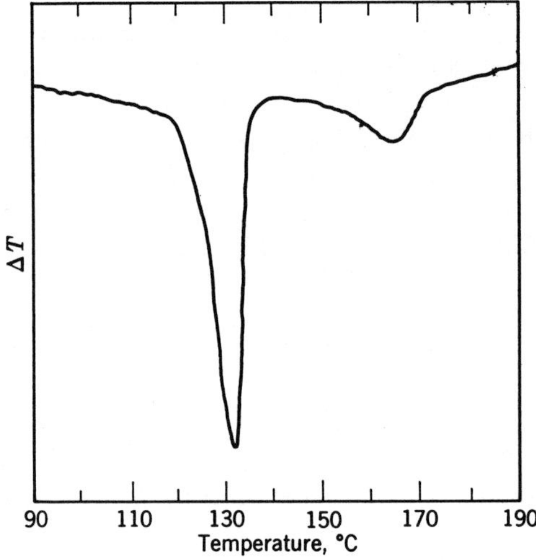

Fig. 13. Thermogram of a physical mixture of high-density polyethylene and isotactic polypropylene (6).

diluent permits estimating the heats and entropies of fusion without knowing the degree of crystallinity of the polymer studied.

A limited number of mixtures of polyethylene with liquid diluents (eg, α-chloronaphthalene and tetralin) have been studied by DTA (21). Within a narrow range of compositions, reproducible melting-point depression was observed. Melting-point depression of both high-density and low-density polyethylenes mixed with several solid diluents have also been studied (21).

Thermograms of high-density polyethylene, and mixtures of it with dotriacontane arranged in the order of increasing volume fraction of the diluent, are shown in Figure 14. Progressive depression of the crystalline melting point occurs with increasing diluent concentrations. Little change in the melting point of dotriacontane was observed with increasing polymer concentration.

The melting-point depression of a polymer by a diluent can be expressed by equation 2, the Flory equation (17), where $T°_m$ and T_m are the melting temperatures of the

$$\left(\frac{1}{T_m} - \frac{1}{T°_m}\right)\Big/\phi_1 = \frac{R}{\Delta H_f}\frac{V_2}{V_1}\left[1 - \left(\frac{\chi V_1}{R}\frac{\phi_1}{T_m}\right)\right] \tag{2}$$

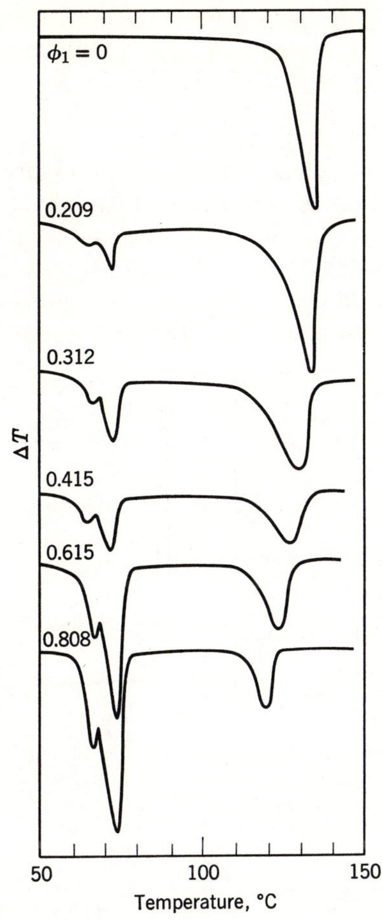

Fig. 14. Thermograms of high-density polyethylene alone and mixed with dotriacontane (21).

pure polymer and the polymer–diluent mixture, ΔH_f is the heat of fusion per polymer repeating unit, R is the gas constant, V_2/V_1 is the ratio of the molar volumes of the polymer repeating unit and the diluent, ϕ_1 is the volume fraction of the diluent, and χ is the interaction-energy density characteristic of the polymer–solvent pair. By plotting $[(1/T_m) - (1/T°_m)]/\phi_1$ vs ϕ_1/T_m, the true heat of fusion and the interaction-energy density can be derived.

Plots of $[(1/T_m) - (1/T°_m)]/\phi_1$ vs ϕ_1/T_m for the high-density polyethylene–dotriacontane mixture and two other polyethylene–solid diluent mixtures have been made. Straight-line plots were obtained in all three cases. From the intercept at $\phi_1 = 0$, the true heats of fusion of the polyethylene repeating unit was estimated to lie between 870 and 970 calories per repeating unit. This range agrees well with the reported value of 940 ± 30 calories (22).

Crystallization

Crystallization in high polymers is not only of theoretical interest for understanding polymer morphology but also of basic importance in such practical operations in plastics fabrication as extrusion and spinning of molten polymers. The mode of crystallization affects the density and crystallinity of the polymer, and consequently its mechanical, thermal, and optical properties.

Polymer crystallization is usually accompanied by evolution of latent heat. This fact can be utilized for following the course of crystallization of polymers by DTA. The heat evolved during crystallization yields exothermic, upward peaks.

Melt Crystallization. Melting and crystallization thermograms of Marlex, DYNH, an isotatic polypropylene, and a mixture of equal amounts, by weight, of a high-density polyethylene and an isotactic polypropylene are shown in Figure 15 (4). Whereas the maximum melting points of polymers are measured by taking the peak position of their melting thermograms, the temperature of the onset of crystallization is measured by the initial point of sharp departure. Crystallization of the high-density polyethylene begins at 123°C; the difference between the maximum melting temperature and the temperature of the onset of crystallization is approximately 13°. The low-density polyethylene, DYNH, has a lower melting point of 110°C, and an initial crystallization temperature of 101°C. Polypropylene shows a high degree of supercooling. Although polypropylene melts near 170°C, its initial crystallization temperature was found to be 129°C—only 6° higher than that for high-density polyethylene. A recent study (23) has shown that the degree of supercooling in polypropylene can be reduced by more than 20° with an appropriate nucleating agent. The resulting polymer showed improved mechanical and optical properties. In the physical mixture with 1:1 weight ratio, at a 2°C/min cooling rate, a single crystallization peak appeared, indicating that the two polymers crystallized together. Simultaneous crystallization can be observed over a broad range of weight proportions.

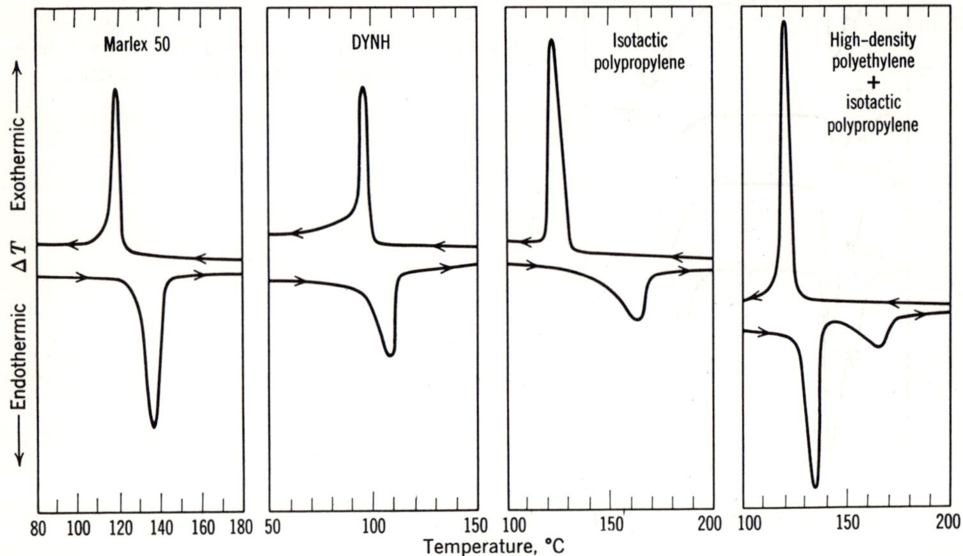

Fig. 15. Melting and crystallization thermograms of some polyolefins (4).

Crystallization thermograms of a physical mixture containing 1:2 parts, by weight, of a high-density polyethylene and isotactic polypropylene and an ethylene–propylene block copolymer are shown in Figure 16. Again, the physical mixture showed simultaneous crystallization. In the block copolymer, crystallization started at about the same temperature as that for the physical mixture. But here an additional crystallization peak appeared, partly overlapping the main peak. To judge by its size, this second peak can be attributed to the polyethylene blocks. The greater degree of supercooling of the polyethylene blocks, which was absent in the physical mixture of the two homopolymers, may also be interpreted as to mean that restrictions exist between the different types of blocks. Barrall and associates (24) recently studied a series of carefully prepared ethylene–propylene block copolymers and confirmed the doublet exothermic peak. These authors made additional correlations between the ethylene content, block length, and degree of randomness.

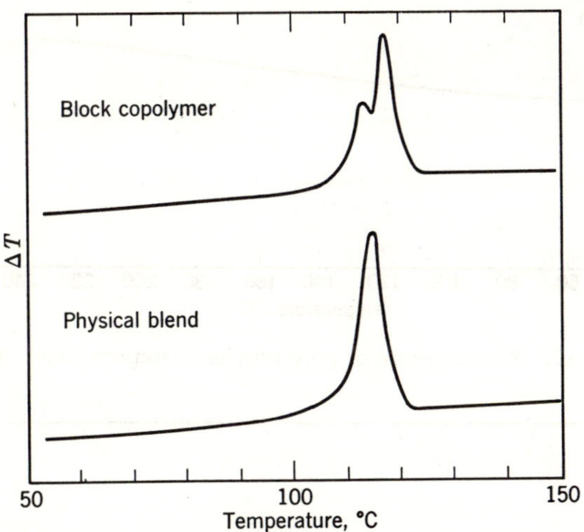

Fig. 16. Crystallization thermograms of an ethylene–propylene block copolymer and a physical mixture of a high-density polyethylene and isotactic polypropylene (13).

Some important factors influencing crystallization, such as the molecular weight of the polymer or the cooling rate, may be conveniently studied by DTA. The rate of crystallization is usually inversely proportional to the molecular weight. The extent of supercooling, namely, the difference between the melting temperature and the temperature for the onset of crystallization, is directly proportional to the cooling rate. For instance, at 1°C/min cooling rate, the supercooling in polypropylene was 35–40°C, and at 7°C/min the supercooling was 50–52°C (25).

In some cases, by measuring the melting and crystallization thermograms consecutively, one may be able to obtain information on how the crystalline nature of the polymer may vary according to crystallization conditions.

Quantitative study of crystallization kinetics is usually made under isothermal conditions. Previous measurements were carried out by the techniques of dilatometry, infrared-absorption spectrophotometry, x-ray diffraction, and polarized microscopy. More recently, several authors successfully adapted DTA to such studies (26–28).

Cold Crystallization. Some high polymers can crystallize, on heating, at temperatures far below the melting point but above the glass-transition temperature. This type of crystallization has been called "cold crystallization," and is assumed to involve the nearest neighboring chain units in the amorphous regions. Such crystallization occurs without molecular rearrangements and leads to production of small crystallites.

When poly(ethylene terephthalate) is quenched from the melt, completely

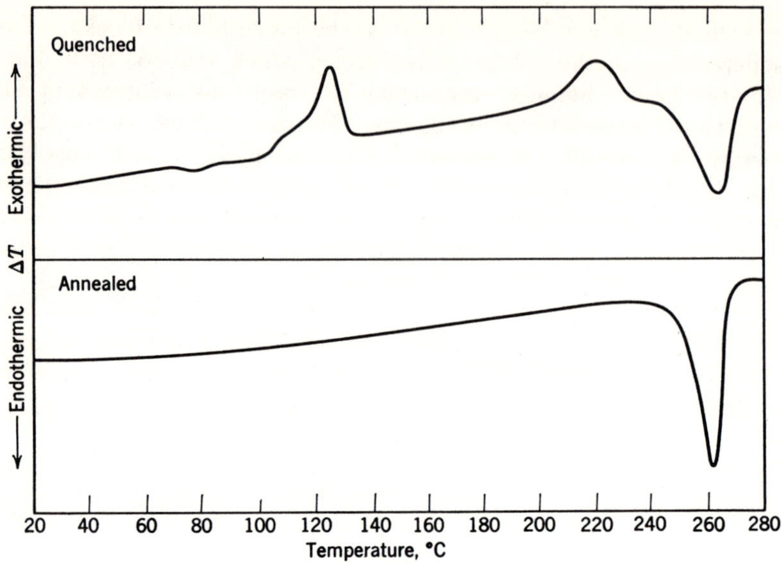

Fig. 17. Thermograms of poly(ethylene terephthalate) (9).

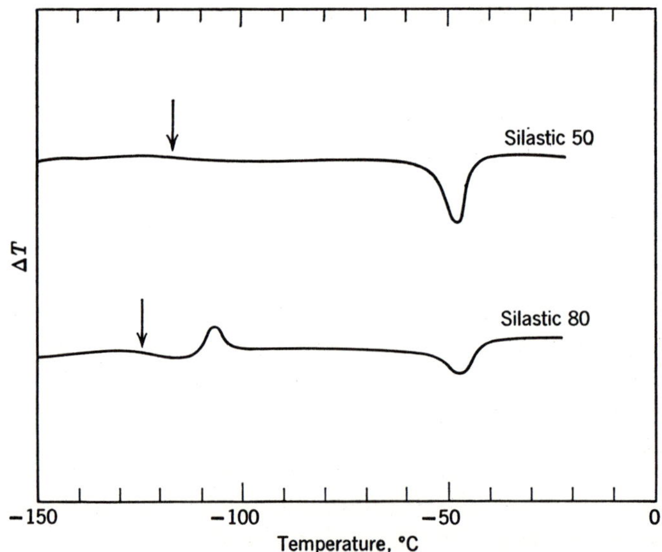

Fig. 18. Thermograms of commercial polysiloxanes (30).

amorphous polymer results. The thermogram for the quenched material is shown at the top of Figure 17; the thermogram for the annealed material is reproduced here for comparison. The deflection near 70°C represents glass transition. An exothermic peak representing cold crystallization starts at 100°C. Premelt crystallization gives an exothermic peak at 220°C. The thermogram shows that melting begins at slightly above 250°C, and ends at 265°C. Similar studies on the transition temperatures and development of crystallinity in poly(ethylene terephthalate) with temperature have recently been made by x-ray diffraction (29); the results agree well with those estimated from the thermogram.

The heats of transition estimated from the various peaks, in calories per gram, are: cold crystallization 7.9, premelt crystallization 6.4, and melting 14.0. Because the sum of the heats of the two crystallizations almost equals the heat of fusion, the original material can be judged as totally amorphous.

In the thermogram for the annealed material, glass transition is barely perceptible, and both crystallization peaks are missing. Annealed poly(ethylene terephthalate) displays a sharper melting peak at 265°C, and a greater heat of fusion, 16.7 cal/g, corresponding to a greater amount of crystallinity developed during annealing (qv).

Cold crystallization can take place at very low temperatures in such polymers as polydimethylsiloxanes. Two commercial polydimethylsiloxane elastomers—Silastic 50 and 80 (Dow Corning)—both vulcanized and containing fillers, were heated at

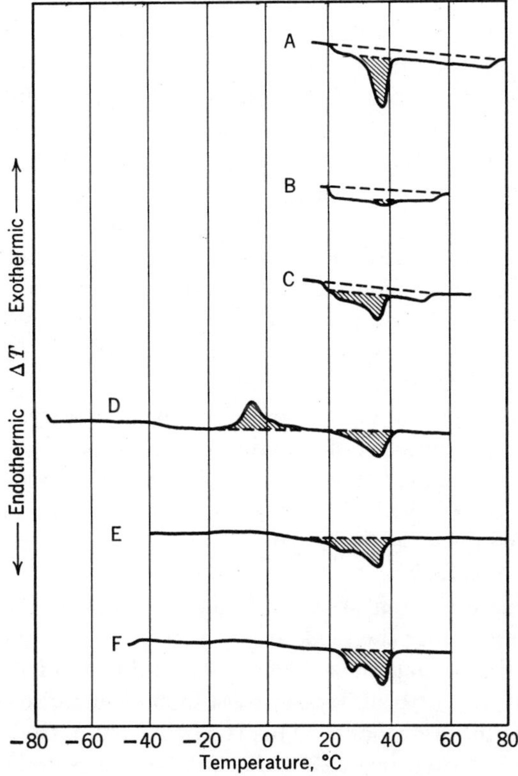

Fig. 19. Melting and crystallization thermograms of a crosslinked polyurethan (31).

2°C/min from −150°C; their thermograms are shown in Figure 18 (30). The numerals 50 and 80 designate the hardness on the durometer scale. Polysiloxane elastomers are known to have very low glass-transition temperatures, which are attributed to the extremely flexible silicon–oxygen backbone chain. Silastic 50 undergoes a glass transition at −112°C, Silastic 80 at −120°C. However, Silastic 80 could be supercooled, as evidenced by the exothermic crystallization peak starting at −108°C upon heating. The supercooling may be related to the manner of quenching, filler content, or state of vulcanization.

In both Silastic 50 and 80, the onset of melting occurs at −55°C, and ends at −44°C. Judging from the size of the melting peaks, it may be inferred that Silastic 50 is more crystalline than Silastic 80.

Another example of cold crystallization occurring at low temperatures is that in "I-rubber," a polyurethan crosslinked with isocyanate (31). This polymer loses its stiffness when heated above 40°C, or when a mechanical stress is applied to it at room temperature. However, the original stiffness can be regained as a result of cold crystallization after long storage.

The thermogram of an isocyanate-linked polyurethan stored for some time, shown in Figure 19A, gives a well-defined melting peak at 38°C. Thermogram B was obtained for the polymer immediately after it was rapidly quenched to 20°C after being held at 60°C for 10 minutes. The melting peak, although located near the same temperature, is only 20% of that in curve A, evidently because there was not enough time for the crystallinity to develop. In fact, the x-ray diffraction pattern of the polymer showed only amorphous rings. If the quenched polymer was stored at room temperature for 24 hours before taking the thermogram, the melting peak became larger (thermogram C). The temperature of the melting peak was only 1°C lower than that of the highly crystalline material. Melting started at a much lower temperature, indicating that much of the polymer was present as smaller crystallites.

Thermogram D (Fig. 19) is for the polymer previously melted and subsequently cooled at 1°C/min to −100°C. The baseline inflection near −40°C represents the glass transition. Cold crystallization started at about −20°C and peaked at −6°C. Melting started at 20°C and peaked at 35°C. The areas of exothermic crystallization peak and the endothermic melting peak are practically equal. When the polymer was cooled to −100°C, as before, but stored at room temperature for 24 hours and then cooled to −40°C before measurement, the thermogram shown in Figure 19E resulted. The exothermic peak due to cold crystallization seen previously in curve D is missing. The melting thermogram consisted of a main peak at 38°C, and a secondary peak at 23°C. If, instead of 24 hours, the supercooled polymer was stored at room temperature for 10 days and then recooled at −40°C before the measurement, the thermogram developed a more pronounced secondary peak at 28°C, as shown in Figure 19F. The primary peak remained at 38°C.

With enough storage time, a sharp melting thermogram like curve A would be obtained again. By measuring the peak areas in curves A, E, and F, almost identical values were obtained, indicating that the total crystallinity of the three samples was the same, except that the crystallite-size distribution changed with time, as small crystallites coalesced into larger ones. This is further supported by the fact that the secondary melting peak shifted from 23°C for the sample stored for 24 hours to 28°C for the sample stored for 10 days. By further analyzing the thermograms more detailed information about the rate of recrystallization could also be obtained (31).

Glass Transition

In the first-order transition of melting, enthalpy undergoes a sudden change at the transition temperature, as illustrated by the many thermograms discussed so far. In glass transition, caused by the onset of motion of chain segments in the amorphous region of a polymer, the change in heat capacity results only in a shift of the thermogram baseline (see Fig. 2). The magnitude of this shift is usually much smaller than that of thermogram peaks involving latent heat. Hence, high stability of the baseline and high sensitivity of the recording instrument are required for measuring glass transitions. To facilitate identification, recording the derivative of the thermogram may be useful (3).

Glass transitions in some polymers have already been shown in many of the thermograms discussed so far: the 45–50°C region for polyamides (Figs. 4 and 11); the 60–70°C region for polyesters (Fig. 17); near −40°C for a crosslinked polyurethan (Fig. 19); and the extremely low-temperature region for polysiloxanes (Fig. 18).

The glass-transition temperature depends on molecular weight, on internal strain in the polymer, and to some extent on the heating rate. The effect of internal strain is clearly shown in the polyesters. The quenched, highly amorphous polymers always display a more pronounced inflection.

The glass-transition temperature depends on molecular weight, on internal strain in the polymer, and to some extent on heating rate. The effect of internal strain is clearly shown in the polyesters. The quenched, highly amorphous polymers always display a more pronounced inflection.

Glass transitions of several amorphous polymers, such as polystyrene, poly(vinyl chloride), poly(methyl methacrylate), vinyl chloride–vinylidene chloride copolymer, and polyacrylonitrile, have been measured by DTA (32). Glass transition in polystyrene was shown previously in Figure 5. The glass-transition temperatures measured by the initial departure of the baseline are in good agreement with values obtained by other techniques. The glass-transition temperature vs molecular-weight relationship has been determined for polyacrylonitrile (32) and polypropylene (30) by DTA. The glass-transition temperature increases rapidly in the region of low molecular weight and levels off at higher molecular weight. Many other polymers, such as polystyrene and polyisobutylene, show similar patterns (33).

Crystal–Crystal Transition

The most well-known crystal–crystal transition is probably that occurring in some higher members of paraffin hydrocarbons. For instance, the thermogram of dotriacontane (6) consists of two closely situated peaks, the first due to chain-rotational transition a few degrees below melting. The premelting peak has been attributed to a crystal–crystal transition from the monoclinic form, in which chain rotation is restricted, to the hexagonal form, in which the molecules can rotate about their chain axes. The heat of fusion is the sum of the heats of rotational transition and normal melting.

Polytetrafluoroethylene (Teflon, Du Pont) displays a unique first-order transition near 20°C, followed by a subsidiary transition at about 30°C, as shown by the thermogram in Figure 20 (30). X-ray studies have indicated that these transitions occurring far below the true melting point (327°C) actually involve a partial disordering of the crystalline region. The disordering has been attributed to either longitudinal translation along, or angular displacement about, the chain axis (34).

An interesting crystal–crystal transition has been reported for isotactic poly-1-butene and has been studied extensively by DTA (35–38). Poly-1-butene exists in three polymorphic forms. The so-called form II (tetragonal) is unstable and is obtained by quenching the polymer rapidly from the molten state. Form II is apparently stable only about 80°C and below −20°C. At room temperature it is transformed into a thermodynamically more stable form I (rhombohedral) simply on long standing. Form III is also a stable form and is prepared only by crystallization from solution.

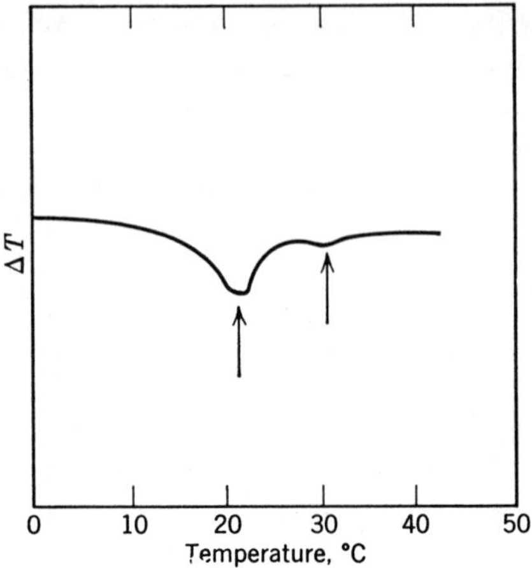

Fig. 20. Thermogram of Teflon (4).

Melting thermograms for poly-1-butene freshly quenched and annealed at room temperature for 16 and 64 hours, respectively, are shown in Figure 21. Pure form II gives a simple melting peak at 118°C. The annealed sample has an additional peak at 131°C caused by the melting of form I. The size of the 131°C peak increases with annealing time; the sample annealed for 64 hours is practically completely converted into form I.

The melting thermogram of form III (curve A, Fig. 22), unlike that of forms I and II, consists of two endothermic peaks, one at 104°C and another at 118°C. The former is presumably characteristic of form III, whereas the latter has already been identified as characteristic of form II. This thermogram appears to suggest that part of the form III sample might have existed as form II. However, if this is the case, then one would also expect conversion of some of form II into form I to have taken place. But, unlike the thermogram of Figure 21, the 131°C peak was absent in the melting thermogram of form III. Clampitt and Hughes (37) suggested that the 104°C peak probably corresponds to a phase transition from form III to II; the latter then melts at 118°C. This was further supported by two experiments (37). A sample of form III was heated to pass the 104°C transition (curve B, Fig. 22) and then immediately quenched. The quenched sample was stored at room temperature for one week before another melting thermogram was taken (curve C). During this heating cycle, the 104°C peak was missing, the 118°C peak due to form II was also small, and the major peak occurred at 132°C. See also CRYSTALLINITY.

Chemical Reactions

Among chemical reactions that can be studied by DTA are polymerizations and various other chemical reactions of interest in the polymer area. Chemical reactions may involve either a polymer reacting with another polymer or with some chemicals, or a transformation caused by some forms of energy such as radiation or heat. Chemical reactions may include oxidation, vulcanization, crosslinking, curing, etc. Effects of radiation may be studied by DTA, especially where crystallinity is involved. Thermal degradation can be most profitably studied in conjunction with gas chromatography.

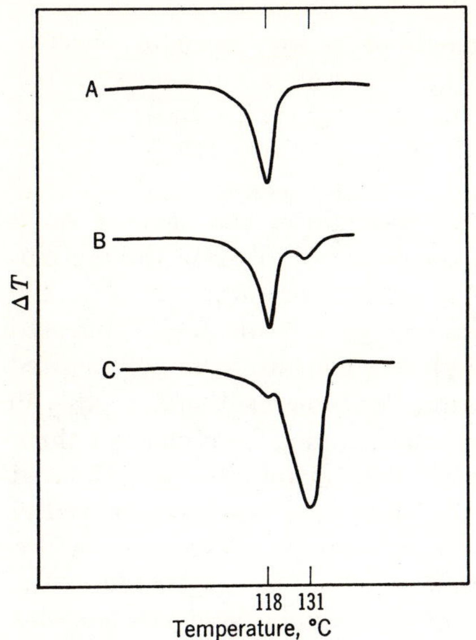

Fig. 21. Thermograms of poly-1-butene freshly quenched from melt (A), and annealed at room temperature for 16 hours (B) and 64 hours (C) (37).

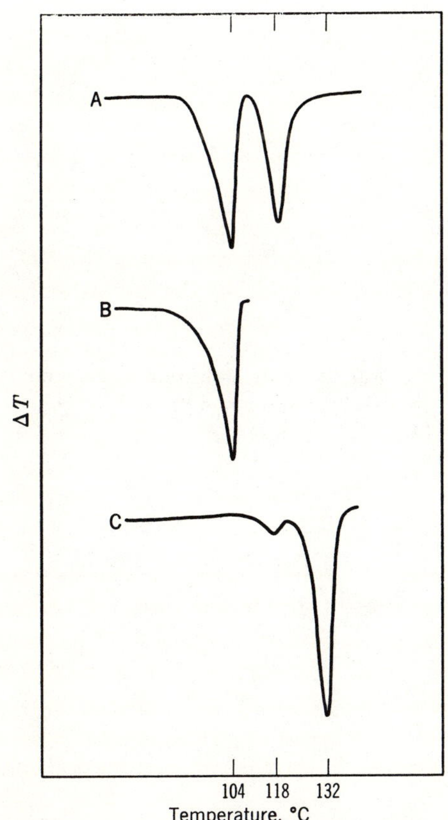

Fig. 22. Thermograms of poly-1-butene crystallized from solution (37).

Polymerization

Polymerizations involving well-defined step reactions accompanied by thermal effects can be studied by DTA. Brasseur and Champetier (39) were the first to use DTA to study the mechanism of polycondensation between such dibasic acids as phthalic acid or its anhydride, and adipic acid with such polyols as ethylene glycol, glycerol, pentaerythritol, and triethanolamine.

More recently, DTA was used to determine the mechanism and heats of polymerization of triallyl cyanurate (TAC) and triallyl isocyanurate (TAIC), the thermograms of which are shown in Figure 23 (40). TAC yields exothermic peaks at 100 and 160–210°C, respectively; TAIC yields only the first exothermic peak. Infrared studies showed that all allyl double bonds in triallyl isocyanurate have polymerized at temperatures beyond the first exotherm. Hence, the first exothermic peak in triallyl isocyanurate can be considered to represent the reaction involving all three double bonds. The size of the first exothermic peak in triallyl cyanurate was estimated to correspond to a reaction of two double bonds. The second exotherm in triallyl cyanurate was then attributed to rearrangement of cyanurate to isocyanurate and a simultaneous polymerization of the remaining double bond. The magnitude of the second exotherm is approximately equal to the heat per double bond reacted plus the difference in heat content between the cyanurate and isocyanurate structure.

The information of thermosetting resins from phenol and formaldehyde has been known for a long time. However, a detailed mechanism is still not completely understood. Polycondensation between phenol and formaldehyde was recently studied

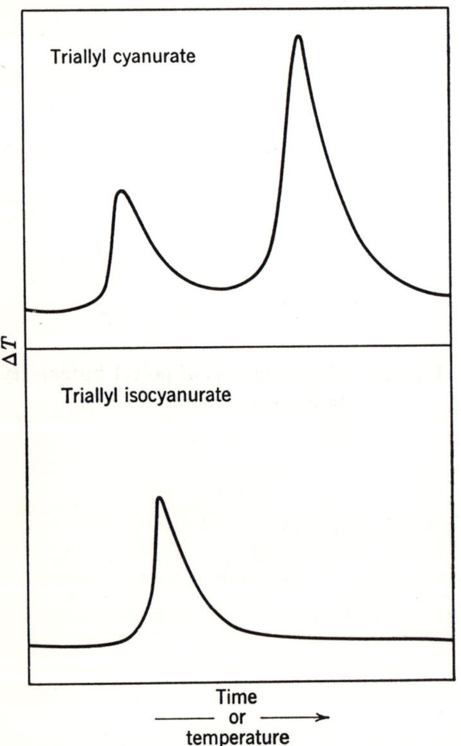

Fig. 23. Polymerization thermograms of triallyl cyanurate and triallyl isocyanurate (40).

by DTA (41,42). The polycondensation reaction is accompanied by an exothermic effect. The thermogram profiles differ considerably, depending on the composition of the mixture and catalyst concentration.

DTA has been applied to the study of Vibrin 135 polyester resin (Naugatuck Chemical) gelled with *tert*-butyl perbenzoate catalyst at room temperature and cured to different degrees; some thermograms are shown in Figure 24 (43). The uncured sample undergoes two exothermic reactions whose mechanism has already been discussed above. For the resin cured for 24 hours at 80°C, the first exotherm decreased considerably; it disappeared completely after postbake for an additional 24 hours at 180°C. The higher temperature exotherm, which involves rearrangement in triallyl cyanurate and polymerization of the remaining allyl bond, is less affected by the curing procedure.

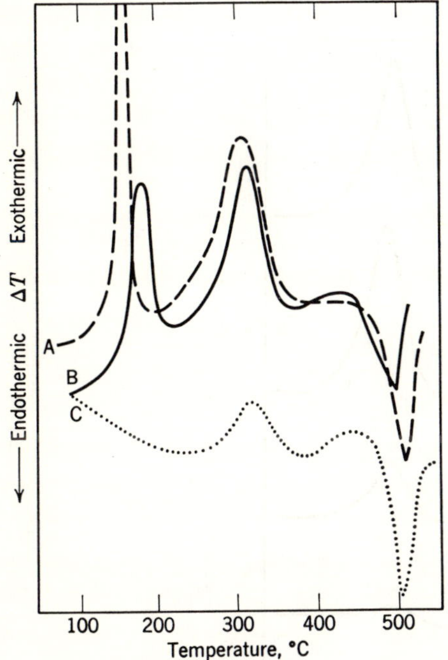

Fig. 24. Effect of curing in Vibrin 135: (A) uncured sample; (B) sample cured at 80°C for 24 hours; (C) sample cured at 180° for 24 hours.

DTA studies have shown that different catalysts can lead to different degrees of curing (43). When benzoyl peroxide was used instead of *tert*-butyl perbenzoate as the catalyst and then cured at 80°C for 24 hours, the 180°C exotherm completely disappeared, indicating that the unsaturation in the polyester and two allyl double bonds in TAC have completely reacted at this low curing temperature. However, the third allyl group in TAC was not affected by the catalyst.

Miscellaneous Chemical Reactions

Very often changes in polymers resulting from a chemical reaction can be determined by DTA. The reaction may involve either another polymer or some micromolecular substance, such as a vulcanizing agent, a crosslinking agent, or oxygen. Chemical reactions caused by such physical agents as radiation and heat should also be amenable to study by DTA.

The only *polymer–polymer reaction* cited in a review article is the unpublished experiment of Doyle (44) on the interaction between nylon and a phenolic resin. Nylon alone displays the usual endothermic melting peak near 270°C. In the nylon–phenolic resin mixture, the melting peak of nylon almost disappeared completely, indicating some sort of interaction had occurred between nylon and the phenolic resin. Also no nylon can be extracted with 98% formic acid from the nylon–phenolic resin composite, although nylon alone dissolves readily in the same solvent.

Crosslinking (qv) may be of different kinds: the polymerization between monomers containing multifunctional groups, reaction between an unsaturated polymer and some multifunctional monomer (eg, curing); intramolecular crosslinking through elimination of some small molecules, as in polyacrylonitrile and polychloroprene; or intermolecular crosslinking through recombination of free radicals. Some examples of the latter two categories will be discussed briefly here.

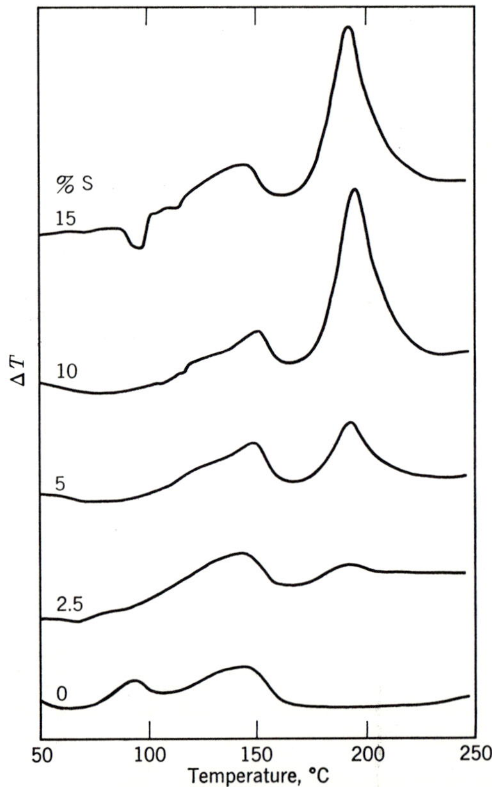

Fig. 25. Thermograms of polybutadiene vulcanized with different amounts of sulfur (4).

Crosslinking of polyethylene is a well-known commercial practice, often used to modify the physical properties of the polymer. Because the interchain crosslinking in polyethylene would inevitably change the crystallinity, DTA should therefore be useful for determining the extent of such a reaction. Ke (4) showed that, when a high-density polyethylene was crosslinked with 1, 3, and 5% of dicumyl peroxide, its crystallinity was lowered by 6, 29, and 50%, respectively.

Crosslinking involving only internal groups of a molecule has been known for polyacrylonitrile and polychloroprene. In polyacrylonitrile, the side groups cyclize to naphthyridene-like rings (see ACRYLONITRILE POLYMERS). Schwenker and Beck (45) studied the crosslinking in polyacrylonitrile (Orlon) and neoprene by taking their thermograms in air and in nitrogen, respectively. An exothermic peak attributable to the crosslinking reaction was observed for each material.

Rubber *vulcanization* is a special case of crosslinking reactions, in which sulfur crosslinks with the unsaturation in polydiene. Because vulcanization is accompanied by a heat change, the reaction can be studied by DTA. Vulcanization of a polybutadiene with sulfur is presented here as an illustration (4).

Thermograms of polybutadiene alone and mixed with different amounts of sulfur are shown in Figure 25. The bottom thermogram is that for pure polybutadiene. Polybutadiene plus 2.5% sulfur yielded a thermogram similar to that of the pure polybutadiene up to 160°C; then a small exothermic band attributable to vulcanization appeared at 195°C. As sulfur content increases, the size of the exotherm in-

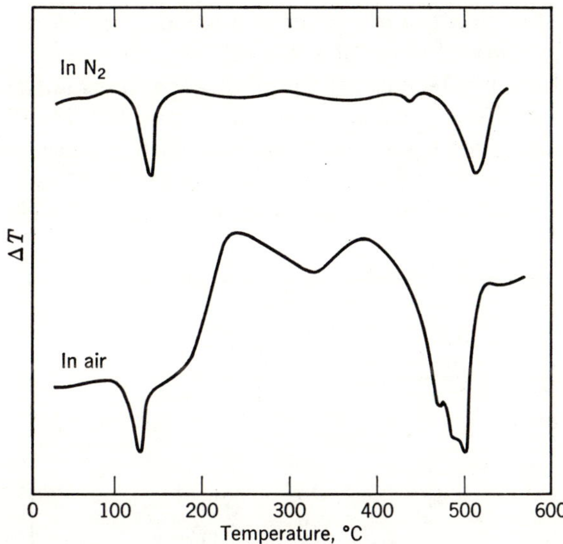

Fig. 26. High-temperature thermograms of polyethylene in air and in nitrogen (25).

creases proportionately. The exotherm area remains constant at sulfur content of 10% or above. This limiting value may be related to the type of sulfur bonds formed in the vulcanizate. For natural rubber the limiting sulfur content is nearly twice as much.

DTA was recently used to examine a rubber–sulfur mixture with a 2:1 weight ratio precured in a press at 153°C for different periods of time (46). As expected, the area of the exotherm was found to decrease with curing time. A dehydrogenation reaction of ebonite by sulfur at high temperatures has also been detected by DTA.

Many polymers undergo *oxidation* at elevated temperatures. The oxidative reaction can usually be established simply by comparing two thermograms, one taken in air and one taken in an inert atmosphere.

Oxidation in high-density polyethylene at elevated temperature is represented by an exotherm ranging from 160 to 450°C, as shown in Figure 26 (25). No such exothermic reaction was visible when the sample was heated under a nitrogen atmosphere. The endothermic peak at 500°C represents thermal depolymerization or degradation of the polymer and is independent of the atmosphere. When heated under nitrogen, a small endothermic peak occurred at about 410°C prior to depolymerization. This peak has been attributed to the rupture of weaker bonds. When heated in the presence of air, the small endothermic peak was overshadowed by the broad oxidation exotherm.

The exothermic reaction during oxidation of polyethylene was recently utilized for the study of oxidation kinetics by DTA (47). The polymer was first melted under nitrogen, producing the regular melting peak near 110°C. The heating block was then brought to the desired temperature at which the rate of oxidation was to be studied. When the temperature became equilibrated, as indicated by a steady baseline, nitrogen was withdrawn, and oxygen was admitted at a slow, steady rate. After an induction period, an exothermic upturn caused by oxidation occurred in the curve. Induction times estimated in this manner and the activation energies calculated from DTA data are in good agreement with literature values obtained by other techniques. The effect of antioxidants on the resistance of a low-density polyethylene to oxidation was also examined by the technique of DTA (47).

Isotactic poly(4-methyl-1-pentene) containing Santonox antioxidant gives a well-defined endothermic melting peak as shown in Figure 27 (7). This same polymer without Santonox also gives a melting peak, but with a considerably smaller area. In addition, a pronounced exothermic peak starting at 178°C and peaking at 186°C appeared, which presumably is caused by oxidation of the polymer. This agrees with

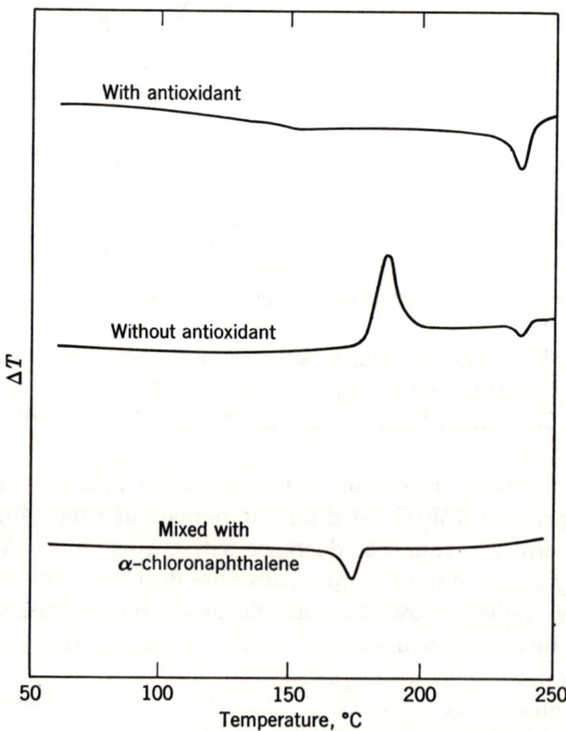

Fig. 27. Thermograms of poly(4-methyl-1-pentene) with and without an antioxidant, and submerged in a solvent (7).

the recent report that poly(4-methyl-1-pentene), among the common polyolefins, has a very high reactivity toward oxygen (48).

There is other evidence that the exothermic peak is associated with oxidation: when the unstabilized polymer was mixed and submerged with a solvent, say, α-chloronaphthalene, the melting thermogram was also free of the exothermic peak, since oxygen was shielded from the polymer by the solvent. The melting peak position was lowered by more than 50°C because of polymer–diluent interaction.

High-energy radiation produces either crosslinking or chain scission in high polymers. Although radiation chemistry of high polymers is a well-established field, the application of DTA to the study of effects of radiation on polymers was reported only recently (49).

Irradiation of poly(vinyl chloride), leads to hydrogen chloride evolution. The nonirradiated poly(vinyl chloride) shows an endothermic peak at 300°C, which is known to be caused by hydrogen chloride evolution. The irradiated material gives a

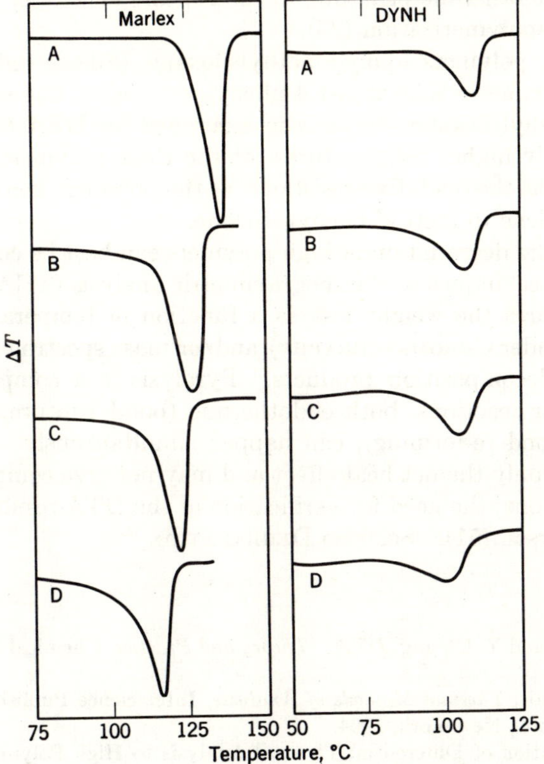

Fig. 28. Thermograms of Marlex and DYNH irradiated with fast electrons: (A) 0 dosage; (B) 100 Mrad; (C) 200 Mrad; (D) 500 Mrad (4).

much smaller endothermic peak, indicating that some hydrogen chloride had already evolved during irradiation (49).

Thermograms of a high-density and a low-density polyethylene, nonirradiated and irradiated with different dosages of electron radiation, are shown in Figure 28 (4). The two thermograms at the top represent nonirradiated Marlex and DYNH.

The three thermograms below are for the polymer irradiated with 100, 200, and 500 Mrad of fast electrons. The lowering of the peak position and the broadening of the peak with irradiation indicate that both crystallite size and crystallite-size distribution were changed. A drastic drop in intrinsic viscosity of the irradiated polymers indicates that chain scission was the primary reaction during irradiation.

Thermal Degradation

Thermal stability of high polymers is of prime importance in the fabrication process. Knowledge gained from studies of polymer degradation may lead to more useful and stable polymers. Application of DTA to the study of thermal degradation is based on the fact that the reaction is always accompanied by pronounced heat changes.

Thermal degradation in polyethylene and polypropylene has already been mentioned. Thermal degradation in both polymers is accompanied by heat absorption, resulting in pronounced endothermic peaks. When polyethylene was heated in nitrogen, a small endothermic peak due to the rupture of weaker bonds could also be detected prior to depolymerization (25).

Three silicone polymers—polydimethylsiloxane (Silicon oil SF96-40, General Electric), poly(tetramethylsiloxanyl-1,4-phenylene), also known as silphenylene, and polymethylphenylsiloxane—have been examined by DTA (50). The exotherm starts at increasingly higher temperatures for the three polymers listed, in the above order, indicating that the oxidative resistance of the polymers increases from dimethyl silicone to silphenylene to methyl phenyl silicone.

Study of thermal degradation of high polymers can best be complemented or corroborated by such techniques as thermogravimetric analysis (TGA) (qv) or derivative TGA, which measures the weight loss as a function of temperature, and gas chromatography (qv under CHROMATOGRAPHY) and/or mass spectrometry, which identify and measure the decomposition products. Pyrolysis is a complex reaction during which various other reactions, both endothermic (bond rupture, volatilization, etc) and exothermic (bond re-forming), can happen simultaneously. In such a case, the thermogram shows only the net heat effect and may not give complete information on all individual reactions; the need for verification of the DTA results by TGA has been suggested by Anderson (51). See also DEGRADATION.

Bibliography

1. W. J. Smothers and Y. Chiang, *DTA: Theory and Practice*, Chemical Rubber Co., Cleveland, 1959.
2. W. W. Wendlandt, *Thermal Methods of Analysis*, Interscience Publishers, a division of John Wiley & Sons, Inc., New York, 1964.
3. B. Ke, "Application of Differential Thermal Analysis to High Polymers," in John Mitchell, Jr., I. M. Kolthoff, E. S. Proskauer, and A. Weissberger, eds., *Organic Analysis*, Vol. 4, Interscience Publishers, Inc., New York, 1960.
4. B. Ke, "Differential Thermal Analysis," in B. Ke, ed., *Newer Methods of Polymer Characterization*, No. 6 in *Polymer Reviews Series*, Interscience Publishers, a division of John Wiley & Sons, Inc., New York, 1964.
5. B. Ke, ed., "Thermal Analysis of High Polymers," *J. Polymer Sci.* [C], Polymer Symposia, No. 6, Interscience Publishers, a division of John Wiley & Sons, Inc., New York, 1964.
6. B. Ke, *J. Polymer Sci.* **42,** 15 (1960).
7. B. Ke, *J. Polymer Sci.* [A] **1,** 1453 (1963).
8. B. Ke and A. W. Sisko, *J. Polymer Sci.,* **50,** 87 (1961).

9. B. Ke, *J. Appl. Polymer Sci.*, **6** (24), 624 (1962).
10. T. R. White, *Nature* **175,** 895 (1955).
11. I. M. Tolchinskii, N. A. Nechitailo, and A. V. Topchiev, *Plasticheskie Massy* **7,** 3 (1960).
12. E. T. Pieski, in A. Renfrew and P. Morgan, eds., *Polythene,* Interscience Publishers, Inc., New York, 1960.
13. B. Ke, *J. Polymer Sci.* **61,** 47 (1962).
14. E. H. Immergut and H. Mark, *Makromol. Chem.* **18/19,** 322 (1956).
15. G. Bier, *Angew. Chem.* **73,** 186 (1961).
16. W. E. Catlin, E. P. Czerwin, and P. H. Wiley, *J. Polymer Sci.* **2,** 412 (1947).
17. P. J. Flory, *Principles of Polymer Chemistry,* Cornell University Press, Ithaca, New York, 1953.
18. M. M. Broubaker, D. D. Coffman, and F. McGrew, U.S. Pat. 2,339,237 (1944).
19. O. B. Edgar and R. Hill, *J. Polymer Sci.* **8,** 1 (1952).
20. P. W. Morgan and S. L. Kwopek, *J. Polymer Sci.* **40,** 299 (1959).
21. B. Ke, *J. Polymer Sci.* **50,** 79 (1961).
22. F. A. Quinn, Jr., and L. Mandelkern, *J. Am. Chem. Soc.* **80,** 3178 (1958).
23. H. N. Beck and H. D. Ledbetter, *Paper, 148th Natl. Meet., Am. Chem. Soc., Chicago, Sept. 1964.*
24. E. M. Barrall, II, R. S. Porter, and J. F. Johnson, *Paper, 148th Natl. Meet., Am. Chem. Soc., Chicago, Sept. 1964.*
25. F. Danusso and G. Palizzotti, *Chim. Ind.* (*Milan*) **44,** 241 (1962).
26. F. H. Müller and H. Martin in Ref. 5, p. 83.
27. H. J. Donald, E. S. Humes, and L. W. White, in Ref. 5, p. 93.
28. J. Chiu, *Anal. Chem.* **36,** 2058 (1964).
29. H. G. Kilian, H. Halboth, and E. Jenckel, *Kolloid-Z.* **172,** 166 (1960).
30. B. Ke, *Polymer Letter* **1,** 167 (1963).
31. F. H. Müller and H. Martin, *Kolloid-Z.* **171,** 119 (1960).
32. J. J. Keavney and E. C. Eberlin, *J. Appl. Polymer Sci.* **3,** 47 (1960).
33. F. Wustlin, in H. A. Stuart, ed., *Die Physik der Hochpolymeren,* Vol. 3, Springer Verlag, Berlin, 1955.
34. C. A. Sperati and H. W. Starkweather, *Fortsch. Hochpolymer.-Forsch.* **2,** 465 (1961).
35. J. Boor and J. C. Mitchell, *J. Polymer Sci.* [A] **1,** 59 (1963).
36. H. Wilski and T. Grewer, in Ref. 5, p. 33.
37. B. H. Clampitt and R. H. Hughes, in Ref. 5, p. 43.
38. C. Geacintov, R. S. Schotland, and R. B. Miles, in Ref. 5, p. 197.
39. P. Brasseur and G. Champetier, *Bull. Soc. Chim. France* **13,** 265 (1946); **14,** 117 (1947); **16,** 793 (1949).
40. B. H. Clampitt, D. E. German, and J. R. Galli, *J. Polymer Sci.* **27,** 515 (1958); **38,** 433 (1959).
41. G. B. Ravich, and A. F. Florova, *Dokl. Akad. Nauk SSSR* **90,** 391 (1953).
42. Y. Nakamura, *Kogyo Kagaku Zasshi* **64,** 392 (1961).
43. C. B. Murphy, A. J. Palm, C. D. Doyle, and E. M. Curtiss, *J. Polymer Sci.* **28,** 447, 453 (1958).
44. C. Doyle, referred to in C. G. Murphy, *Anal. Chem.* **32,** 168R (1960); and private communication.
45. R. F. Schwenker, Jr. and L. R. Beck, Jr., *Textile Res. J.* **30,** 624 (1960).
46. M. L. Bhaumik, A. K. Sircar, and D. Banerjee, *J. Appl. Polymer Sci.* **4,** 366 (1960).
47. A. Rudin, H. P. Schreiber, and M. H. Waldman, *Ind. Eng. Chem.* **53,** 137 (1961).
48. F. H. Winslow and W. Matreyek, *Paper, 141st Natl. Meet., Am. Chem. Soc., Washington, D.C., March 1962.*
49. C. B. Murphy and J. A. Hill, *Nucleonics* **18,** 78 (1960).
50. G. P. Brown, J. A. Hill, and C. B. Murphy, *J. Polymer Sci.* **55,** 119 (1961).
51. H. C. Anderson, *Nature* **191,** 1088 (1961).

Bacon Ke
Charles F. Kettering Research Laboratory

THERMOGRAVIMETRIC ANALYSIS

The term "thermobalance" seems to have been introduced by Honda in 1915 to describe an instrument that he constructed which continuously measured weight changes of a substance at gradually varying temperatures (1,2). In 1926, Saito (3) described an improved balance, and in succeeding years various other Japanese workers made further improvements which allowed hundreds of pyrolysis curves to be obtained (2). From 1923 to 1938, French workers, eg, Guichard and his students, developed thermobalance instrumentation and studied the scope and limitations of the method (1,2). In 1943, Chevenard developed a thermobalance, which bears his name, that was both rugged and sensitive enough to be employed in an industrial laboratory. In 1946, Duval and his co-workers began studies of the pyrolysis of analytical precipitates employing the Chevenard thermobalance, and in the following four years they recorded the pyrolysis curves of several hundred inorganic precipitates.

Most of the early work using the thermobalance involved inorganic materials and did not involve estimations of kinetic parameters in pyrolysis, such as frequency factor, reaction order, and activation energy. One of the early attempts to estimate these parameters for the pyrolysis of organic compounds such as coal and polystyrene was made in 1951 by Van Krevelen et al. (4). However, the methods proposed were tedious and involved many assumptions. In 1958, Freeman and Carroll (5) developed a widely used method for the determination of reaction kinetics using the thermobalance. During the past few years, much progress has been made in developing more suitable methods for the determination of kinetic parameters, and some of these methods will be discussed in this article, along with the application of TGA to specific materials.

Primarily, dynamic thermogravimetric analysis (TGA) and some associated complementary methods in polymer degradation studies will be discussed. We will define TGA as a continuous process that involves the measurement of sample weight as the reaction temperature is changed by means of a programmed rate of heating. Isothermal and essentially isothermal methods will not be covered. However, results obtained by these methods will be compared, where possible, with results obtained by TGA.

In the discussion that follows, a more or less systematic presentation of methods involved in TGA based upon their "exactness" and versatility is given. This is followed by a description of qualitative and arbitrary quantitative estimations of thermal stabilities of polymers based upon their thermograms. Finally, the application of TGA methods to studies of the behavior of some polymeric materials undergoing pyrolysis is covered. For a fuller treatment see Ref. 6.

Instrumentation

Figures 1 and 2 show block schematic diagrams of two of the many commercially available thermobalances, ie, the Thermo-Grav and the Chevenard recording thermo-

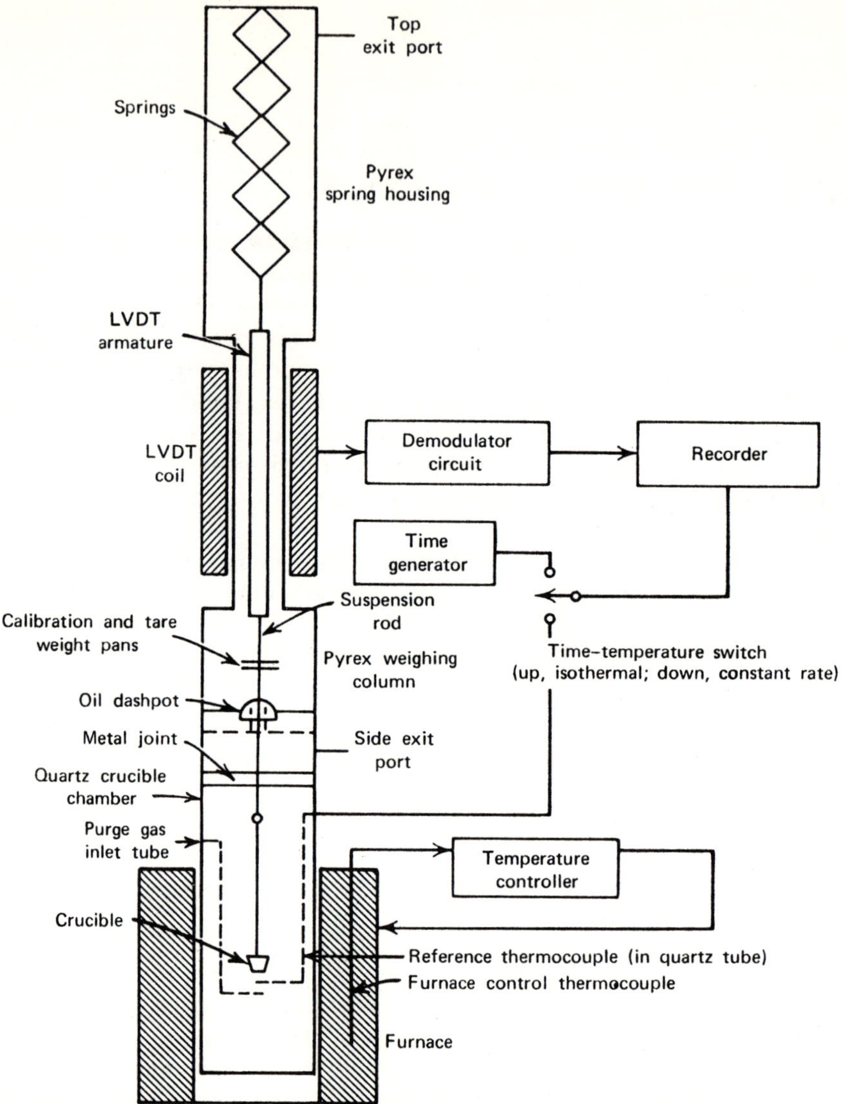

Fig. 1. Schematic diagram of Thermo-Grav recording thermobalance.

balances. (For fuller coverage, see Refs. 7–9.) The Thermo-Grav is manufactured by the American Instrument Co., Inc., Silver Spring, Md.; the Chevenard is manufactured by the Société A.D.A.M.E.L., Paris, France.

To use the Thermo-Grav balance, a preweighed sample is placed in a crucible holder suspended from precision springs. The armature of a linear variable differential transformer (LVDT) which is mounted on the suspension rod actuates the coil as the sample changes weight during heating. The corresponding electrical signal is demodulated, amplified, and used as the input for the vertical Y axis of the recorder. The horizontal X axis of the recorder is used to indicate temperature (up to 1000°C) or time for either TGA or isothermal experiments, respectively. In TGA, a chromel–alumel thermocouple near the crucible records the sample temperature. Various

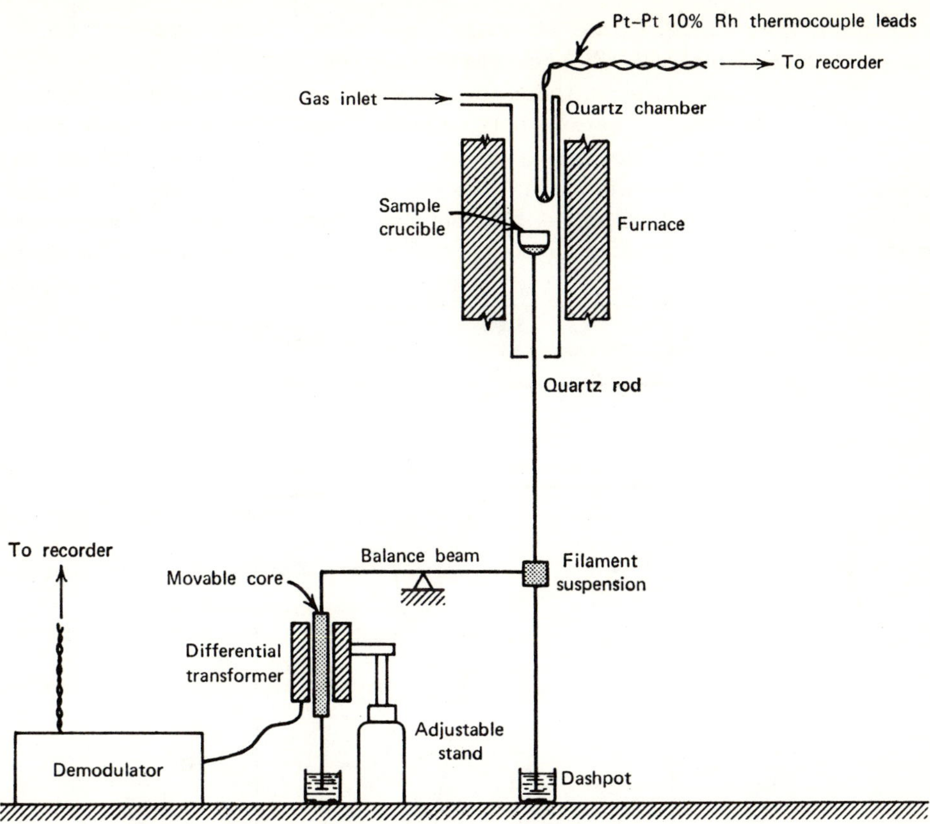

Fig. 2. Schematic diagram of Chevenard recording thermobalance.

nominal furnace heating rates may be chosen by means of a selector switch, and a periodic electrical signal results in a spike on the TGA thermogram. The exact heating rate can be estimated from these spikes. The weighing column may be evacuated for runs in vacuo, or various gases may be introduced into the system during a run at atmospheric pressure. A major disadvantage of this balance is that high-boiling volatile material may deposit on portions of the weighing system.

The Chevenard balance tends to avoid such deposits by an arrangement whereby the weighing system is situated below the sample to be pyrolyzed. This balance is also a deflection-type instrument containing a crucible which is supported by a quartz rod with an end ring. A balance beam is also suspended from the rod, and, by means of a tungsten filament, beam movements actuate an LVDT which emits electrical signals corresponding to weight changes. As in the Thermo-Grav, the electrical signals are demodulated and used to actuate a strip chart recorder. A two-channel recorder plots furnace thermocouple temperature (up to 900°C) and sample weight versus time. A wide range of heating rates may be employed and, as in the case of the Thermo-Grav, oil dashpots are used to damp out parasitic oscillations. The vacuum model can be used for controlled-atmosphere operation as well as for vacuum operation.

It may also be mentioned here that thermoanalyzers are available which can simultaneously record temperature and sample weight loss along with derivative thermogravimetric (DTG) (rate of weight loss versus temperature or time) and differential thermal analysis (DTA) curves (Fig. 3) (10).

As the reader may surmise, there are many experimental variables that may influence the determination of various parameters obtained from pyrolyses of polymers using TGA techniques. Thus, activation energy associated with a pyrolysis may be affected by the shape of the crucible; the location and arrangement of temperature measuring devices; the size, shape, and physical characteristics of the sample (eg, particle size); the furnace arrangement; and the surrounding sample atmosphere.

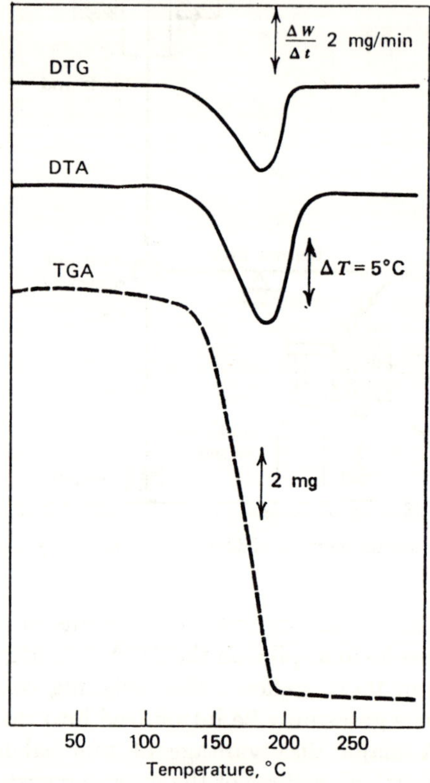

Fig. 3. Simultaneous TGA/DTG/DTA of the dehydration of calcium oxalate monohydrate (10). Heating rate is 10°C/min, sample weight 93.3 mg.

Nevertheless, valid and consistent results can be obtained by observing such precautions as use of a sample small enough to ensure temperature uniformity during decomposition as well as direct sample-temperature measurement; use of samples with a uniform sample size and which are uniformly packed in the crucible; and adjustment of the gas flow, pressure, and sample shape to reduce the effects of effluent gases from the sample.

Determination of Kinetic Parameters

In many polymer pyrolyses the TGA curve follows a relatively simple sigmoidal path. Thus the sample weight decreases slowly as reaction begins, then decreases rapidly over a comparatively narrow temperature range, and finally levels off as the reactant becomes spent. The shape of the curve depends primarily upon the kinetic parameters involved, ie, upon reaction order (n), frequency factor (A), and activation energy (E). The values of these parameters can be of major importance in the elucidation of mechanisms involved in polymer degradation (11,12) and in the estimation of thermal stability (13). However, it should be realized that the expressions utilized to evaluate these parameters are generally valid for fluid systems but are of questionable validity in solid-state reactions. Therefore, too much significance should not be given to the values of these parameters without substantiating evidence.

Thermogravimetric curves may be more complex than described above. Thus, if a material degrades by a multistep mechanism which involves rate-controlling steps of similar order, and if the activation energies of the rate-controlling steps are of a similar magnitude, a relatively simple trace may be obtained which provides an overall activation energy for the sample degradation. However, if the values of E of the rate-controlling steps differ sufficiently, the TGA trace may involve two or more sigmoids, and if the reaction orders for the various rate-controlling steps have values greater than zero, two or more inflection points may be observed. The separate sigmoidal traces may be individually analyzed for values of E, n, and A by methods similar to those employed for TGA curves that possess one sigmoid. However, if values of E for the rate-controlling steps are not sufficiently different, then one reaction may overlap another and analyses of the TGA curves may be difficult, if not impossible. TGA studies give values of overall kinetic parameters which, in themselves, may often shed little light on the mechanism involved in any particular pyrolysis. It is therefore often necessary to complement TGA studies with differential thermal analysis and with chromatographic, infrared, and mass spectrographic methods.

There are several advantages to using TGA methods rather than isothermal methods in the determination of kinetic parameters (14,15). These include the following:

1. Considerably fewer data are required. The temperature dependence of the volatilization rate may be determined over various temperature ranges from the results of a single experiment, whereas several separate experiments are required for each temperature range if isothermal methods are employed.

2. The continuous recording of weight loss versus temperature ensures that no features of the pyrolysis kinetics are overlooked.

3. A single sample is used for the entire TGA trace, thereby avoiding a possible source of variation in the estimation of kinetic parameters.

4. In the isothermal method, a sample may undergo premature reaction and this may make the subsequent kinetic data difficult, if not impossible, to analyze properly.

A major disadvantage is that precise temperature control for kinetic experiments is much more difficult with TGA than with isothermal methods.

Several so-called exact methods have been proposed for estimating kinetic parameters from TGA curves. These methods are based upon the assumptions that thermal and diffusion barriers are negligible and that the Arrhenius equation is valid. The former is reasonable since, in general, small quantities of powdered samples are employed in TGA studies. In the following discussion, the first four methods described allow the estimation of E and n (and A). Succeeding methods described involve the determination of only a single kinetic parameter. Finally, methods are presented that involve additional assumptions and/or approximations.

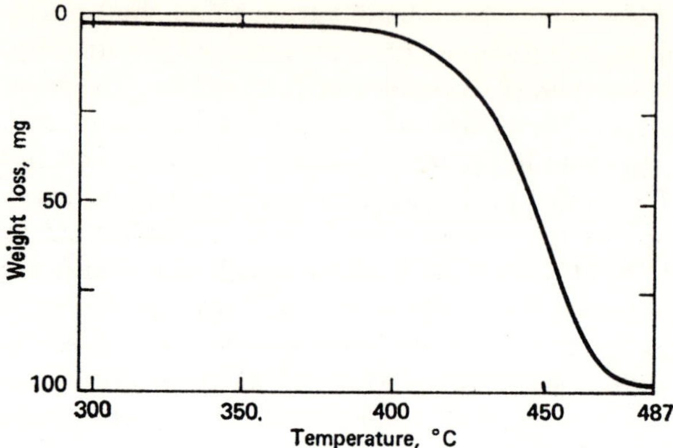

Fig. 4. Thermogravimetric curve for the degradation of polyethylene at 1 mm Hg pressure (16). Sample weight is 100 mg, heating rate 5°C/min.

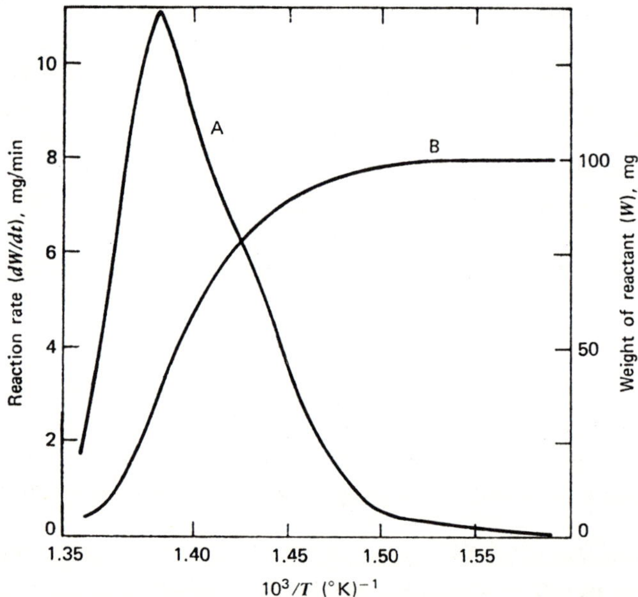

Fig. 5. Graph of the first derivative of the thermogravimetric curve (*dW/dt*) (curve A) and the weight of the reactant (curve B) as a function of reciprocal absolute temperature for the degradation of polyethylene in vacuum (16).

Method of Freeman and Carroll (5,16). For reactions which involve a decomposition of the type

$$\text{A (solid)} \rightarrow \text{B (solid)} + \text{C (gas)}$$

a general rate expression may be written as equation 1, where W is weight of active

$$R_T = -dW/dT = (A/RH)(e^{-E/RT}W^n) \tag{1}$$

material remaining for a particular reaction and RH is the rate of heating. When equation 1 is applied at two different temperatures and the resulting expressions are subtracted from one another (RH is constant), equation 2 is obtained where R_t =

$$\left.\begin{matrix}\Delta \log R_T \\ \text{or} \\ \Delta \log R_t\end{matrix}\right\} = n\,\Delta \log W - (E/2.303R)\,\Delta(1/T) \qquad (2)$$

$RH(R_T)$. From equation 2 it can be seen that values of E and n may be calculated from a single TGA curve. Thus, $\Delta \log R_t$ should be linear with $\Delta \log W$ when $\Delta(1/T)$ is held constant. The slope of the resulting linear curve will give a value for n, whereas the intercept will give a value for E.

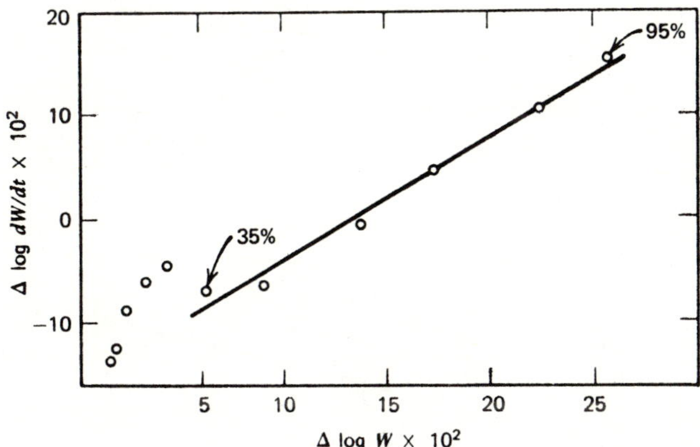

Fig. 6. Kinetics of the thermal degradation of polyethylene in vacuum (14). Data are taken from Figure 3 (16).

Figure 4 shows a primary thermogram for a sample of polyethylene. In Figure 5 the curves obtained using data from Figure 4 are shown. Values of R_t and W at equally spaced intervals of $1/T$ may thus be obtained. From such values, equation 2 may be plotted as shown in Figure 6. Above 35% reaction it was found that n was essentially unity and that E had a value of 67 ± 5 kcal/mole. These values appear to agree favorably with values reported by Madorsky (17), who used isothermal procedures. Other dependencies were found for the polyethylene degradation in vacuum below 35% conversion and are indicated in Figure 7.

These results obtained for polyethylene degradation illustrate advantages for procedures such as the method of Freeman and Carroll. Comparatively little data are required and the kinetics can be studied continuously over an entire temperature range. This can be particularly important in cases of polymer pyrolyses where the kinetic parameters change with conversion.

An important disadvantage of this method lies in the need for obtaining slopes from steep portions of the primary thermogram. As a result, there is often sufficient

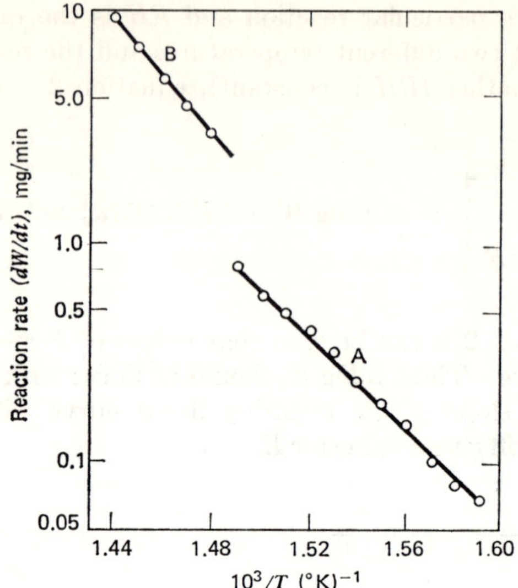

Fig. 7. Temperature dependence of the low-temperature thermal degradation of polyethylene in vacuum: curve A, up to 3% degradation; curve B, from 3 to 15% degradation (16).

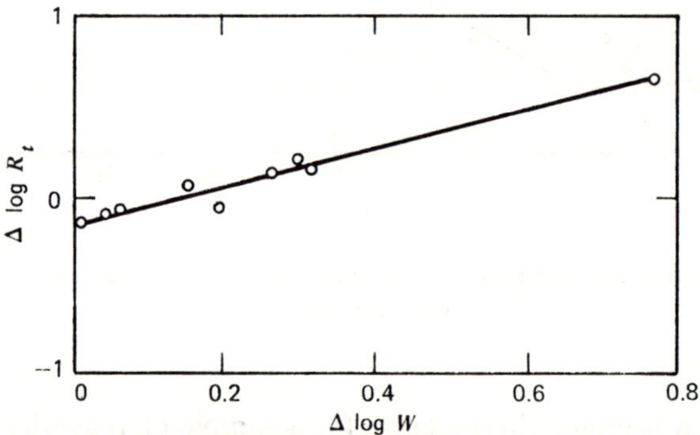

Fig. 8. Δ log R_t plotted against Δ log W for the thermal degradation of polytetrafluoroethylene (Teflon, Du Pont) in vacuum (11).

scatter in the plot to make accurate evaluation of kinetic parameters difficult. The case of the thermal degradation of polytetrafluoroethylene in vacuum (Fig. 8) is an example of this. There is often enough scatter for small changes in the slope of the line to produce relatively large changes in E and n. It was recently found (18) that the method of Freeman and Carroll could be applied to the second step of the thermal decomposition of thorium 8-quinolinol chelate but that a linear relationship could not be obtained by this method for the first step of the degradation (18).

Methods Involving Maximization of Rate (19,20). A method for evaluating kinetic parameters that emphasizes the position of the inflection point on the primary thermogram has been reported by Reich and co-workers (19) and by Fuoss and co-workers (20). If equation 1 is differentiated with respect to T and the result is set equal to zero, equation 3 is obtained after rearranging (19,20). In this equation

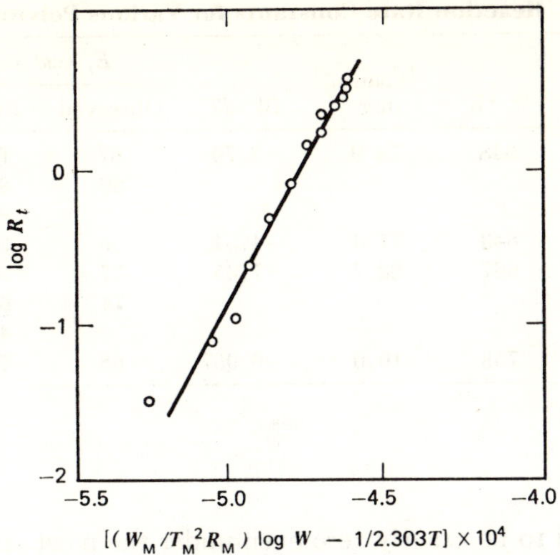

Fig. 9. Log R_t plotted against $(W_M/T_M^2R_M)$ log $W - (1/2.303T) \times 10^4$ for polytetrafluoroethylene at a heating rate of 6°C/min (24).

$$n = (E/R)(W_M/R_MT_M^2) \tag{3}$$

W_M, R_M, and T_M are the weight of active material remaining, the slope, and the temperature, respectively, at the inflection point on the primary thermogram. If equation 1 is converted into a logarithmic expression and equation 3 is substituted into this expression, equation 4 is obtained (19).

$$\log R_t = \log A + (E/R)[(W_M/R_MT_M^2) \log W - (1/2.303T)] \tag{4}$$

In a plot of log R_t from equation 4, it can be readily seen that the slope of the resulting linear relation will give the value of E and that the intercept will give the value of A (Fig. 9). After the value of E has been obtained, the value of n may be obtained from equation 3. Values of kinetic parameters obtained by this method for the pyrolysis of polytetrafluoroethylene in vacuum were found to agree well with values in the literature.

Fuoss and co-workers (20) presumed the reaction order in order to estimate E from equation 3. Apparently, they were unaware of equation 4. The value of E was estimated from a single slope at the inflection point of the primary thermogram. This procedure may be difficult to apply to samples whose TGA curves become very steep. Also, if the value of n is not integral, as is usually assumed, but fractional, the value of E may be in considerable error. Nevertheless, Fuoss and co-workers were able to obtain values for E and A by their procedure which appear to be in good agreement with values in the literature (Table 1).

Equation 4 resembles equation 2 in various respects, and it might be anticipated that both methods would have similar advantages and disadvantages. It may appear in theory that the former expression is limited in use since it can be applied only to reaction orders with values larger than zero. However, in practice, the value of n

Table 1. Reaction Rate Constants for Various Polymers (20)

Polymer	T, °K	W_{initial}, mg	dW/dT	E, kcal/mole Observed	E, kcal/mole Reported	A
polytetrafluoroethylene[a]	848	78.9	−3.70	67	67	2.4×10^{16}
				69	80	
					80.5	
acrylic resin[b]	643	71.3	−4.84	56	32–55	2.0×10^{18}
polystyrene	667	83.7	−7.35	77	58	5.0×10^{24}
				74	60 ± 5	
					45	
$CaC_2O_4 \rightarrow CaCO_3 + CO$	758	16.0	−0.957	68	74	8.2×10^{18}

[a] Teflon, Du Pont.
[b] Lucite, Du Pont.

would not be expected to be exactly zero even under the most stringent experimental techniques employed. Therefore, this method should have wider applicability to pyrolysis kinetics than might appear on first inspection.

Method of Multiple Heating Rates (21–23). If the value of the constant heating rate is changed from run to run, other conditions remaining the same, different TGA curves will be obtained (Fig. 10). Equations can be readily obtained from equation 1.

$$\ln R_t = \ln A - E/RT + n \ln W \tag{5}$$

When W is held constant, a plot of $\ln R_t$ versus $1/T$, employing data from the different TGA curves, should give a linear relation whose slope will give the value of E and whose intercept the value of A. It may be advisable to carry out a series of such plots at various (constant) values of W, as shown in Figure 11. In this manner, average

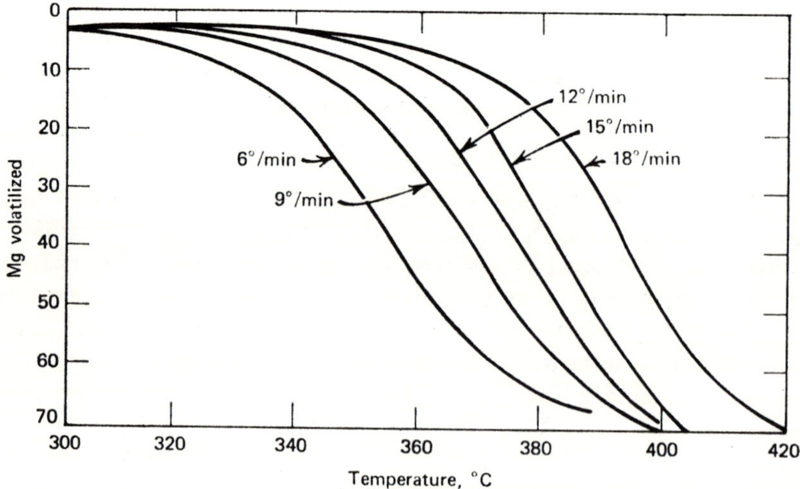

Fig. 10. Primary thermograms for *m*-phenylenediamine-cured halogenated epoxides at various heating rates (11).

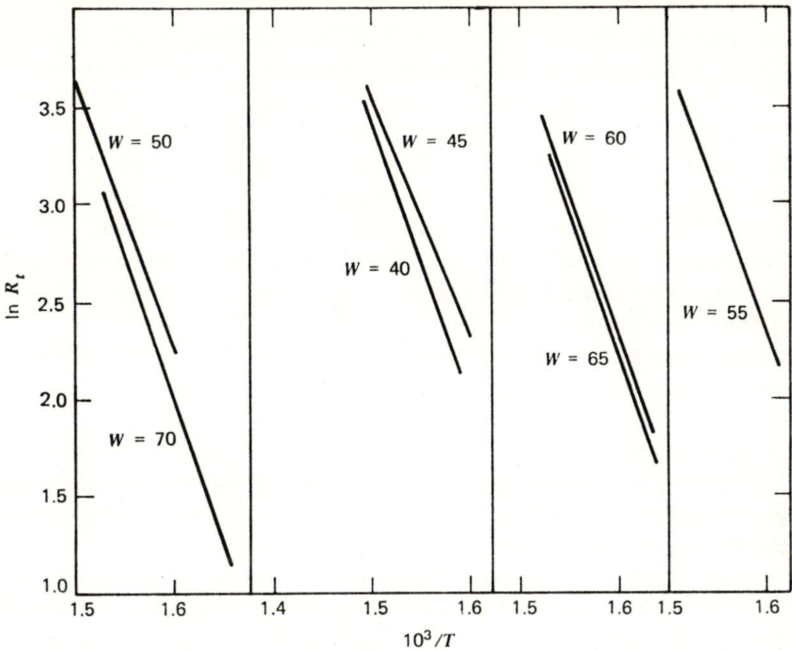

Fig. 11. Ln R_t plotted against $1/T$ for *m*-phenylenediamine-cured halogenated epoxide at the indicated values of W (11).

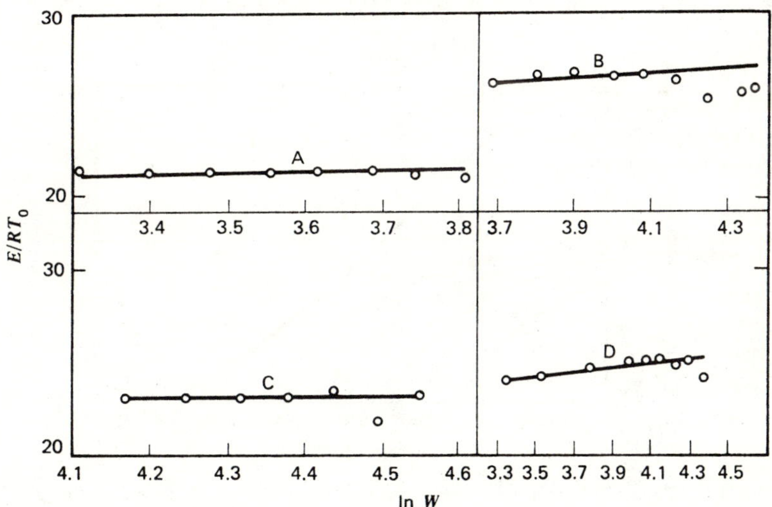

Fig. 12. Determination of reaction order for epoxy resins (11). Key: curve A, uncured epoxy resin based on epoxide of the formula

$$\overset{\displaystyle O}{\overbrace{CH_2-CH}}CH_2O-C_6H_4-\underset{CF_3}{\overset{CF_3}{\overset{|}{\underset{|}{C}}}}-C_6H_4-OCH_2\overset{\displaystyle O}{\overbrace{CH-CH_2}}$$

(6F); curve B, 6F cured with *m*-phenylenediamine; curve C, uncured Epon 820 (Shell Chemical Co.); curve D, Epon 820 cured with *m*-phenylenediamine.

values of E and A may be obtained over a range of conversion. The series of curves obtained can indicate the conversion at which the pyrolysis kinetics begin to vary.

In order to evaluate n, we may employ equation 6, which applies at $\ln R_t = 0$.

$$E/RT_0 = \ln A + n \ln W \qquad (6)$$

In this case, a plot of E/RT_0 versus $\ln W$ should give a linear relation whose slope will be n (see Fig. 12).

Anderson (22,23) has employed a variation of the above method. He carried out TGA experiments at three different rates of heating and subsequently solved three simultaneous expressions, each of the form of equation 5, for values of E, n, and A by means of a computer.

The above method has been tried using a number of constant values of W and several rates of heating, and it appears to give satisfactory values for overall activation energy. Although more data are required than in other procedures, this very fact tends to enhance confidence in the results obtained. Changes in kinetics (and mechanism) may also be detected rather satisfactorily by this method. However, scatter in the plot for estimating n may make the determination of this quantity somewhat less precise than the evaluation of activation energy.

Method of Variable Heating Rate for a Single Thermogram (24). A method involving a different type of thermogram has recently been proposed (24). In one variation of this method the polymer sample is heated at a given rate, eg, 6°C/min, and after about 30% decomposition, the heat input is raised so that the heating rate approaches a higher value (15°C/min in Figure 13). In another variation (see Figure 14), the heating cycle is reversed. An initial high heating rate of about

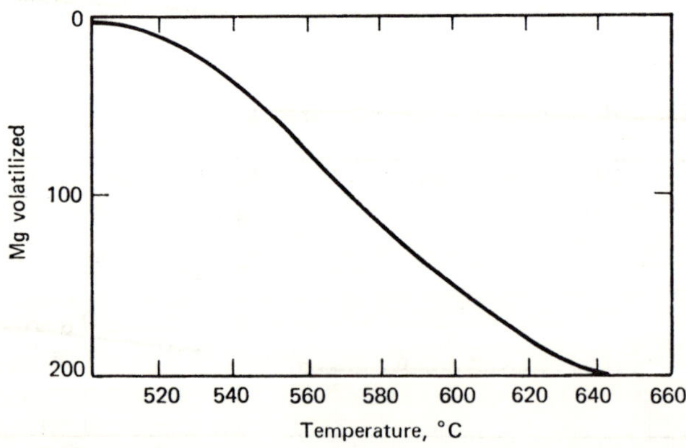

Fig. 13. Variable heating rate thermogram for polytetrafluoroethylene (24).

16°C/min was employed in obtaining this curve, and after about 30% decomposition the heat input was drastically reduced so that decomposition occurred while the material was actually cooling.

For either of the cases described above, equations 7 and 7a may be written. When R_T is held constant

$$\Delta \log (RH)/\Delta(1/T) = n[\Delta \log W/\Delta(1/T)] - (E/2.303R) \qquad (7)$$

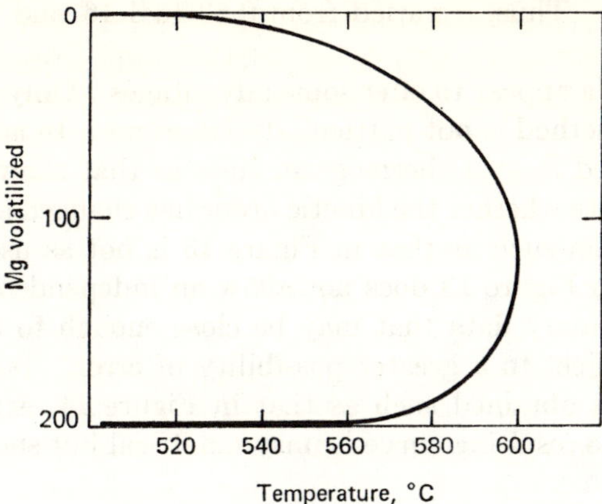

Fig. 14. Variable heating rate thermogram for polytetrafluoroethylene (24).

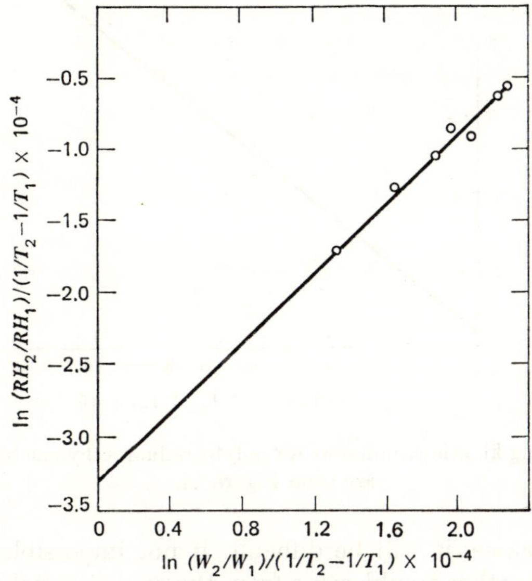

Fig. 15. Plot for obtaining kinetic parameters for polytetrafluoroethylene by equation 7 (24). Data are from Figure 13.

and when temperature is held constant

$$n = (\Delta \log R_t)/(\Delta \log W) \tag{7a}$$

Equations 7 and 7a are readily obtained from equation 1.

In using equation 7, data from thermograms such as those shown in Figures 13 and 14 were employed. The method consisted of obtaining pairs of values of RH, T, and W from sections of these curves which gave equal values of $R_T(dW/dT)$. Then, appropriate plots, as illustrated in Figures 15 and 16, gave lines whose slopes yielded values for n and whose intercepts gave E. The value of n was checked independently by constructing isotherms on the original thermogram (Figure 14) and by substituting values into equation 7a. Values of kinetic parameters obtained by this method for the pyrolysis of polytetrafluoroethylene (Teflon) in vacuum were in satisfactory agreement with literature values. Thus, n varied from 0.83 to 1.16 and E ranged from 66 to 74 kcal/mole.

Equations 7 and 7a appear to offer some advantages. Only one thermogram need be obtained and the method is not particularly laborious. In addition, the value of n may be readily checked from a thermogram such as that shown in Figure 14. The method can also indicate whether the kinetic order has changed at various conversions. However, a thermogram such as that in Figure 13 is not as useful as that shown in Figure 14. The one in Figure 13 does not allow an independent check of n and also requires the use of primary data that may be close enough to the extremities of the thermogram to be subject to a greater possibility of error. It should be noted that when a thermogram is obtained such as that in Figure 14, experimental conditions should be such that the resulting curve is unsymmetrical but shows distinct curvature

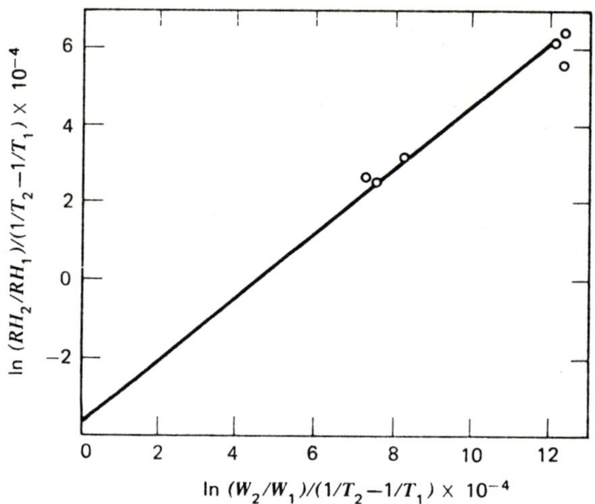

Fig. 16. Plot for obtaining kinetic parameters for polytetrafluoroethylene by equation 7 (24). Data are from Figure 14.

on both sides. Otherwise, it will be difficult, if not impossible, to use equation 7. This condition can be rather troublesome from the experimental point of view.

Methods Involving Approximate Integration of the Rate Equation. Various such methods are discussed here.

Methods of Reich (25,26). If equation 1 is expanded in an asymptotic series (27,28), and it is assumed that $(2RT/E) \ll 1$ (which is usually valid in polymer pyrolyses), equation 1 becomes equation 8.

$$-\int_{W_0}^{W} dW/W^n = [A/(RH)](RT^2/E)e^{-E/RT} \tag{8}$$

Using two thermograms for the same material having different heating rates, we may write equations 9 or 9a for any particular active residual weight, W.

$$[A/(RH)_1](RT_1^2/E)e^{-E/RT_1} = [A/(RH)_2](RT_2^2/E)e^{-E/RT_2} = \text{constant} \tag{9}$$

or

$$\ln\left[\frac{(RH)_2}{(RH)_1}\left(\frac{T_1}{T_2}\right)^2\right] = (E/R)\left(\frac{1}{T_1} - \frac{1}{T_2}\right) \tag{9a}$$

When more than two thermograms are employed, it is more convenient to express equation 9 as equation 9b.

$$\ln\,[(RH)/T^2] = -E/RT + \ln\left[\frac{AR}{E(\text{constant})}\right] \tag{9b}$$

When various specific values of W are chosen and $\ln\,[(RH)/T^2]$ is plotted versus $1/T$, a series of nearly parallel lines should be obtained whose slopes will give values of E.

Some advantages of the above method are: (a) no prior knowledge of reaction order or kinetic process is required; (b) only primary data are used from the thermogram; (c) the method consumes relatively little time; (d) no curve fitting or laborious plots are necessary; (e) values of E may be obtained at various conversions thereby determining whether any change in mechanism is occurring as the conversion changes. Some disadvantages of the method are: (a) the reaction order is indeterminate; (b) at least two thermograms are required; (c) it is sensitive to changes in temperature.

In order to illustrate the applicability of the method, results obtained by equation 9 were compared with those obtained by theoretically exact methods for an uncured and a cured epoxy resin. The results are listed in Table 2. From this table, it can be seen that the values of E obtained by equation 9 and standard methods are in good agreement irrespective of reaction order. However, the limitations of the method presented must not be overlooked.

Table 2. Application of Equation 9 (25)

RH, °C/min	Reaction order (standard methods)	E, kcal/mole		W/W_0	
		Eq. 9	Standard method	This method	Standard methods
		Uncured epoxy resin			
10.5, 14.0	0	17	17	0.50	0.60–0.95
11.0, 14.0		15		0.70	
14.0, 19.5		15		0.70	
		Cured epoxy resin			
6.0, 12.0	1	26	28	0.50	0.50–0.95
8.5, 12.0		26		0.50	

Reich (26) has presented another method which involves a combination of equations 1 and 9. When these expressions are combined, eliminating the term $A/(RH)$, and generalizing by allowing for an inactive residue, equations 10a and 10b are obtained

$$E/R = S/W_c \ln\,(W_{0,c}/W_0) \qquad \text{for } n = 1 \tag{10a}$$

$$E/R = S(1-n)/W_c^n(W_{0,c}^{1-n} - W_c^{1-n}) \qquad \text{for } n \neq 1 \tag{10b}$$

where $S = dW_c/d(1/T)$

$W_{0,c} = W_0 - W_R$

$W_c = W - W_R$

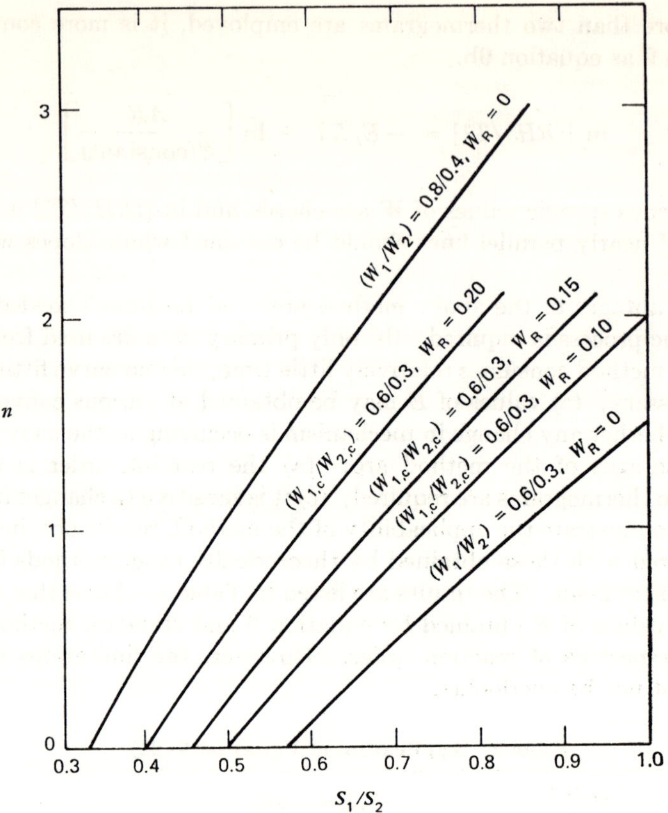

Fig. 17. Theoretical plots of reaction order, n, against S_1/S_2 for various values of W_1/W_2 or $W_{1,c}/W_{2,c}$ and W_R (26).

W is the weight fraction of material remaining at time t and W_R is the weight fraction of inactive material remaining after a pyrolysis. From equations 10a and 10b the following expressions may be obtained, respectively:

$$S_1/S_2 = (W_{1,c}/W_{2,c}) \log (W_{0,c}/W_{1,c})/\log (W_{0,c}/W_{2,c}) \qquad \text{for } n = 1 \qquad (10c)$$

and

$$S_1/S_2 = (W_{1,c}/W_{2,c})^n \left(\frac{1 - (W_{1,c}/W_{0,c})^{1-n}}{1 - (W_{2,c}/W_{0,c})^{1-n}}\right) \qquad \text{for } n \neq 1 \qquad (10d)$$

Prior to determining the overall activation energy of a pyrolysis, it is necessary to estimate n. From equations 10c and 10d it can be seen that for a particular, arbitrarily selected ratio of $(W_{1,c}/W_{2,c})$ and value of W_R the corresponding ratio S_1/S_2 may be calculated for various values of n. Such calculated values were used to construct the theoretical curves shown in Figure 17. In this figure, various values of (W_1/W_2) or $(W_{1,c}/W_{2,c})$ and of W_R are given, from which the ratio S_1/S_2 could be calculated. The curves in Figure 17 can be used to estimate n from experimental TGA curves. After the value of n has been determined, it may be substituted into equations 17a and 17b to obtain the value of E. After E and n have been estimated, the value of A may then be obtained by means of an expression such as equation 1.

When this method was applied to TGA curves for polytetrafluoroethylene and polyethylene the kinetic parameters obtained were in satisfactory agreement with values given in the literature.

It may be of interest to note here that equations 10c and 10d may be employed to obtain a simple expression involving n. Thus, at values of W equal to 0.8 and 0.1, values of S_1/S_2 (designated as $S_{0.8}/S_{0.1}$) were calculated for various values of n, with $W_R = 0$. From these values the following approximate expression may be written for values of n from about $\frac{1}{4}$ to 2

$$(S_{0.8}/S_{0.1}) \approx D(T_{0.8}/T_{0.1})^2 \approx 0.8n \tag{10e}$$

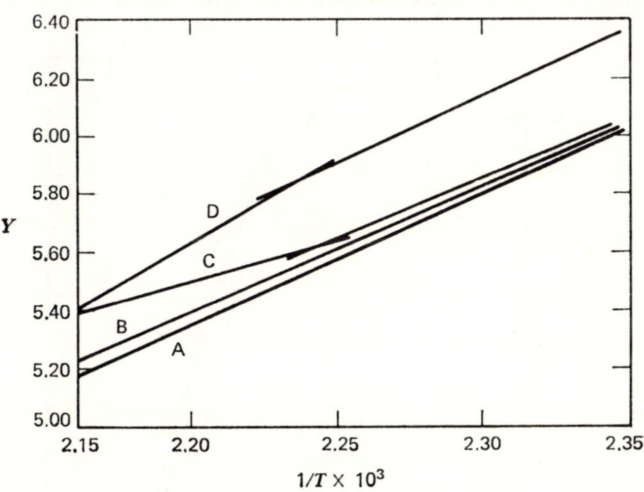

Fig. 18. Water loss from calcium oxalate monohydrate. $Y = -\log [(1 - W^{1-n})]/[T^2(1 - n)$ for $n = 0$, 1/2, and 2/3, and $-\log(-\log W/T^2)$ for $n = 1$. Key: curve A, $n = 2/3$; curve B $n = 1/2$; curve C, $n = 0$; and curve D, $n = 1$ (29).

where $D \equiv (dW/dT)_{0.8}/(dW/dT)_{0.1}$. Since the value of $(T_{0.8}/T_{0.1})$ is generally a little less than unity, equation 10e may be written as equation 10f. Equation 10f

$$D \approx n \tag{10f}$$

may be used when it is desired to obtain a rapid, approximate estimation of n from a primary thermogram which possesses a value of $W_R = 0$. From experimental thermograms obtained in the authors' laboratory for degradation of polytetrafluoroethylene in vacuum at different heating rates, values of D were determined which were consistently near unity, as anticipated.

Method of Coats and Redfern (29). Coats and Redfern (29) developed a method for estimating E by use of an integrated form of the rate equation which was similar to equation 8. Thus, when $(2RT/E) \ll 1$ and the logarithm of both sides of equation 8 is taken, equation 11 is obtained, where I denotes the term involving the integral.

$$\ln\left[\frac{\int_W^{W_0} dW/W^n}{T^2}\right] = \ln\frac{I}{T^2} = \ln\frac{AR}{(RH)E} - \frac{E}{RT} \tag{11}$$

The value of I may be readily evaluated once n is known or assumed. In the latter case, when a plot of ln (I/T^2) against $1/T$ gives a linear relation, then it is considered that the correct value of n was assumed. When $(2RT/E)$ is not neglected, the logarithmic term on the righthand side of equation 11 becomes ln $[AR/(RH)E]$ $[1 - (2RT/E)]$. Coats and Redfern indicate that for most values of E and for the temperature range over which decomposition reactions generally occur, the preceding logarithmic term is essentially constant.

Figure 18 illustrates the trial and error procedure employed to study the loss of water from calcium oxalate monohydrate. From this figure it can be seen that the results best fit a ⅔ order assumption, and the corresponding value of E = 21.7 kcal/mole. These results are in close agreement with other reported values.

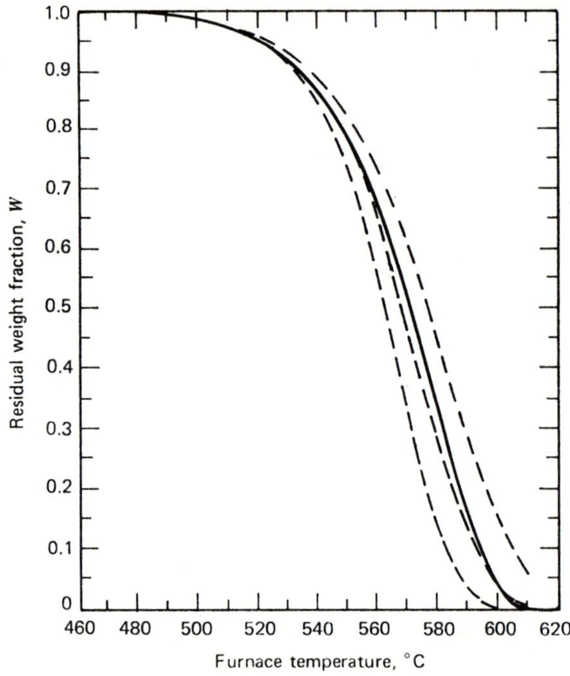

Fig. 19. Thermogravimetric analysis of pulverized polytetrafluoroethylene at 180°C/hr in dry nitrogen (27). Solid curve is experimental; lower, middle, and upper dashed curves are calculated with E = 80, 70, and 60 kcal/mole, respectively.

Method of Doyle (27). Doyle (27) attempted to evaluate E and A as constants of the equation of the thermogram rather than of the rate equation. He obtained equation 12 for the equation of the thermogram. In equation 12 U, having values x,

$$g(h) = \frac{EA}{(RH)R}\left(\frac{e^{-x}}{x} - \int_x^\infty \frac{e^{-U}}{U}\,dU\right) \tag{12}$$

replaces E/RT. Equation 12 may also be written as equation 13, where $p(x)$ is the term in parentheses in equation 12.

$$g(h) = \frac{EA}{(RH)R}\,p(x) \tag{13}$$

Common logarithmic values of $p(x)$ are tabulated (27), as well as first differences in log $p(x)$ for use in interpolating. Before E can be evaluated, it is necessary to determine x_a, the value of x at T_a (subscript a denotes point functions); E can then be found by using equation 14.

$$E = RT_a x_a \tag{14}$$

Methods are illustrated by Doyle for estimating x_a and hence E by this equation for the thermogram. In each case determination of E (and A) requires successive approximations or curve fitting. Figure 19 shows an experimental curve and several calculated curves for polytetrafluoroethylene using trial values of E.

Although this method appears to be workable, it is a very laborious one. Also, considering the approximations involved in the derivation, eg, reaction order must be presumed, it does not seem likely that it will find wide use in studies of polymer degradation.

Method of Flynn and Wall (30). Equation 13a can be obtained from equations 1 and 13.

$$\int_0^W \frac{-dW}{W^n} \equiv g(h) = \frac{EA}{(RH)R}\, p(E/RT) \tag{13a}$$

In logarithmic form, equation 13a becomes equation 13b.

$$\log g(h) = \log (EA/R) - \log (RH) + \log p(E/RT) \tag{13b}$$

For values of $E/RT \geqslant 20$, $\log p\ (E/RT)$ may be closely approximated by $-2.315 - 0.457E/RT$, and equation 13b becomes equation 13c.

$$\log g(h) \approx \log (EA/R) - \log (RH) - 2.315 - 0.457E/RT \tag{13c}$$

Upon differentiating equation 13c at constant degree of conversion equation 13d

$$-d \log (RH)/d(1/T) \approx 0.457E/R \tag{13d}$$

is obtained. From equation 13d it can be seen that from the slope of a plot of log (RH) against $1/T$, the activation energy may be calculated. This procedure may be repeated at various degrees of conversion, thereby testing the constancy of E with respect to conversion and temperature.

It may also be noted here that Ozawa (31) also developed expressions similar to equation 13d and set up appropriate theoretical master curves of conversion against $\log (AE/(RH)R)p(E/RT)$ for simple nth order reactions. By means of such master curves, various kinetic parameters could be estimated from the value of the displacement of an experimental curve along the abscissa that was required to overlap an appropriate master curve.

Miscellaneous Methods. In addition to the foregoing, several other attempts have been made to estimate kinetic parameters based on simplifying assumptions and approximations. Some of these are briefly outlined below.

Newkirk (32) and Smith (15) assumed that thermal decompositions may be fitted by first-order kinetics with respect to weight loss. Calculations were made of first-order rate coefficients at each of a large number of temperatures on the continuous weight-loss curve. This was followed by construction of the appropriate Arrhenius plot, which was then examined for linearity. Linearity or near linearity over a wide

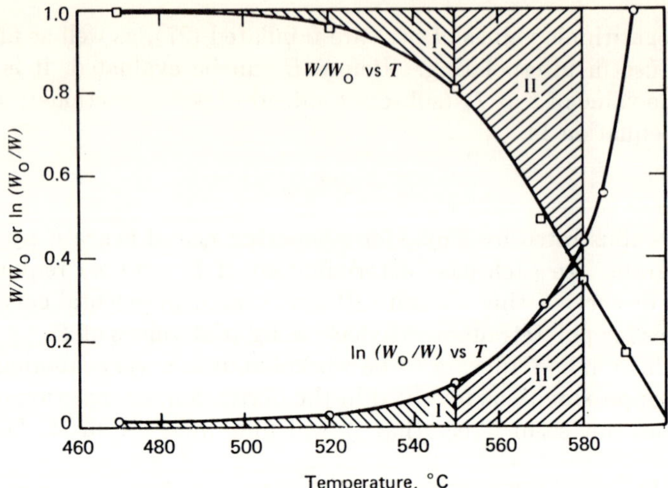

Fig. 20. Zero- and first-order rate plots (13). Areas I and II are discussed in the text.

range of conversion was considered as evidence for a first-order kinetic form of the rate-determining reaction.

Moiseev (33) integrated the rate equation. However, for data treatment he used the rate equation as such, employing graphical differentiation to obtain rates. By selecting values of n, he was able to find the best fit to the experimental data by means of the criterion of linearity.

In conjunction with developing arbitrary thermal stability indexes for various polymers undergoing pyrolysis, Reich and Levi (13) employed an approximate, rapid method for obtaining E. This method involved the measurement of areas, A_{I} and A_{II}, as shown in Figure 20. Two arbitrary temperatures were selected on the TGA curve and E was evaluated from expression 14a. The reaction order was arbitrarily assumed

$$E = \frac{\log\left(\dfrac{A_{(\mathrm{I}+\mathrm{II})}}{A_{\mathrm{I}}}\right) 4.6\, T_1 T_2}{T_2 - T_1} \tag{14a}$$

to be zero in order to calculate overall values of E from equation 14. This was justified from the fact that for many polymeric degradations it is often difficult to distinguish clearly between zero- and first-order kinetics, up to about 50% conversion in constant-temperature experiments. (It is interesting to note that Smith (15) has indicated that an incorrect assumption of the value of n may not necessarily alter the value of E very much.)

Because of an error in the derivation of equation 14, which does not affect the validity of this expression (25,34), the authors feel that an outline of the derivation should be given. Assuming that for relatively low conversions zero-order kinetics apply, we may write, from equation 1

$$\int_1^W -\frac{dW}{T^2} = \left(\frac{A}{RH}\right)\int_0^T \frac{e^{-E/RT}\,dT}{T^2} \tag{15}$$

or

$$(1/T_{\mathrm{av}})^2(1 - W) = [AR/(RH)E]e^{-E/RT} \tag{16}$$

where

$$(1/T_{\mathrm{av}})^2 = \int dW/T^2/\int dW \approx \text{constant}$$

Equation 16 may further be converted into expression 17

$$\left(\frac{1}{T_{av}}\right)^2 \left(\frac{1}{T'_{av}}\right)^2 \int_0^T (1 - W)dT = [A/(RH)] \left(\frac{R}{E}\right)^2 e^{-E/RT} \qquad (17)$$

where

$$\left(\frac{1}{T'_{av}}\right)^2 = \int \frac{(1 - W)}{T^2}\, dT/\int (1 - W)\, dT \approx \text{constant}$$

The constancy of the average temperatures, as defined above, was checked by means of a TGA curve for the degradation of a cured epoxy resin in vacuum. Up to about 40% conversion the following values were obtained, employing 10% conversion intervals:

$$(1/T_{av})^2 = (2.2 \pm 0.05) \times 10^{-6}$$

and

$$(1/T'_{av})^2 = (2.1 \pm 0.02) \times 10^{-6}\ (°K)^{-2}$$

In equation 17 the integral represents the area of the upper portion of the TGA curve bounded by the zero-reaction horizontal line ($W/W_0 = 1$) and extending from temperature T to the temperature at which the curve and the zero-reaction horizontal line meet (see Fig. 20). If this area is denoted by A_{II}, equation 17 may be transformed, for two different temperatures T_1 and T_2 into equation 14. Similar considerations may be applied to first-order kinetics. In this respect, it may be mentioned that this method gave values of E for polytetrafluoroethylene of 68 and 69 kcal/mole when zero- and first-order expressions, respectively, were used. This method will be referred to again in the next section.

Estimation of Thermal Stability from TGA Curves

Many factors can affect the thermal stability of polymers, including melting or softening points, bond strengths, activation energies, crosslinking, and the presence of low-molecular-weight volatile material (or impurities), "weak" links, and groups which are readily affected by heat, etc (see also DEGRADATION; DEPOLYMERIZATION; HEAT-RESISTANT POLYMERS). In order properly to assess the thermal stability of polymers, many of the preceding factors must be simultaneously considered. Since this is virtually impossible, various investigators have devised arbitrary qualitative and quantitative methods. Some of these methods, which do not involve TGA curves, are briefly described in this section.

Madorsky (35) suggested that for polymers which volatilize completely below 600°C, the relative thermal stability may be estimated by heating the polymers under equivalent conditions and comparing the amount of vaporization that occurs in a vacuum pyrolysis for ½ hr at various temperatures. Another procedure involves the comparison of temperatures (T_R) at which 50% of the weight of a given polymer is vaporized under standard pyrolysis conditions. Wright (36) also presented a method which involves a comparison of temperatures at which polymers undergo a 25% weight loss after 2 hr at an elevated temperature. Two disadvantages of this method are: (*1*) The actual percentage weight loss may not be significant unless it can be related to the degradation process, eg, a dimethylsilicone polymer can be completely degraded to silica with only a 20% weight loss. (*2*) Certain polymers, eg, poly(vinyl chloride), may crosslink during pyrolysis and this may decrease the rate of volatiliza-

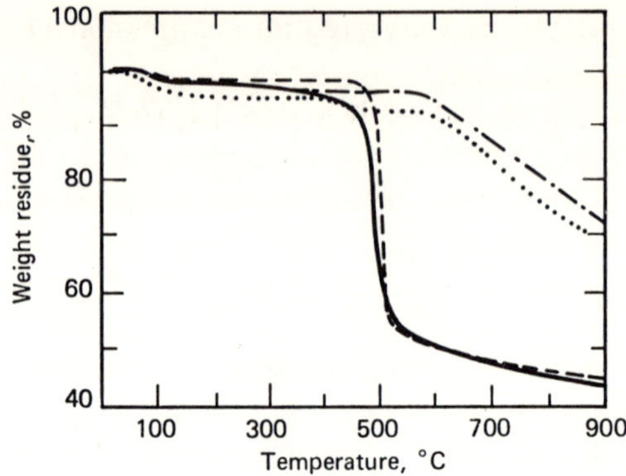

Fig. 21. Thermographic analysis of four polybenzimidazoles (41). T = 150°C/hr. Key:

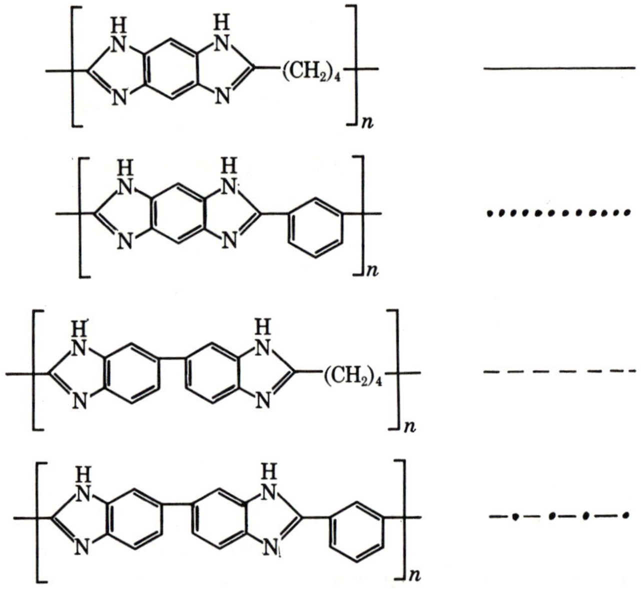

tion. Madorsky (37) proposed that for carbon-chain polymers, thermal stability might be better defined as the tendency of such polymers to yield a more or less carbonized residue.

Thermomechanical procedures have also been employed to evaluate thermal stability. It has been suggested (38) that a polymeric material may retain its "usefulness" when it retains 50% of its strength after 1 hr of exposure to a specified temperature and that the limit is reached after a 10% weight loss has occurred. It has also been specified that secondary effects such as cracking and crazing do not occur. However, this limit depends upon many assumptions and may vary widely for similar materials. Jurkov (39) studied the heat resistance of polymers by measuring changes in their moduli of elasticity caused by increasing temperature. Others (40) suggested that the use of moduli of elasticity to determine heat resistance is more significant than deformation measurements of polymers under linear compression, which have been carried out by various workers. However, in selecting the thermomechanical procedures to be used, the choice of the method of measurement, the accuracy of the measurements, as well as the complexity and availability of the apparatus must be considered. See also MECHANICAL PROPERTIES.

The following sections discuss methods for estimating thermal stability of polymers which are based essentially on TGA measurements. Some of these methods are essentially qualitative in character, others are semiquantitative in nature, and a few rest on a firm theoretical basis.

Qualitative Methods. It is often possible, merely by visual observation, to ascertain the relative thermal stabilities of various polymers. Thus, Marvel (41) has used TGA thermograms to compare stabilities of four polybenzimidazoles (qv under POLYIMIDAZOLES) in nitrogen (Fig. 21). From Figure 21 it may be seen that when an alkyl group was replaced by a phenyl, the initial decomposition temperature rose from about 400 to 500°C, the percentage residue at 900°C increased from about 45 to about 70%, and the decomposition rate decreased considerably. Many similar qualitative analyses have been reported but only a few will be described here.

Recently, Jeffreys (42) discussed the thermal stability of various phenolic resins in air. Results obtained from a static test were compared with thermobalance results (Table 3). In order to establish a criterion for evaluating the resin decomposition,

Table 3. Test Results for Evaluating Thermal Stability of Phenolic Resins (42)

		Static test, loss at 300°C after 1 hr, %	Thermobalance results			
					Weight-loss curve maxima	
Type	Resin		10% DT, °C	50% DT, °C	Temperature, °C	Value, mg/min
1	phenol–formaldehyde	7.0	430	830	415	≥0.7
					530	≥1.4
2	*m*-cresol–formaldehyde	9.5	395	650	430	≥3.0
3	*m*-isopropylphenol–formaldehyde	12.7	360	610	450	≥2.4
4	*m-tert*-butylphenol–formaldehyde	26.4	335	490	355	≥3.2
5	*p-tert*-octylphenol–formaldehyde	7.15[a]	360	490	470	≥3.5
6	*p*-dodecylphenol–formaldehyde	19.4[a]	345	460	440	≥5.4
7	cardanol–formaldehyde	19.6	340	485	445	≥5.4
8	*p*-octadecylphenol–formaldehyde	22.05[a]	340	480	470	≥4.5
9	phenol–benzaldehyde	11.4	360	740	broad peak at about 550°C (1.2 mg/min)	
10	phenol–furfural	14.2	320	640	broad peak at about 410°C (1.4 mg/min)	
11	*m*-isopropylphenol–furfural	35.0[a]	290	520	300	≥1.6
					410	≥1.7
12	*m-tert*-butylphenol–furfural	27.0[a]	270	440	330	≥2.6
13	*p-tert*-octylphenol–furfural	36.0[a]	315	490	390	≥2.0
					485	≥3.0
14	*p*-dodecylphenol–furfural	28.75[a]	350	510	475 (broad)	≥2.9
15	cardanol–furfural	22.2	340	485	445	≥5.0
16	phenol–formaldehyde (stearic acid modified)	13.0	375	710	455	≥2.0
17	phenol–formaldehyde (oleic acid modified)	13.1	390	665	455	≥1.8
18	phenol–formaldehyde (linseed fatty acids modified)	11.9	380	655	455	≥2.4

[a] Resins in which skin effect was observed or in which caking effect occurred.

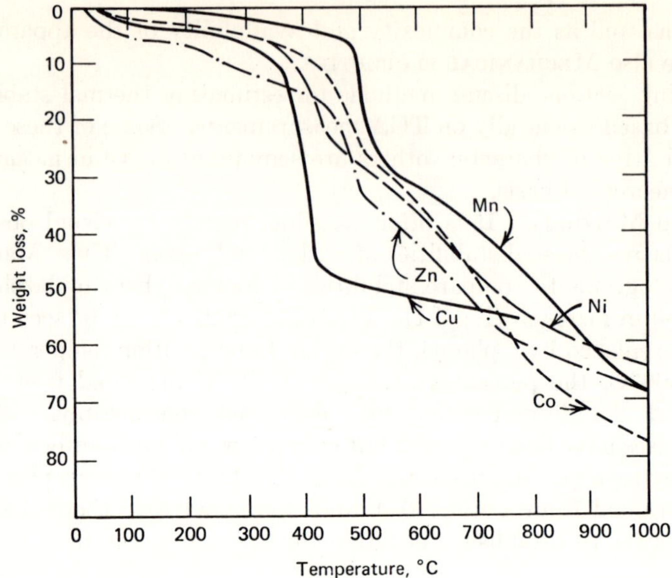

Fig. 22. Thermograms of Mn(II), Co(II), Ni(II), and Zn(II) 5,5′-[methylene bis(*p*-phenylenenitrilomethylidyne)]di-8-quinolinol coordination polymers (44). Heating rate in vacuum, 2.5°C/min.

the temperatures at which 10% decomposition (10% DT) and 50% decomposition (50% DT) had occurred were noted. Temperatures were also recorded at which there occurred maximum rates of decomposition. From Table 3 it can be seen that, based upon resin types *1*, *2*, *3*, *4*, and *7*, the thermal stability of the resins decreases with increasing molecular weight of the meta-substituted phenol, ie, stability decreases in the order, phenol > *m*-cresol > *m*-isopropylphenol > cardanol > *m*-*tert*-butylphenol. The anomalous position of the *m*-*tert*-butylphenol indicates that branching of the side

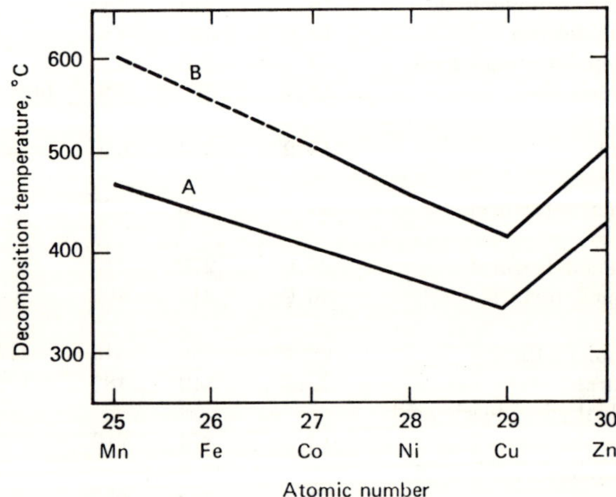

Fig. 23. Plot of decomposition temperatures of polymers against atomic number of coordinated metal. Curve A, 5,5′-[methylenebis(*p*-phenylenenitrilomethylidyne)]di-8-quinolinol polymers; curve B, bis(8-hydroxy-5-quinolyl)methane polymers (44).

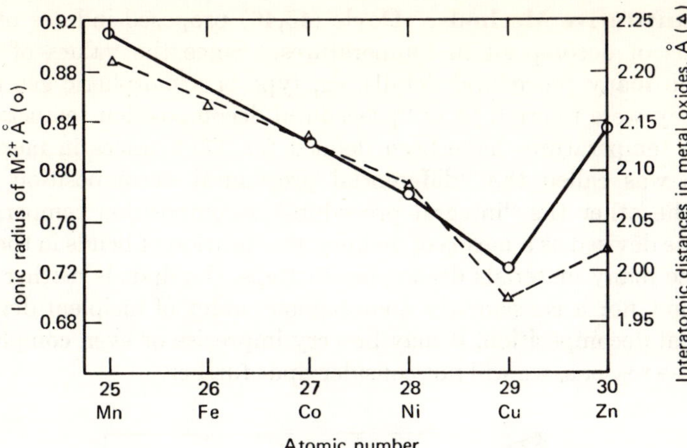

Fig. 24. Plot of ionic radius (O) and interatomic distances in metal oxides (Δ) against atomic number of metal (44).

chain has a significant effect, especially when branching occurs from the α-alkyl carbon atom which is attached to the phenolic nucleus. Other generalizations have been made by Jeffreys.

Recently, Horowitz and Perros (43,44) discussed the thermal stability of coordination polymers (qv). The thermal stability of the Mn(II), Co(II), Ni(II), and Zn(II) 5,5′-[methylenebis(*p*-phenylenenitrilomethylidyne)]di-8-quinolinol coordination polymers was studied under vacuum by means of a thermobalance. The data obtained are shown in Figure 22. After an initial weight loss, a rather sharp break occurred in each of the thermograms, indicating the onset of a decomposition process involving a rapid loss of volatile fragments. As a measure of thermal stability, decomposition temperatures of the polymers were determined. This involved drawing straight lines through the experimental points on the expanded thermogram before and after the initial sharp break in the curve that occurred at moderately low weight losses. The intersection of these two lines was then projected on the temperature axis and this temperature was arbitrarily defined as the decomposition temperature. These temperatures are plotted for each run as a function of the atomic number of the metal contained in the backbone of the polymers in Figure 23, curve A. Curve B shows a similar relationship obtained for coordination polymers of bis(8-hydroxy-5-quinolyl)-methane polymers. The following conclusions were deduced from curves A and B: the substitution of a large organic bridge for the CH_2 group in the 5,5′ position of the ligand lowers the apparent decomposition temperature of the polymers by 80 to 100°C; the decomposition mechanism, near the temperature at which accelerated weight loss occurs, appears to involve metal–ligand scissions in both polymer systems and is dependent on the nature of the metal which unites the ligands in the polymer. For the latter case, it is interesting to note that when the metal–oxygen interatomic distances for the divalent metal oxides are plotted against the atomic numbers of the metals, a relationship as shown in Figure 24 is obtained. The relationship between ionic radii of the metals and their atomic numbers is also shown in this figure. The shape of both of these plots bears a strong resemblance to those shown in Figure 23 (curves A and B) where the decomposition temperature has been plotted as a function of the atomic numbers of the metals in the polymers.

Semiquantitative Methods. Doyle (45,46) proposed indexes of thermal stability in terms of decomposition temperatures. Since the values of these indexes depended upon many procedural details, eg, type of atmospheric gas used and rate of heating, they were referred to as "procedural decomposition temperatures." Two types of such temperatures have been defined for TGA traces in inert atmosphere. One of these was called the "differential procedural decomposition temperature" (dpdt), and the other the "integral procedural decomposition temperature" (ipdt). The former was devised as a means of defining the location of bends in the TGA curves. However, since many materials decompose in steps, the dpdt is neither a unique empirical end point nor a consistently unambiguous index of incipient degradation. In cases of gradual decomposition, it may be very imprecise or even completely unavailable. For these reasons, we will not consider dpdt further.

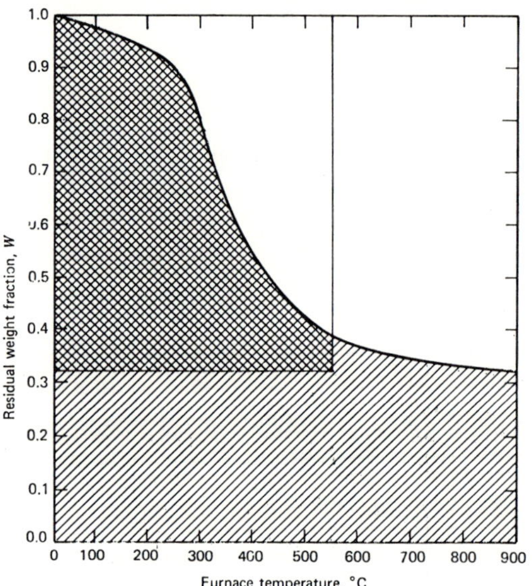

Fig. 25. Thermogram areas, A^* (crosshatched) and K^* (doubly crosshatched) (46).

The second type of index, ipdt, was devised as a way of summing up the whole shape of the "normalized" data curve. As such, it is readily available and is highly reproducible. In order to place materials on an equal procedural basis, the total experimentally accessible temperature range from 25 to 900°C was used. In Figure 25 the area of all the crosshatched region divided by the area of the total rectangular plotting region was denoted as A^*. From the value of A^*, the value of T^* may be obtained from expression 18. In this expression, it was assumed that all materials volatilize completely below 900°C and do so at a single temperature.

$$T^* = 875A^* + 25 \tag{18}$$

The value of T^* serves as a crude measure of refractoriness; however, many materials which possess high refractory weight fractions up to 900°C begin to decompose at much lower temperatures. Thus, a second area ratio, K^*, was defined as the doubly crosshatched region in Figure 25. The boundaries of the regions involved are established by the temperature T^*, by the residual weight fraction at the arbitrary temperature of 900°C, and by the horizontal zero-loss line. The ipdt may now be calculated from equation 19.

$$\text{ipdt} = 875A^*K^* + 25 \tag{19}$$

Table 4 shows values of ipdt for several polymers. Values of ipdt have been used to compare thermal stabilities of a homologous series of polymers containing azulene (47).

Table 4. Integral Procedural Decomposition Temperatures of Some Polymers (46)

Polymer	ipdt, °C
polystyrene	395
epoxy resin cured with maleic anhydride	405
poly(methyl methacrylate)[a]	345
nylon-6,6	419
polytetrafluoroethylene[b]	555
vinylidene fluoride–hexafluoropropylene copolymer[c]	460

[a] Plexiglas, Rohm & Haas Co.
[b] Teflon, Du Pont.
[c] Viton A, Du Pont.

Other thermal stability indexes have been reported (13). As mentioned previously, these indexes were devised in conjunction with equation 14. From this equation, overall activation energies for about thirty polymers were rapidly estimated, and attempts were made to correlate these values with various arbitrary parameters.

The energies obtained were correlated with the so-called 10% tangential temperature (TTN). The tangential temperature (TN) was obtained by drawing a tangent at the break in the TGA curve prior to the maximum rate portion. The temperature at the point of intersection of the tangent and the zero-reaction line is the TN temperature. The TTN is then obtained by multiplying the TN by a ratio of areas. The latter is obtained by observing where the TGA curve intersects the 10% reaction line. The total pertinent rectangular area is bounded by the zero-reaction line and extends from 25°C to the temperature obtained at the intersection of the 10% reaction line and the TGA curve. The portion of the area under the curve divided by this total area is the ratio of areas which when multiplied by TN will afford the TTN.

Another parameter referred to as the sigma temperature, ΣT, was also used. This parameter denotes the temperature corresponding to the intersection of a horizontal line with the TGA curve. This horizontal line is drawn at the ordinate whose value is determined by adding unity to the ordinate value of W/W_0 at 900°C and dividing the sum by two.

Since the decomposition of the polymer itself, and not secondary effects of the reaction, was of primary interest it was reasoned that the TN would give some measure of the onset of appreciable decomposition. This in conjunction with a measure of initial stages of volatilization, which leads to TTN, might thus be expected to represent some measure of thermal stability. For the purpose of including some effects of later stages of reaction, the ipdt was also considered. In Figurc 26, values of overall energies, E_0, are plotted against two parameters, ipdt and TTN. From this figure, it may be observed that values of ipdt are relatively insensitive to changes in E_0. However, the TTN values showed a surprisingly good linear correlation with E_0 for values above 155°C. The linear correlation coefficient for the latter parameter was

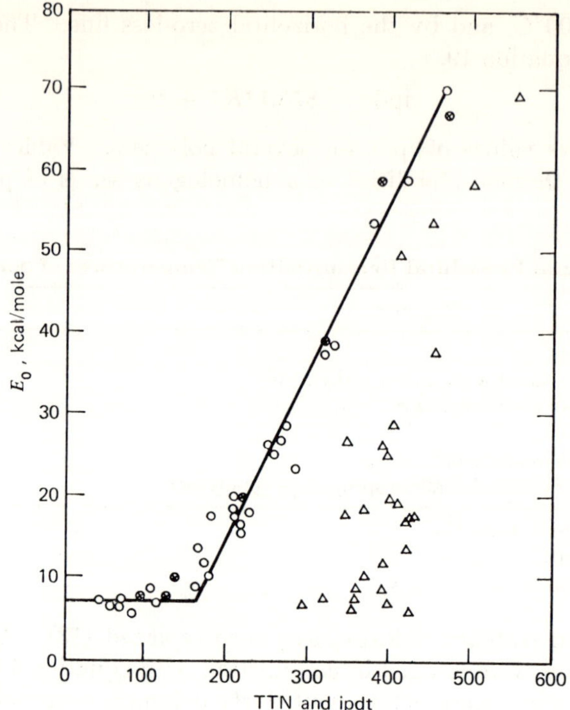

Fig. 26. E_0 plotted against the parameter TTN (heating rate 3°C/min (O); 2.5°C/min (⊗), and the parameter ipdt (Δ) (13).

estimated to be 0.98 for values of E_0 obtained at 3°C/min. Below 155°C, the values of E_0 leveled off at an average value of 7 for the range of TTN values investigated. When E_0 was plotted against the sum (TTN + idpt), no improvement in the linearity was observed. Smith (13) has also compared values of ipdt with overall activation energies for several polymeric systems. He also found that values of E_0 did not correlate well with values of ipdt. Figure 27 shows a plot of E_0 against a parameter which was obtained by dividing TTN by the difference between ΣT and the value of TTN. If the points represented by the dotted circles were neglected, the correlation coefficient had a value of 0.98. In contrast to other plots, there was no sharp break in the line below certain parameter values.

The values of E_0 obtained from the TGA curves were strongly dependent upon the initial portions of the curves as well as upon later portions. Therefore, a TTN parameter rather than the TN should give a better correlation with E_0, as has been observed. Since the ipdt places much emphasis on the late portions of the TGA curve, it would be anticipated that the ipdt would be relatively insensitive to changes in E_0.

A correlation between E_0 and the thermal stability of a polymer is probably always quite complex. It is presumed that E_0 gives an overall picture of the pyrolysis energetics. It probably involves energies associated with diffusion processes, bond ruptures, volatilization, etc. Prior to designating any literal interpretation to the magnitude of E_0, it should be realized that knowledge of the processes involved in the degradation must be available. Any correlation presented in terms of E_0 is at best semiquantitative and should not be used for the determination of precise activation

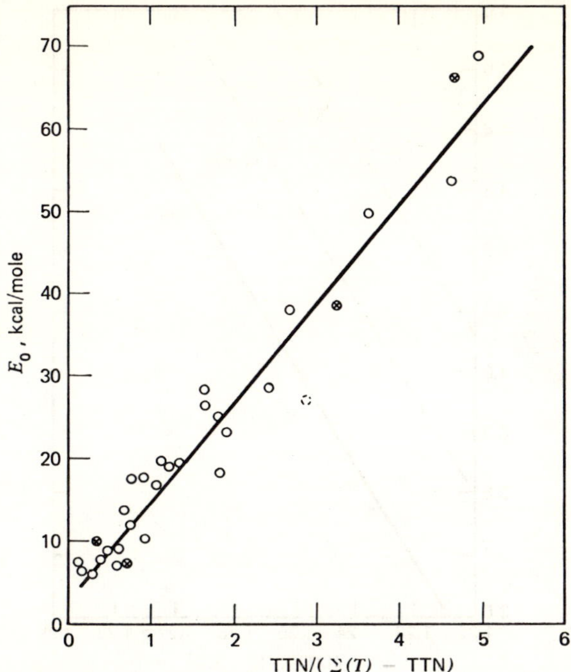

Fig. 27. E_0 plotted against the parameter TTN/Σ(T) − TTN; heating rate 3°C/min (○); 2.5°C/min (⊗).

energies at the expense of the more elaborate methods. However, such rapid correlations may be of considerable value if the thermal stability of a rather large number of experimental polymers must be compared.

Quantitative Methods. The isothermal "life" of a polymer may be quantitatively correlated with TGA curves. Thus, Doyle (48) has obtained equation 20,

$$\log t_i \approx -E/2.3RT + \text{constant} \tag{20}$$

based upon various simplifying assumptions. In this equation t_i is isothermal aging time, and T is absolute thermogravimetric analysis temperature corresponding to the equivalent aging time t_i. Based upon isothermal experiments and TGA curves for polytetrafluoroethylene in a nitrogen atmosphere, a value of E of 67 kcal/mole was obtained.

An expression similar to equation 20 may be obtained by the use of equation 8. When this expression is used in conjunction with the isothermal rate (eq. 21) expression equation 22 is obtained, for any value of n.

$$\int \frac{dW}{W^n} = kt_i \tag{21}$$

$$\log (t_i/T^2) = -E/2.3RT + \log \frac{AR}{(RH)Ek} \tag{22}$$

From equation 22 it can be seen that from a single TGA curve and from a single isothermal degradation experiment the value of E may be readily obtained, irrespective of reaction order. Equation 22 has been plotted for the degradation of polytetra-

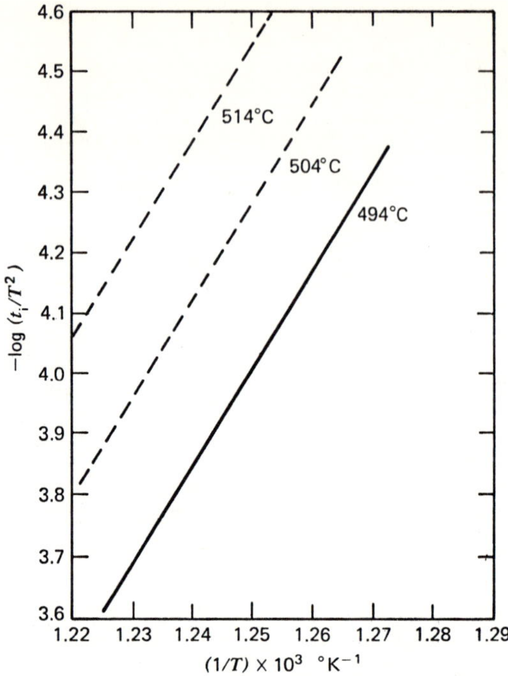

Fig. 28. Log (t_i/T^2) plotted against $1/T$ for degradation of polytetrafluoroethylene.

fluoroethylene in vacuum, as shown in Figure 28, using an isothermal temperature of 494°C. From the slope of the linear relation obtained, a value of E of 74 kcal/mole was obtained. Once the value of E has been obtained, the isothermal stability of the material in question may be quantitatively estimated for other temperatures (other than 494°C). Thus, plots for other temperatures should involve a series of parallel lines whose vertical separation, Δ, may be ascertained from equation 23, where, T_i' and

$$\Delta = (E/2.3R)(1/T_i' - 1/T_i) \tag{23}$$

T_i are the particular isothermal temperatures involved. In this manner, the dotted lines in Figure 28 were constructed for various temperatures, based upon the line obtained at 494°C.

Thermal Degradation Behavior of Some Polymers by TGA Methods

In this section some studies of polymer degradation which are relatively thorough and which involve primarily TGA methods are discussed. Auxiliary work carried out simultaneously with these methods is also mentioned. Where possible, results from isothermal studies of polymer degradation are compared with TGA results.

Styrenated Polyester. Thermal properties of a styrenated polyester (Laminac 4116, American Cyanamid Co.) were studied by Anderson and Freeman (49). The polymer was synthesized by condensation of a glycol and two dicarboxylic acids, one of which was unsaturated. Crosslinking with styrene, used both as solvent and copolymer, was effected by the use of free-radical initiators (see also POLYESTERS, UNSATURATED). TGA, DTA, infrared, and mass spectrometer techniques were used to study the thermal degradation of this polymer both in air and in argon. Based upon infrared analysis, the unit basic structure of the polyester was taken to be (**1**).

$$
\begin{array}{l}
\quad\quad\sim\!\!\sim\!-\mathrm{CH}-\!\sim\!\!\sim \\
\quad\quad\quad\quad\;| \\
\quad\quad\quad\quad\mathrm{CH_2} \\
\quad\quad\quad\quad\;| \\
\mathrm{C_6H_5}-\mathrm{CH} \qquad\qquad\quad \mathrm{CH_3} \\
\quad\quad\quad\quad\;| \qquad\qquad\qquad\quad | \\
\mathrm{H(OOCCHCHCOOCHCH_2OOCC_6H_4COO)_nH} \\
\quad\quad\quad\quad\quad\;\;| \\
\quad\quad\mathrm{C_6H_5}-\mathrm{CH} \\
\quad\quad\quad\quad\quad\;\;| \\
\quad\quad\quad\quad\quad\mathrm{CH_2} \\
\quad\quad\quad\quad\quad\;\;| \\
\quad\quad\quad\quad\quad\mathrm{CH_2} \\
\quad\quad\quad\quad\quad\;\;| \\
\quad\quad\sim\!\!\sim\!-\mathrm{CH}-\!\sim\!\!\sim
\end{array}
$$

(**1**)

Figure 29 shows TGA-derived curves for the thermal degradation of polyester (**1**) in both air and argon. From this figure it can be seen that for both types of atmospheres, the polymer began to degrade at about 200°C. However, in argon, the reaction was complete at about 450°C whereas, in the case of air, there was another reaction stage from about 450 to 550°C. In air, four reaction stages appeared, ie, from 200 to 260°C; from 260 to 360°C; from 360 to 450°C; and from 450 to 550°C. There also appeared to be extensive overlap between the reactions involved in the second and third stages. In argon, degradation occurred in two stages with extensive overlap. The first stage occurred between 200 and 365°C and the second between 365 and 450°C. Figure 30 shows a comparison between the thermal degradation of polyester

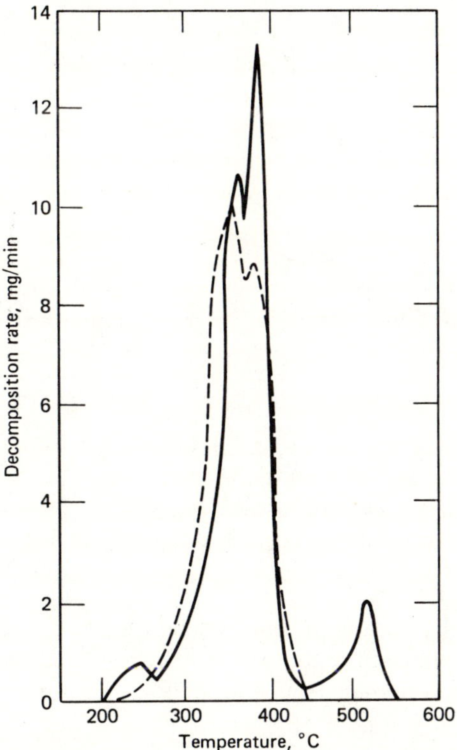

Fig. 29. Derivative thermogravimetric curves for the thermal degradation of a styrenated polyester (Laminac 4116, American Cyanamid Co.) (49). Solid curve, in air; dashed curve, in argon (49).

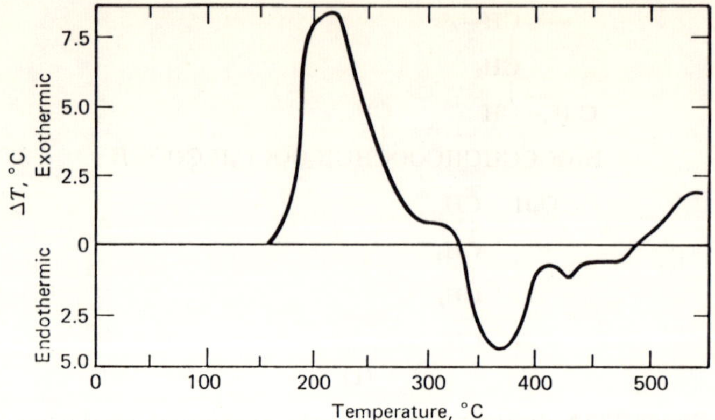

Fig. 30. Differences in temperature differential from DTA between a styrenated polyester (Laminac 4116) heated in air and in argon as a function of sample temperature ($T_{air} - T_A$) (49).

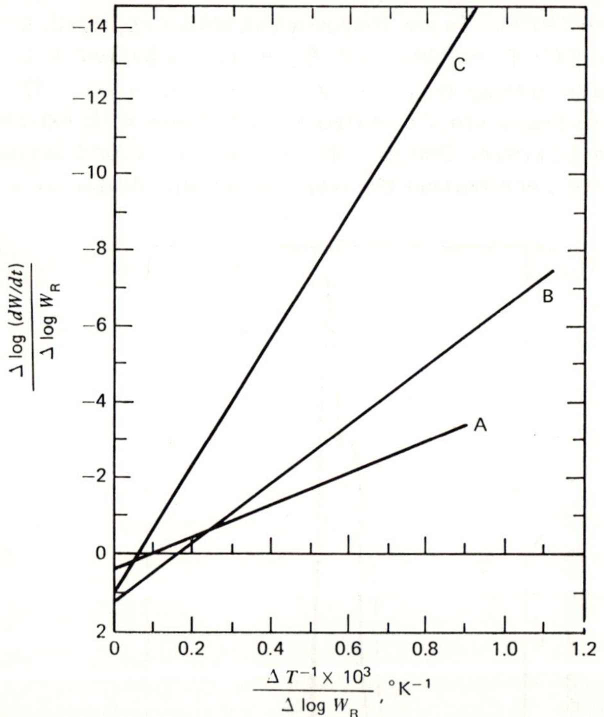

Fig. 31. Kinetics of the thermal degradation of a styrenated polyester (Laminac 4116) in air (49). Key: curve A, initial stage of reaction; curve B, second and third stages of reaction; curve C, fourth stage of reaction.

(1) in air and in argon by means of a difference plot of the DTA curves obtained in these atmospheres (the curve in argon was subtracted from that in air). Two regions of relative exotherm are apparent, from 150 to 290°C and from 470 to 550°C.

Also, degradation in air over the temperature range of 290 to 413°C is more endothermal than decomposition in argon. Upon heating the polyester in air from room temperature to about 500°C, the following noncondensable compounds were

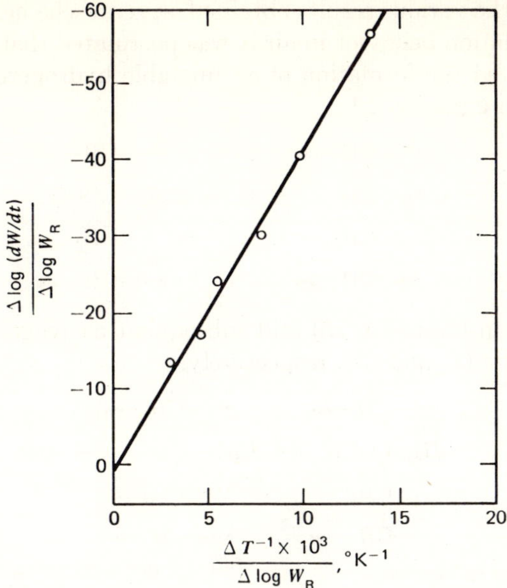

Fig. 32. Kinetics of the thermal degradation of a styrenated polyester (Laminac 4116) in argon (49).

obtained: carbon dioxide, hydrogen, methane, and propylene. The condensable compounds consisted of benzaldehyde and unsaturated hydroxy esters (between 200 and 300°C); phthalic anhydride and an oily liquid containing hydroxy esters (300–400°C); and a mixture of phthalic acid, phthalic anhydride, and a liquid containing low-molecular-weight esters of propylene glycol (400–500°C).

The degradation kinetics were determined by the method of Freeman and Carroll (5). Figure 31 gives plots for the degradation of a styrenated polyester in air. Three linear relationships were obtained which represented the initial stage of the reaction, the combination of second and third stages, and the fourth (and last) stage of the reaction. The corresponding values of E, n, and A are listed in Table 5. Figure 32 shows a similar graph for the reaction of a styrenated polyester in argon. The linear relationship obtained represented the combined first and second stages of reaction. Values of E, n, and A in this atmosphere are also given in Table 5.

Table 5. Kinetic Parameters for Thermal Degradation of a Styrenated Polyester[a] (49)

Stage of reaction	Temp range, °C	Order of reaction	Energy of activation, kcal/mole	First-order frequency factor, sec^{-1}
		In air		
1	200–260	0.4	19	1.3×10^5
2, 3	260–450	1.2	35	2.7×10^9
4	450–550	1.0	79	4.2×10^{19}
		In argon		
1, 2	200–450	1.0	20	4.8×10^3

[a] Laminac 4116, American Cyanamid Co.

On the basis of the various results obtained, several schemes were postulated (49). Thus, for the degradation behavior in air it was postulated that the initial exothermal reaction stage involved the formation of an unstable hydroperoxide intermediate, for example, as shown in equation 24.

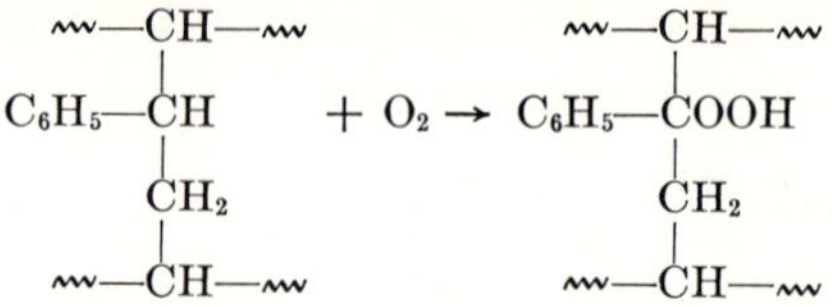

$$\begin{array}{c} \sim\!\!\sim\text{—CH—}\sim\!\!\sim \\ | \\ C_6H_5\text{—CH} \\ | \\ CH_2 \\ | \\ \sim\!\!\sim\text{—CH—}\sim\!\!\sim \end{array} + O_2 \rightarrow \begin{array}{c} \sim\!\!\sim\text{—CH—}\sim\!\!\sim \\ | \\ C_6H_5\text{—COOH} \\ | \\ CH_2 \\ | \\ \sim\!\!\sim\text{—CH—}\sim\!\!\sim \end{array} \qquad (24)$$

Radical formation (equation 25) and subsequent cleavage (equations 26 and 27) could lead to products (**2**) and (**3**), respectively.

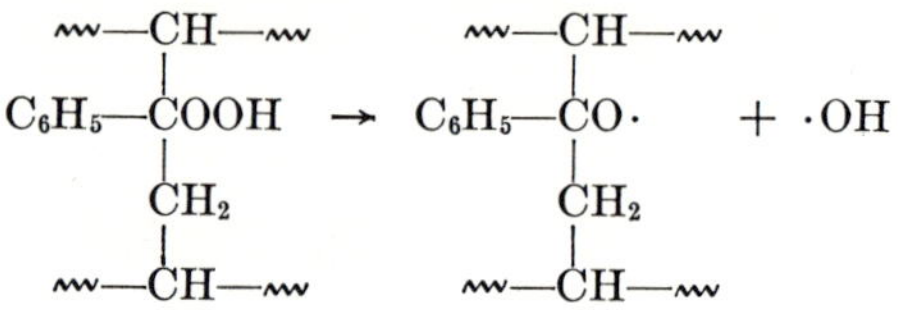

$$\begin{array}{c} \sim\!\!\sim\text{—CH—}\sim\!\!\sim \\ | \\ C_6H_5\text{—COOH} \\ | \\ CH_2 \\ | \\ \sim\!\!\sim\text{—CH—}\sim\!\!\sim \end{array} \rightarrow \begin{array}{c} \sim\!\!\sim\text{—CH—}\sim\!\!\sim \\ | \\ C_6H_5\text{—CO}\cdot \\ | \\ CH_2 \\ | \\ \sim\!\!\sim\text{—CH—}\sim\!\!\sim \end{array} + \cdot OH \qquad (25)$$

$$\begin{array}{c} \sim\!\!\sim\text{—CH—}\sim\!\!\sim \\ | \\ C_6H_5\text{—CO}\cdot \\ | \\ CH_2 \\ | \\ \sim\!\!\sim\text{—CH—}\sim\!\!\sim \end{array} + \cdot OH \rightarrow \underset{(\mathbf{2})}{\begin{array}{c} O \\ /\!/ \\ C_6H_5\text{—C} \\ | \\ CH_2 \\ | \\ \sim\!\!\sim\text{—CH—}\sim\!\!\sim \end{array} + \begin{array}{c} \sim\!\!\sim\text{—CH—} \\ | \\ OH \end{array}} \qquad (26)$$

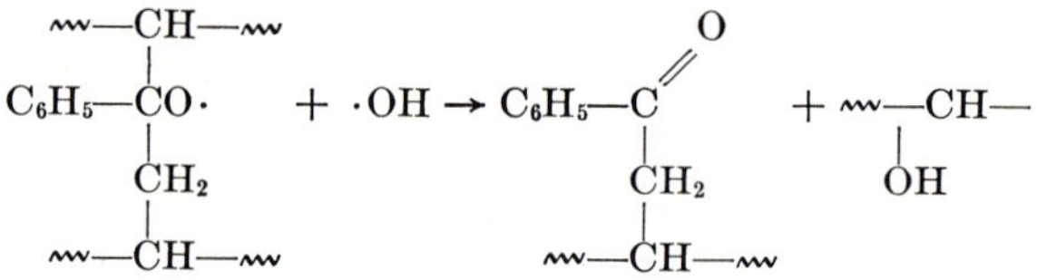

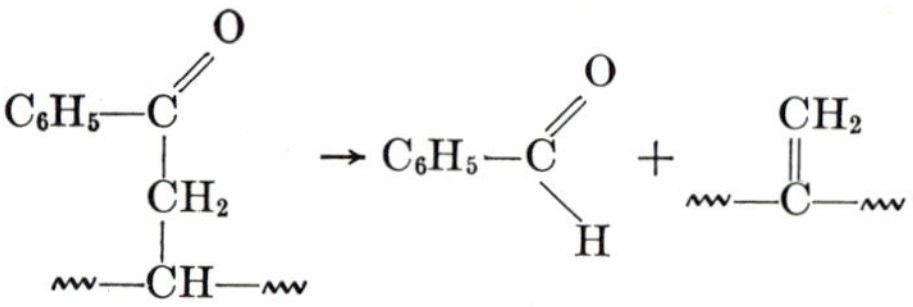

$$\begin{array}{c} O \\ /\!/ \\ C_6H_5\text{—C} \\ | \\ CH_2 \\ | \\ \sim\!\!\sim\text{—CH—}\sim\!\!\sim \end{array} \rightarrow \underset{(\mathbf{3})}{\begin{array}{c} O \\ /\!/ \\ C_6H_5\text{—C} \\ \diagdown \\ H \end{array} + \begin{array}{c} CH_2 \\ \| \\ \sim\!\!\sim\text{—C—}\sim\!\!\sim \end{array}} \qquad (27)$$

The value of E of 19 kcal/mole for the initial degradation phase in air falls in the range of values reported for hydroperoxide formation.

Since the second and third reaction stages in air were endothermal, bond rupture was indicated. Values of E, n, and A for these stages were of the order expected for the rupture of chemical bonds (50–52). Based upon the gaseous products obtained from these stages, the following free-radical mechanism was suggested (eqs. 28–30). Cleavage occurs at the carboxyl oxygen atoms to form phthalic anhydride (eq. 28).

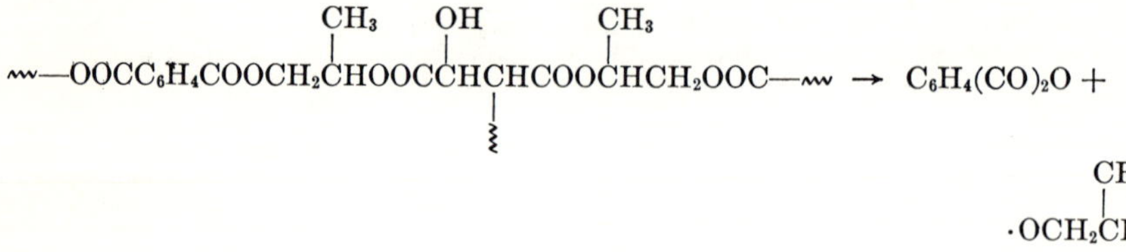

$$\sim\!\!\sim\text{—OOCC}_6H_4COOCH_2\overset{CH_3}{\overset{|}{C}}HOOC\underset{\sim\!\sim}{\underset{|}{\overset{OH}{\overset{|}{C}}}}HCHCOO\overset{CH_3}{\overset{|}{C}}HCH_2OOC\text{—}\sim\!\!\sim \rightarrow C_6H_4(CO)_2O + \cdot OCH_2\overset{CH_3}{\overset{|}{C}}HOOC\text{—}\sim\!\!\sim \qquad (28)$$

The hydroxy ester may then be formed by reaction with hydrogen (eq. 29).

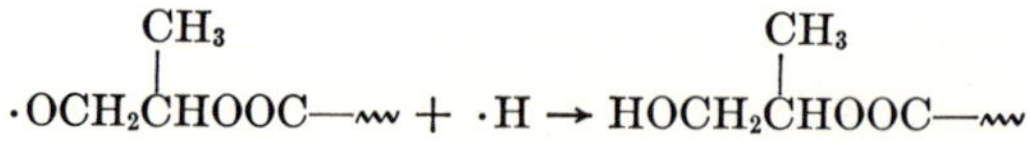

$$\cdot OCH_2\overset{CH_3}{\overset{|}{C}}HOOC\text{—}\sim\!\!\sim + \cdot H \rightarrow HOCH_2\overset{CH_3}{\overset{|}{C}}HOOC\text{—}\sim\!\!\sim \qquad (29)$$

Decarboxylation of the polyester (eq. 30) should occur readily and accounts for the formation of propylene and carbon dioxide.

$$\sim\!\!\sim\!\!-COOCH_2\overset{\overset{\displaystyle CH_3}{|}}{C}HOOC-\!\!\sim\!\!\sim \rightarrow 2\,CO_2 + CH_2{=}CHCH_3 \qquad (30)$$

In the fourth degradation stage (450–550°C) the exothermal trend and the high value of E of 79 kcal/mole indicate that oxidation of carbon, formed in the third reaction stage, occurs. This is supported by reported values of E for the reaction between carbon and oxygen (80 kcal/mole). Furthermore, a residue of carbon is found after degradation is complete in argon.

Polytetrafluorethylene. The isothermal decomposition of polytetrafluoroethylene (Teflon) in vacuum was reported by Madorsky and co-workers (53) to involve first-order kinetics. However, Wall and Michaelsen (54) later indicated, by the use of isothermal methods in a nitrogen atmosphere, that first-order kinetics obtain above about 510°C, whereas below 480°C the polymer degrades by a zero-order law. Subsequently, Madorsky and Straus (55) carried out another series of isothermal experiments in vacuum at lower temperatures than they had previously employed (below 485°C). They concluded that the degradation of polytetrafluoroethylene by heat was a first-order reaction through the temperature range of 425 to 513°C, and obtained a value of E of 80.5 kcal/mole. The preceding discordance of reaction order obtained by isothermal methods was the topic of several succeeding studies that employed TGA methods. See also TETRAFLUOROETHYLENE POLYMERS.

Anderson (56) studied the pyrolysis of polytetrafluoroethylene in vacuum, by means of TGA techniques, over the temperature range of 450 to 550°C. Employing the method of Freeman and Carroll (5) of estimating kinetic parameters, he found that, based upon eleven replicate TGA experiments, the values of E and n were 74.8 ± 3.9 kcal/mole and 1.02 ± 0.07, respectively. Others (11,19) have also used the method of Freeman and Carroll with the maximization method and isothermal techniques and have found values of E and n of a similar order of magnitude, even for temperatures below 480°C. Anderson (22,57) also employed thermogravimetric cycling experiments to determine whether the kinetics and mechanism are consistent as the polymer is degraded. Cycling experiments were carried out between 25°C and a fixed higher temperature; the ratio W_R/W_i for each cycle should follow a curve expressed by equation 31, where W_R is the weight of polymer remaining after cycle M,

$$(W_R/W_i) = (1 - W/W_i)^M \qquad (31)$$

W_i is initial weight, and W is weight loss during the first cycle. From Figure 33 it can be seen that good agreement is obtained for polytetrafluoroethylene between theoretical and experimental cycling values. From this figure it can also be observed that the agreement between theoretical and experimental cycling values is unsatisfactory for a copolymer of tetrafluoroethylene and hexafluoropropylene. TGA curves indicate that there were two stages in the decomposition of this copolymer. Values of E varied from about 35 to 60 kcal/mole as the conversion rose from 10 to 30%. The cycling experiments (up to 500°C, which involved the initial stage) also indicate that the initial decomposition stage involves chain segments of lowered stability (see DEGRADATION). Wall and Straus (58) have reported that a similar copolymer produced certain volatile products during decomposition which were more abundant in the early pyrolysis stages. When the cycling experiments were carried up

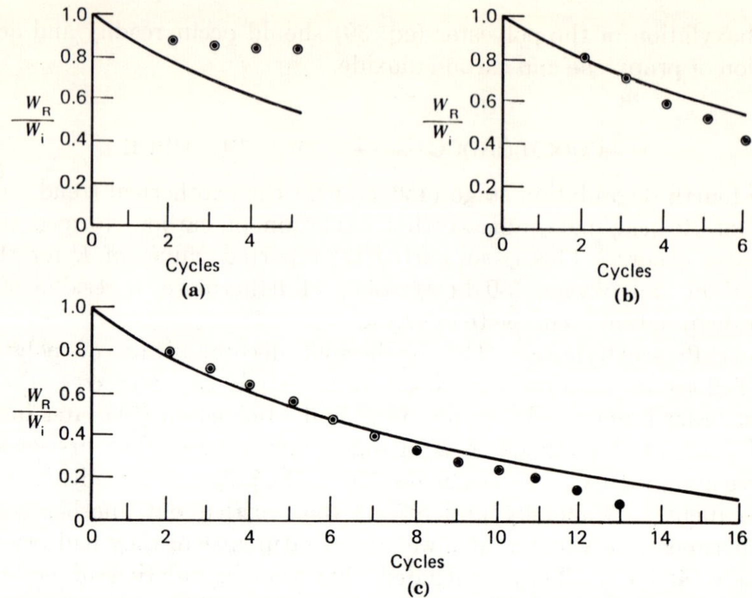

Fig. 33. Thermogravimetric cycling curves: W_i, initial weight at 25°C, W_R, residual weight at cycling temperature. The solid line represents the theoretical curve (57). Key: (**a**) fluorinated ethylene–propylene copolymer at 500°C; (**b**) fluorinated ethylene–propylene copolymer at 540°C; (**c**) polytetrafluoroethylene at 540°C.

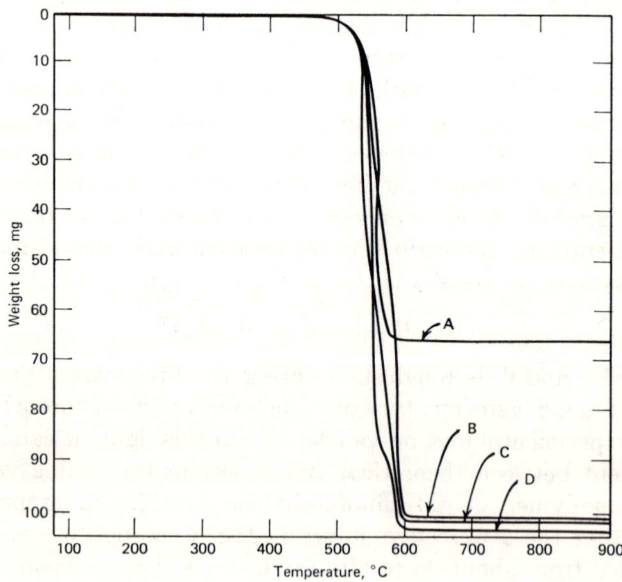

Fig. 34. Thermogravimetric analysis curves of polytetrafluoroethylene and polytetrafluoroethylene–silica mixtures in air (59). Curve A, 50% polytetrafluoroethylene, 50% silicon dioxide (100.0 mg sample); curve B, 90% polytetrafluoroethylene, 10% silicon dioxide (101.0 mg sample); curve C, polytetrafluoroethylene (102.0 mg sample); curve D, 75% polytetrafluoroethylene, 25% silicon dioxide (104.0 mg sample).

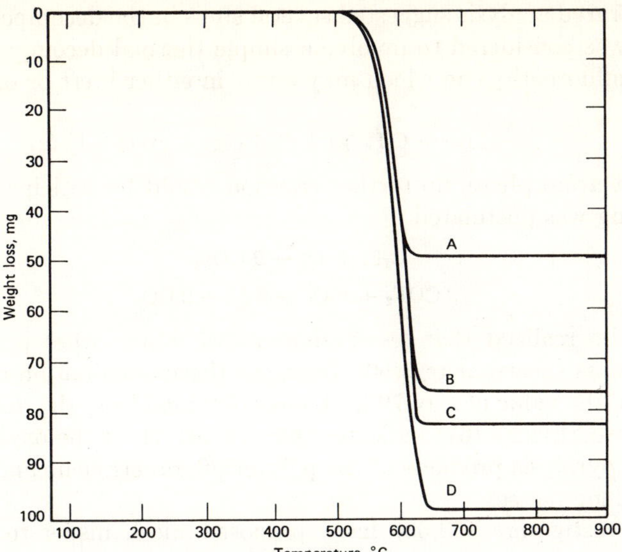

Fig. 35. Thermogravimetric analysis curves of polytetrafluoroethylene and polytetrafluoroethylene–silica mixtures in helium (59). Curve A, 50% polytetrafluoroethylene, 50% silicon dioxide (100.0 mg sample); curve B, 75% polytetrafluoroethylene, 25% silicon dioxide (103.0 mg sample); curve C, 90% polytetrafluoroethylene, 10% silicon dioxide (92.0 mg sample); curve D, polytetrafluoroethylene (100.0 mg sample).

to 540°C (which involved the second stage), the corresponding results indicate that the pyrolysis products consist of other volatile products during the later degradation stages. Thus, it has been found that at 10% conversion, the light volatiles contained 85% of C_3F_6 and 9% of C_2F_4, whereas at 92% decomposition, the volatiles contained 19% of C_3F_6 and 68% C_2F_4.

Light and co-workers (59) recently reported TGA methods for the analysis of silica-filled polytetrafluoroethylene. Figures 34 and 35 present TGA curves in air and helium, respectively, for pure powdered polytetrafluoroethylene and three mixtures of polytetrafluoroethylene with colloidal silica. In helium, the decomposition temperature range for all the samples studied was from about 500 to 650°C. At the completion of decomposition, all of the polytetrafluoroethylene volatilized, leaving a percentage of silica residue which agreed well with the theoretical value (assuming no interaction between polytetrafluoroethylene and silica). However, in air, the decomposition temperature range was from about 500 to 600°C, and based upon the amount of silica remaining it was apparent that interaction had occurred between silica and oxidation products of polytetrafluoroethylene. In order to account for the results obtained in air, the following stoichiometric equation (eq. 32) was proposed.

$$\left(C_2F_4\right)_n + SiO_2 + 2\,O_2 \rightarrow 2\,COF_2 + SiF_4 + 2\,CO_2 + (n-2)\,C_2F_4 + C_2F_4 \text{ (recombination products)} \tag{32}$$

From equation 32 it may be calculated that polytetrafluoroethylene–silica mixtures containing up to 23% of the silica will be completely volatilized (see Fig. 34). For a mixture containing 50% silica it can be calculated that there should be a total weight loss of 65%. An actual loss of 66% was observed (see Fig. 34). The stoichiometry of the reaction, the need for air to be present, and the identification of pyrolysis products by infrared analysis suggested several steps in the decomposition mechanism. The first step was considered to involve a simple thermal decomposition (unzipping) of the polytetrafluoroethylene which may occur in either inert or oxygen atmosphere (eq. 33).

$$-(C_2F_4)-_n\ (s) \rightarrow C_2F_4\ (g) + [C_3F_6\ (g) + \text{cyclo-}C_4F_8\ (g)] \quad (33)$$

In an inert atmosphere, no further reaction would be anticipated. However, in air, the following was postulated

$$C_2F_4 + O_2 \rightarrow 2\ COF_2 \quad (34)$$

and

$$2\ COF_2 + SiO_2 \rightarrow SiF_4 + 2\ CO_2 \quad (35)$$

It should be realized that lesser amounts of many other products have been identified, by mass spectrometry (60), from the thermal oxidation of polytetrafluoroethylene. Also, the value of E of 79 kcal/mole obtained from the isothermal oxidation of polytetrafluoroethylene (61) indicates that the attack of the oxygen occurs mainly on the gaseous pyrolysis products of the polytetrafluoroethylene and not on the polytetrafluoroethylene per se.

Several investigators (62,63) have proposed mechanisms to account for the degradation behavior of polytetrafluoroethylene in the absence of air. Thus, from TGA, isothermal, molecular-weight, etc, techniques, the following experimental observations must be accounted for: the pyrolysis product is mainly monomer; the pyrolysis rate is independent of molecular weight; the pyrolysis follows first-order kinetics over the entire conversion range; and the molecular weight decreases during vacuum pyrolysis. The following plausible scheme involves random initiation followed by depropagation and bimolecular termination (by disproportionation):

$$\sim\sim \xrightarrow{k_i} \sim\sim\cdot + \cdot\sim\sim$$

$$\sim\sim\cdot \xrightarrow{k_d} \sim\sim\cdot + C_2F_4$$

$$\sim\sim\cdot + \cdot\sim\sim \xrightarrow{k_t} \sim\sim CF{=}CF_2 + CF_3{-}CF_2{-}\sim\sim$$

From the above scheme, we may write equations 36–39, in which R_i = initiation

$$R_i = 2\,k_i\,\bar{N}_n[\mathrm{P}] \quad (36)$$

$$d[\mathrm{M}]/dt = k_d[\mathrm{R}\cdot]V = -(1/m)dW/dt \quad (37)$$

$$-d[\mathrm{R}\cdot]/dt = k_t[\mathrm{R}\cdot]^2 \quad (38)$$

$$[\mathrm{P}] = W/Vm\bar{N}_n = \rho/m\bar{N}_n \quad (39)$$

rate; $\bar{N}_n$ = number-average degree of polymerization; [R] and [P] denote the concentration of radicals and polymer molecules, respectively; [M] is the number of moles of monomer; V = volume of polymer; and m denotes the monomer molecular weight. Furthermore, we may write equation 40 for the rate of change in the number of polymer molecules (N).

$$dN/V\,dt = -k_i\bar{N}_n[\mathrm{P}] + k_t[\mathrm{R}\cdot]^2 \quad (40)$$

From the definition of number-average molecular weight, $\bar{M}_n = W/N$, equations 41 and 42 are obtained.

$$d\bar{M}_n/dt = (\bar{M}_n/W)\,dW/dt - (\bar{M}_n^2/W)\,dN/dt \quad (41)$$

$$d\bar{M}_n/dt = -(\bar{M}_n/W)\,m\,k_d[\mathrm{R}\cdot]V + (\bar{M}_n^2/W)k_i\,\bar{N}_n[\mathrm{P}]V - (\bar{M}_n^2/W)\,k_t[\mathrm{R}\cdot]^2V \quad (42)$$

From steady-state considerations,

$$k_t[\mathrm{R}\cdot]^2 = 2\,k_i\bar{N}_n[\mathrm{P}]$$

or

$$[\mathrm{R}\cdot] = 2^{1/2}(k_i\,\bar{N}_n[\mathrm{P}]/k_t)^{1/2} \tag{43}$$

The kinetic chain length (L) may be defined by equation 44.

$$L = k_d/k_t[\mathrm{R}\cdot] = (k_d/k_t)(k_t^{1/2}/2k_i\bar{N}_n[\mathrm{P}]^{1/2}) = k_d(m/2\rho k_i k_t)^{1/2} \tag{44}$$

Inserting the expression for L (eq. 44) into the expression for $d\bar{M}_n/dt$ (eq. 41) gives equation 45.

$$d\bar{M}_n/dt = -\bar{M}_n k_i\left(\frac{\bar{M}_n}{m} + 2L\right) \tag{45}$$

When $L \ll \bar{M}_n/m$,

$$\frac{1}{\bar{M}_n} - \frac{1}{M_{n,0}} = \frac{k_i}{m}t \tag{46}$$

Furthermore, from equation 37 and equations 39 and 43, the following equation can be obtained:

$$-dW/dt = 2^{1/2}k_d\left(\frac{mk_i}{\rho k_t}\right)^{1/2} W \tag{47}$$

From the above expression, equation 48 can be obtained.

$$k(\mathrm{exptl}) = 2k_i L \tag{48}$$

Finally, when the expression for t from equation 47 is substituted into equation 46, equation 49 is obtained.

$$\frac{\bar{M}_{n,0} - \bar{M}_n}{\bar{M}_n\bar{M}_{n,0}} = (1/2mL)\ln\frac{W_0}{W} \tag{49}$$

From equations 44, 47, and 48 it can be seen that the rate of weight loss of the polytetrafluoroethylene should be first-order in W and should be independent of molecular weight, as observed. Also, from equation 49, the value of $\bar{M}_n$ should decrease with conversion, as observed. It should be noted here that if end groups influenced the pyrolysis rate and terminal initiation was the predominant initiation step, then the pyrolysis rate would be a function of the molecular weight. This is not observed experimentally.

From equation 47 the following approximate expression may be written for the activation energies for the various steps in the mechanism (assuming that $E_t \approx 0$)

$$E(\mathrm{exptl}) \approx E_d + \tfrac{1}{2}E_i \tag{50}$$

Since the observed activation energy possesses a value of about 80 kcal/mole and assuming that the E_i = 74 kcal/mole (calculated from thermodynamic data for the bond energy of the C–C bond in polytetrafluoroethylene (64)), the value of E_d should be about 43 kcal/mole. It is interesting to note that the activation enthalpy for depropagation of polytetrafluoroethylene at 480°C was estimated to be 44 kcal/mole (63).

Bibliography

1. C. Duval, *Inorganic Thermogravimetric Analysis,* 1st ed., Elsevier Publishing Co., New York, 1953.
2. A. E. Newkirk and E. L. Simons, *Rept. No. 63-RL-3498C,* General Electric Co., Nov. 1963.
3. H. Saito, *Bull. Imper. Acad. Japan* **2,** 58 (1926).
4. D. W. Van Krevelen, C. Van Heerden, and F. J. Huntjens, *Fuel* **30,** 253 (1951).
5. E. S. Freeman and B. Carroll, *J. Phys. Chem.* **62,** 394 (1958).
6. L. Reich and D. W. Levi, *Macromol. Rev.* **1,** pp. 173 ff. (1967).
7. W. W. Wendlandt, *Thermal Methods of Analysis,* Interscience Publishers, a division of John Wiley & Sons, Inc., New York, 1964.
8. P. D. Garn, *Thermoanalytical Methods of Investigation,* Academic Press, Inc., New York, 1965.
9. P. E. Slade, Jr., and L. T. Jenkins, eds., *Techniques and Methods of Polymer Evaluation,* Marcel Dekker, Inc., New York, Vol. 1, 1966.
10. *Tech. Bull. T-102,* Mettler Instrument Corp., Princeton, N.J.
11. D. W. Levi, L. Reich, and H. T. Lee, *Polymer Eng. Sci.* **5,** 135 (1965).
12. H. L. Friedman, *U.S. Dept. Commerce, Office Tech. Serv., PB Rept.* **145,** 182 (1959).
13. L. Reich and D. W. Levi, *Makromol. Chem.* **66,** 102 (1963).
14. A. W. Coats and J. P. Redfern, *Analyst* **88,** 906 (1963).
15. D. A. Smith, *Rubber Chem. Technol.* **37,** 937 (1964).
16. D. A. Anderson and E. S. Freeman, *J. Polymer Sci.* **54,** 253 (1961).
17. S. L. Madorsky, *J. Polymer Sci.* **9,** 133 (1952).
18. J. H. Van Tassel and W. W. Wendlandt, *J. Am. Chem. Soc.* **81,** 813 (1959).
19. L. Reich, H. T. Lee, and D. W. Levi, *J. Polymer Sci.* [B] **1,** 535 (1963).
20. R. M. Fuoss, I. O. Salyer, and H. S. Wilson, *J. Polymer Sci.* [A] **2,** 3147 (1964), errata p. 5027.
21. H. L. Friedman, *Paper, Am. Chem. Soc. Meet., Div. Polymer Chem., New York, Sept. 1963; Polymer Preprints* **4** (2), p. 662; in *Thermal Analysis of High Polymers,* B. Ke, ed., Interscience Publishers, a division of John Wiley & Sons, Inc., New York, 1964, p. 183.
22. H. C. Anderson, *Paper, Am. Chem. Soc. Meet., Div. Polymer Chem., New York, Sept. 1963; Polymer Preprints* **4** (2), p. 655.
23. H. C. Anderson, *J. Polymer Sci.* [B] **2,** 115 (1964).
24. L. Reich, H. T. Lee, and D. W. Levi, *J. Appl. Polymer Sci.* **9,** 351 (1965).
25. L. Reich, *J. Polymer Sci.* [B] **2,** 621 (1964).
26. L. Reich, *J. Appl. Polymer Sci.* **9,** 3033 (1965).
27. C. D. Doyle, *J. Appl. Polymer Sci.* **5,** 285 (1961).
28. L. A. Dudina and N. S. Yenikolopyan, *Vysokomolekul. Soedin.* **5,** 986 (1963).
29. A. W. Coats and J. P. Redfern, *Nature* **201,** 68 (1964).
30. J. H. Flynn and L. A. Wall, *J. Polymer Sci.* [B] **4,** 323 (1966).
31. T. Ozawa, *Bull. Chem. Soc. Japan* **38,** 1881 (1965).
32. A. E. Newkirk, *Anal. Chem.* **32,** 1558 (1960).
33. B. M. Moiseev, *Russ. J. Phys. Chem.* **37,** 357 (1963).
34. C. D. Doyle, *Makromol. Chem.* **80,** 220 (1964).
35. S. L. Madorsky and S. Straus, *J. Res. Natl. Bur. Std.* **55,** 223 (1955).
36. W. W. Wright, in *Thermal Degradation of Polymers,* Monograph No. 13, Society of Chemical Industry, London, 1961, p. 248.
37. S. L. Madorsky and S. Straus, p. 60, Ref. 36.
38. F. R. Eirich and H. F. Mark, p. 43, Ref. 36.
39. C. N. Jurkov, *Dokl. Akad. Nauk SSSR* **47,** 493 (1946).
40. I. F. Kanavetz and L. G. Batalov, *SPE Trans.* **1,** 63 (1961).
41. C. S. Marvel, *SPE J.* **20,** 220 (1964).
42. K. D. Jeffreys, *Brit. Plastics* **36,** 188 (1963).
43. E. Horowitz and T. P. Perros, *J. Inorg. Nucl. Chem.* **26,** 139 (1964).
44. E. Horowitz and T. P. Perros, *J. Res. Natl. Bur. Std.* **69A,** 53 (1965).
45. C. D. Doyle, *WADD Tech. Rept. 60-283,* June 1960.
46. C. D. Doyle, *Anal. Chem.* **33,** 77 (1961).
47. S. S. Stivala, G. R. Sacco, and L. Reich, *J. Polymer Sci.* [B] **2,** 943 (1964).
48. C. D. Doyle, *J. Appl. Polymer Sci.* **6,** 639 (1962).
49. D. A. Anderson and E. S. Freeman, *J. Appl. Polymer Sci.* **1,** 192 (1959).

50. I. Marshall and A. Todd, *Trans. Faraday Soc.* **49,** 67 (1953).
51. S. Straus and L. A. Wall, *J. Res. Natl. Bur. Std.* **60,** 39 (1958).
52. G. Blyholder and H. Eyring, *J. Phys. Chem.* **61,** 682 (1957).
53. S. L. Madorsky, V. E. Hart, S. Straus, and V. A. Sedlak, *J. Res. Natl. Bur. Std.* **51,** 327 (1953).
54. L. A. Wall and J. D. Michaelsen, *J. Res. Natl. Bur. Std.* **56,** 27 (1956).
55. S. L. Madorsky and S. Straus, *J. Res. Natl. Bur. Std.* **64A,** 513 (1960).
56. H. C. Anderson, *Makromol. Chem.* **51,** 233 (1962).
57. H. C. Anderson, in *Thermal Analysis of High Polymers,* B. Ke, ed., Interscience Publishers, a division of John Wiley & Sons, Inc., New York, 1964, p. 175.
58. L. A. Wall and S. Straus, *J. Res. Natl. Bur. Std.* **65A,** 227 (1961).
59. T. S. Light, L. F. Fitzpatrick, and J. P. Phaneuf, *Anal. Chem.* **37,** 79 (1965).
60. R. E. Kupel, M. Nolan, R. G. Keenan, M. Hite, and L. D. Scheel, *Anal. Chem.* **36,** 386 (1964).
61. J. M. Cox, B. A. Wright, and W. W. Wright, *J. Appl. Polymer Sci.* **8,** 2951 (1964).
62. H. L. Friedman, *Aerophys. Res. Memo 37,* Tech. Info. Series R59SD385, General Electric Co., June 19, 1959.
63. J. C. Siegle, L. T. Muus, T.-P. Lin, and H. A. Larsen, *J. Polymer Sci.* [A] **2,** 391 (1964).
64. C. R. Patrick, *Tetrahedron* **4,** 26 (1958).

Leo Reich and David W. Levi
Picatinny Arsenal

INDEX